**Optical and Wireless Convergence
for 5G Networks**

Optical and Wireless Convergence for 5G Networks

Edited by

Abdelgader M. Abdalla
Instituto de Telecomunicações, Aveiro, Portugal

Jonathan Rodriguez
Instituto de Telecomunicações, Aveiro, Portugal
University of South Wales, Pontypridd, Wales
UK

Issa Elfergani
Instituto de Telecomunicações, Aveiro, Portugal

Antonio Teixeira
University of Aveiro, Portugal
Instituto de Telecomunicações, Aveiro, Portugal

Registered Offices
John Wiley & Sons, Inc., 111 River Street, Hoboken, NJ 07030, USA
John Wiley & Sons Ltd, The Atrium, Southern Gate, Chichester, West Sussex, PO19 8SQ, UK

Editorial Office
The Atrium, Southern Gate, Chichester, West Sussex, PO19 8SQ, UK

For details of our global editorial offices, customer services, and more information about Wiley products visit us at www.wiley.com.

Wiley also publishes its books in a variety of electronic formats and by print-on-demand. Some content that appears in standard print versions of this book may not be available in other formats.

Library of Congress Cataloging-in-Publication Data

Names: Abdalla, Abdelgader M., 1970- editor. | Rodriguez, Jonathan., editor.
    | Elfergani, Issa, editor. | Teixeira, Antonio, editor.
Title: Optical and wireless convergence for 5G networks / edited by:
    Abdelgader M. Abdalla, editor, Instituto de Telecomunicacoes Jonathan
    Rodriguez, editor, Instituto de Telecomunicacoes, University of South
    Wales, Pontypridd, Wales, UK, Issa Elfergani, editor, Instituto de
    Telecomunicacoes, Antonio Teixeira, editor, Instituto de Telecomunicacoes.
Description: Hoboken, NJ, USA : John Wiley & Sons, Inc., [2020] | Includes
    bibliographical references and index. |
Identifiers: LCCN 2019003146 (print) | LCCN 2019007370 (ebook) | ISBN
    9781119491613 (Adobe PDF) | ISBN 9781119491606 (ePub) | ISBN 9781119491583
    (hardcover)
Subjects: LCSH: Mobile communication systems. | Optical
    communications–Technological innovations.
Classification: LCC TK5103.25 (ebook) | LCC TK5103.25 .O68 2019 (print) | DDC
    621.3845/6–dc23
LC record available at https://lccn.loc.gov/2019003146

Cover Design: Wiley
Cover Image: © Iaremenko / Getty Images

Set in 10/12pt WarnockPro by SPi Global, Chennai, India
Printed and bound in Singapore by Markono Print Media Pte Ltd

10 9 8 7 6 5 4 3 2 1

Contents

About the Editors

Abdelgader Mahmoud Abdalla received his PhD in Telecommunications Engineering from MAP-Tele Doctoral Programme, Universidades de Minho, Aveiro, Porto (MAP-Tele), Portugal in October 2014. From 2010 to 2014, during his PhD study, he was involved in several national and international projects that included EURO-FOS Network, TOMAR-PON, BONE, NGPON2, and PANORAMA II. In December 2014, he joined Instituto de Telecomunicações (IT), Aveiro as a Senior Researcher. Recently, he finished the European research project ENIAC-THING2DO successfully, whilst acting as work package leader of Design Enablement that involved 17 partners. Currently, he is acting as a task leader of Design Methodology and Automation in European research OCEAN12 project, ECSEL-JU-Call 2017 within H2020. His main research interests include low-power nanoscale integrated circuit design, systems-on-a-chip design for optical and wireless communications, DSP-enhanced high-simulation runtime mixed-signal integrated circuit design, FPGA/ASIC design of high-simulation runtime digital and optical communication devices, as well as nonlinear modeling for nanotechnologies. He is the author of several journals and conference publications. He is an active IEEE member, acting as TPC member and reviewer for a number of respected conferences, journals, and magazines.

Jonathan Rodriguez received his MSc and PhD degrees in Electronic and Electrical Engineering from the University of Surrey, United Kingdom in 1998 and 2004, respectively. In 2005, he became a researcher at the Instituto de Telecomunicações, Aveiro (Portugal), and a member of the Wireless Communications Scientific Area. In 2008, he became a Senior Researcher and was granted an independent researcher role where he established the Mobile Systems Research Group with key interests in 5G networking, radio-optical convergence, and security. He has served as project coordinator for major international research projects, including Celtic Eureka LOOP and GREEN-T, and FP7 C2POWER, whilst serving as technical manager for FP7 COGEU and FP7 SALUS. He is currently leading the H2020-ETN SECRET project, a European Training Network on 5G communications. In 2008, he became an Invited Assistant Professor at the University of Aveiro (Portugal), attaining Associate status in 2015. He has also served as General Chair for the ACM sponsored MOBIMEDIA 2010 (6th International Mobile Multimedia Communications Conference), Co-Chair for the EAI sponsored WICON 2014 (8th International Wireless Internet Conference), and TPC co-chair for BroadNets 2018, as well as serving as workshop chair on 17 occasions in major international conferences that include IEEE Globecom and IEEE ICC, among others. He is the author of more than 450 scientific works, that include over 100 peer-reviewed international journals and 11 edited books. His professional affiliations include Senior Member of the IEEE, Chartered Engineer (CEng) and Fellow of the IET (2015). In 2017, he became Professor of Mobile Communications at the University of South Wales (UK).

Issa Elfergani received his MSc and PhD in Electrical and Electronic Engineering from the University of Bradford (UK) in 2008 and 2012, respectively, with a specialization in tunable antenna design for mobile handset and UWB applications. He is now a Senior Researcher at the Instituto de Telecomunicações, Aveiro (Portugal), working with several national and international research funded projects, while serving as technical manager for ENIAC ARTEMIS (2011–2014), EUREKA BENEFIC (2014–2017), CORTIF (2014–2017), GREEN-T (2011–2014), VALUE (2016–2016) and H2020-SECRET Innovative Training Network (2017–2020). In 2014 Issa received a prestigious FCT fellowship for his post-doctoral research. He is an IEEE and American Association for Science and Technology (AASCIT) member. Since his PhD graduation, he has successfully completed the supervision of several Master and PhD students. He is the author of around 93 high-impact publications in academic journals and international conferences; in addition, he is the author of two books and nine book chapters. He has been a reviewer for several good ranked journals such as IEEE Antennas and Wireless Propagation Letters, IEEE Transactions on Vehicular Technology, IET Microwaves, Antennas and Propagation, IEEE Access, Transactions on Emerging Telecommunications Technologies, Radio Engineering Journal, IET-SMT, and IET Journal of Engineering. He was the chair of both the 4th and 5th International Workshops on Energy Efficient and Reconfigurable Transceivers (EERT). He has been on the technical program committee of a large number of IEEE conferences. Issa has several years of experience in 3G/4G and 5G radio frequency systems research with particular expertise on several and different antenna structures along with novel approaches in accomplishing size reduction, low cost, improved bandwidth, gain and efficiency. His expertise includes research in various antenna designs such as MIMO, UWB, balanced and unbalanced mobile phone antennas, RF tunable filter technologies and power amplifier designs.

 **Antonio Luís Teixeira** got his PhD from University of Aveiro, Portugal, in 1999, partly developed at the University of Rochester. He holds an EC in Management and Leadership from MIT Sloan School and a post-graduation qualification in Quality Management in the field of Higher Education. He has been a professor at the University of Aveiro since 1999, as an Associate Professor with Agregação. He worked from 2009 to 2013 in Nokia Siemens Networks and in Coriant (2013–2014) as a standardization expert in the field of optical access (in FSAN, ITU-T, IEEE 802.3). Since 2014, he has been the Dean of the University of Aveiro Doctoral School aggregating 50 PhD programs and 1300 students. He has published more than 400 papers (more than 130 in journals), has edited a book, and contributed to several others. He holds 11 patents, and tutored successfully more than 70 MScs and 14 PhDs, having participated in more than 35 projects (national, European and international). In 2014 he co-founded PICadvanced, a startup focused on providing solutions based on optical assemblies targeting biotech and optical networking (including access networks). He served the ECOC TPC from 2008 to 2015 in the SC for subsystems, having chaired it in 2010, 2011 and 2015. He served the access subcommittee in OFC from 2011 to 2014, and was the General Chair of ICTON 09, Networks 2014. He is a Senior Member of OSA and a member of IEEE and the IEEE standards association.

List of Contributors

Professor Ghassemlooy received a BSc(Hons) in Electrical and Electronics Engineering from Manchester Metropolitan University, UK in 1981, and an MSc (1984) and PhD (1987) from University of Manchester, UK. From 1987 to 1988 he was a Post-Doctoral Research Fellow, City University, UK. In 1988 he joined Sheffield Hallam University as a Lecturer, becoming a Professor in 1997. In 2004 he joined University of Northumbria, Newcastle as an Associate Dean (AD) for research in the School of Engineering, and from 2012 to 2014 was AD for Research and Innovation, Faculty of Engineering, where he currently is Head of the Optical Communications Research Group. In 2001 he was awarded the Tan Chin Tuan Fellowship in Engineering from Nanyang Technological University, Singapore. In 2016 he was a Research Fellow and in 2015 a Distinguished Professor at Chinese Academy of Sciences, Quanzhou, China. He was a Visiting Professor at University Tun Hussein Onn Malaysia (2013–17), and Huaqiao University, China (2017–18). He has published over 770 papers (305 journals and 6 books), over 92 keynote/invited talks, and supervised over 60 PhDs. Research interests include optical wireless communications, free space optics, and visible light communications. He was the Vice-Chair of EU Cost Action IC1101 (2011–16). He is the Chief Editor of the British Journal of Applied Science and Technology and the International Journal of Optics and Applications. He is the fellow of the IET, a senior member of IEEE, a senior member of OSA, and Charted Engineer. He is a co-author of a CRC book on *Optical Wireless Communications – Systems and Channel Modelling with Matlab* (2012); and co-editor of four books including a Springer book on *Optical Wireless Communications – An Emerging Technology* (2016), a CRC book *Visible Light Communications Theory and Applications*, CRC June 2017, an IGI Global book *Intelligent Systems for Optical Networks Design Advancing Techniques*, 31 Mar 2013, and an IET book *Analogue Optical Fibre Communications, IEE Telecommunication Series 32*, 1995. From 2004 to 2006 he was the IEEE UK/IR Communications Chapter Secretary, then the Vice-Chairman (2004–2008), and the Chairman (2008–2011), and Chairman of the IET Northumbria Network (Oct 2011–2015). Web site http//soe .northumbria.ac.uk/ocr/people/ghassemlooy/

Paulo P. Monteiro received a diploma "Licenciatura" in Electronics and Telecommunications Engineering from the University of Aveiro in 1988, an MSc in Electronic Engineering from the University of Wales UK, in 1990 and a PhD in Electrical Engineering, from the University of Aveiro in 1999. Presently, he is Associate Professor at the University of Aveiro and Researcher at the Instituto de Telecomunicações. From October 2002 until March 2007 he was at Siemens SA, Portugal as a Head of Research of Optical Networks. From April 2007 until December 2009 he was at Nokia Siemens Networks (NSN) Portugal as a Research Manager at Transport, Aggregation and Fixed Access, and from January 2010 until June 2012 as a R&D manager of the Network Optimization unit at NSN. From July 2012 until May 2013 he was Research Manager at NSN Portugal. In May 2013, the Optical Networks business unit of Nokia Siemens Networks began operating as a new company Coriant and he left in June 2013. In 1992, he joined the Department of Electronic and Telecommunications Engineering of University of Aveiro and the Optical Communications Group of Institute for Telecommunications as an Assistant Professor and Researcher, respectively. In 1999, he became an Auxiliary Professor at the University of Aveiro and he was promoted to Associate Professor in 2005. His main research interests include Optical Communication Networks, Microwave Photonic and Electronic Subsystems. He has tutored and co-tutored successfully more than 14 PhDs, having participated in more than 28 projects (national, and international). He was the coordinator of a CELTIC project OPTRONET and a large-scale integrating project FUTON (FP7 ICT-2007–215533). He has authored/co-authored more than 18 patent applications and over 100 papers in journals and 360 conference contributions. He is a member of ECOC 2018 Sub-Committee 5 on Photonic and Microwave Photonic Subsystems and Senior Member of IEEE.

José Capmany was born in Madrid, Spain, on December 15 1962. He received the Ingeniero de Telecomunicacion degree from the Universidad Politécnica de Madrid (UPM) in 1987 and the Licenciado en Ciencias Físicas in 2009. He holds a PhD in Electrical Engineering from UPM and a PhD in Quantum Physics from the Universidad de Vigo. Since 1991 he has been with the Departamento de Comunicaciones, Universitat Politécnica de València (UPV), where he started the activities on optical communications and photonics, founding the Photonics Research Labs (www.prl.upv.es). He was an Associate Professor from 1992 to 1996, and Full Professor in optical communications, systems, and networks since 1996. In parallel, he has been Telecommunications Engineering Faculty Vice-Dean from 1991 to 1996, and Deputy Head of the Communications Department since 1996. From 2002 to 2016, he was the Director of the Institute of Telecommunications and Multimedia (iTEAM) at UPVa (www.iteam.upv.es). His research activities and interests cover a wide range of subjects related to optical communications including microwave photonics (MWP), integrated optics, optical signal processing, fiber Bragg gratings, and more recently quantum cryptography and quantum information processing using photonics. He has published over 520 papers in international refereed journals and conferences and has been a member of the Technical Program Committees of the European Conference on Optical Communications (ECOC), the Optical Fiber Conference (OFC), the Integrated Optics and Optical Communications Conference (IOOC), CLEO Europe, and the Optoelectronics and Communications Conference (OECC). Professor Capmany has also carried out activities related to professional bodies and is the Founder and current Chairman of the IPS Spanish Chapter, and a Fellow of the Institute of Electrical and Electronics Engineers (IEEE), the Optical Society of America (OSA) and the Institution of Electrical Engineers (IEE). He has acted as a reviewer for over 35 SCI journals in the field of photonics and telecommunications. He is also a founder and chief innovation officer of the spin-off company VLC Photonics (www.vlcphotonics.com) dedicated to the design of photonic integrated circuits and EPHHOX (www.ephoox.com) dedicated to MWP instrumentation.

Professor Capmany is the 2012 King James I Prize Laureate on novel technologies, the highest scientific distinction in Spain, for his outstanding contributions to the field of microwaves.

Professor Marco Ruffini received his MEng in Telecommunications in 2002 from Polytechnic University of Marche, Italy. After working as a research scientist for Philips in Germany, he joined Trinity College Dublin in 2005, where he received his PhD in 2007.

Since 2010 he has been Assistant Professor (tenured 2014) at TCD. He is Principal Investigator (PI) at both the CONNECT Telecommunications Research Centre at TCD, and the IPIC photonics integration center headquartered at the national Tyndall institute. Professor Ruffini is currently involved in several Science Foundation Ireland (SFI) and H2020 projects and leads the Optical Network Architecture group at Trinity College Dublin. His main research is in the area of 5G optical networks, where he carries out pioneering work on the convergence of fixed-mobile and access-metro networks, and on the virtualization of next generation networks. Professor Ruffini and has been invited to share his vision through several keynotes and talks at major international conferences across the world. He has authored over 100 international publications, over 10 patents and contributes to the BroadBand Forum (BBF) standardization body.

Professor Stanislav Zvanovec received his MSc and PhD degrees from the Faculty of Electrical Engineering, Czech Technical University (CTU) in Prague in 2002 and 2006, respectively. He works as a full professor and deputy head of the Department of Electromagnetic Field and chairperson of PhD branch at CTU. His current research interests include free space optical and fiber optical systems, visible light communications, OLED and RF over optics. He is the head of the Free-Space and Fiber Optics Group at CTU (optics.elmag.org). Until 2014 he was a chair of the Joint MTT/AP/ED/EMC chapter of the IEEE Czechoslovakia Section. He is holder of projects national TACR and GACR agencies and participates within the frame of international projects Visible Light Based Interoperability and Networking (ViSioN) European ITN MSCA project, EU COST projects IC1101 OPTICWISE (Optical Wireless Communications – An Emerging Technology), IC0802, IC0603, CA1622, etc. He is author of two books (co-author of a recent book *Visible Light Communications Theory and Applications*), several book chapters and more than 250 journal and conference papers.

Ifiok Otung is Professor of Satellite Communications at the University of South Wales. He is author of more than 100 scientific works in the area of mobile and satellite communication systems. In particular, he has led earthspace radio wave propagation research for many years, supervising 16 PhD and over 100 MSc projects and participating in two EPSRC grants (GR/M40356/01, GR/K96601/01) as well as KTP (TCS 3661) and ERDF (WEFO 80247) funding and many EU COST programs and collaborative projects with industry. He has also been active in the general area of communication systems where he has authored two textbooks targeted at post-graduate students and researchers, including *Digital Communications and Broadband Satellite-Integrated Network Design*. His research includes mobile and satellite propagation channels, prediction modeling, and RF design.

Martin Maier is a full professor with the Institut National de la Recherche Scientifique (INRS), Montréal, Canada. He received MSc and PhD degrees, both with distinctions (summa cum laude), in Electrical Engineering from the Technical University Berlin, Germany in 1998 and 2003, respectively. In the summer of 2003, he was a post-doctoral fellow at the Massachusetts Institute of Technology (MIT), Cambridge, MA. He was a visiting professor at Stanford University, Stanford, CA, October 2006 through March 2007. Further, he was a co-recipient of the 2009 IEEE Communications Society Best Tutorial Paper Award. He was a Marie Curie International Incoming Fellow (IIF) of the European Commission from March 2014 through February 2015. More recently, in March 2017, he received the prestigious Friedrich Wilhelm Bessel Research Award from the Alexander von Humboldt (AvH) Foundation in recognition of his outstanding accomplishments in research on FiWi enhanced networks for realizing the Tactile Internet vision for the year 2020. In May 2017, he was also awarded as one of the three most promising scientists in the category "Contribution to a Better Society" of the prestigious Marie Skłodowska-Curie Actions (MSCA) 2017 Prize Awards of the European Commission.

Raed A. Abd-Alhameed is Professor of Electromagnetic and Radio Frequency Engineering at the University of Bradford, UK. He received BSc and MSc degrees from Basrah University, in 1982 and 1985, respectively, and a PhD degree from the University of Bradford, UK in 1997, all in electrical engineering. He has many years' research experience over 25 years in the areas of radio frequency, antennas and electromagnetic computational techniques, and has published over 500 academic journal and conference papers; in addition he is co-author of three books and several book chapters. He is the senior academic responsible for electromagnetics research in the communications research group, in which a new antenna design configurations and computational techniques were developed and several patents were considered and filed. Jointly with Professor Excell (now the Dean of the Engineering School in Wrexham University), he has developed the principle of the "hybrid" method for electromagnetic field computation, which is able to combine the most appropriate method for differing regions of a problem (e.g. the human head and a mobile telephone). This method is recognized as being a leading area of research in bioelectromagnetic field computation. He also investigates the reduction of the size of antennas for personal mobile communications. The development of this kind of antenna is under active investigation (three patent applications submitted). He has also developed the mathematical tools needed for the simulation of non-linear circuits, including energy-storing devices. Moreover, he has written three different programs for electromagnetic scattering problems (wire antenna design; dielectrically-loaded antennas and microstrip wire antennas) and one code for analysis of nonlinear circuits using Volterra series. Interest has been shown by publishing houses in finding ways of disseminating this work.

Abubakar Sadiq Hussaini, MSc, PhD, Head of Programmes/Director at Commonwealth ITU Group (CIG), actively participates in the ITU activities of the radiocommunication, telecommunication standardization and telecommunication development sections. He is Senior Researcher and Project Development Manager with the 4TELL research group at Instituto de Telecomunicações, Aveiro, Portugal; Visiting Researcher at the University of Bradford, UK; Assistant Professor at the American University of Nigeria; and Visiting Assistant Professor at the Modibbo Adama University of Technology, Yola, Nigeria. He was microwave radio transmission operation and maintenance senior engineer with Nigerian Telecommunication Limited (NITEL), Abuja, Nigeria (10 years). He was a member of the Senate Committee of the University of Bradford, UK, and received his MSc in Radio Frequency Communication Engineering from the University of Bradford, UK in 2007 and his PhD in Telecommunication Engineering from

University of Bradford, UK in 2012. He is a member of IEEE, IET, and Optical Society of America, has contributed to numerous publications, and is involved in European and CELTIC research projects. His research interests include radiofrequency system design and high-performance RF-MEMS tunable filters with specific emphasis on energy efficiency and linearity, and his achievements comprise the participation in the design of energy efficient power amplifier at 3.5 GHz (mobile WiMAX frequency); the design of high-performance RF-MEMS tunable filter with tuning range from GSM 1.8 GHz to LTE 2.6 GHz; the design and development of a "radio over fiber" optical transmitter and receiver (1550 nm wavelength) in which the frequency limitations of quantum well laser in direct RF to light transponding was investigated. He successfully attained several European projects among which are MOBILIA (2009-2011), ARTEMOS (2011-2014) and THINGS2DO (2014-2018). He has participated actively in events, conferences and seminars organized by the Information Society Technology (IST) research program, Institute of Electrical and Electronics Engineers (IEEE), Institution of Engineering and Technology (IET), and European NanoElectronics Forum. He participated in several annual European project reviews. His collective role is to defend the project objectives and results to a panel of examiners that are considered European experts in their respective technical fields. He has served as a workshop organizer and a workshop chair. He is a TPC member and reviewer for many international conferences and journals. He is a guest editor for an IET Science, Measurement and Technology special issue.

Andreas Stöhr was born on 7 May 1965. He received the Dipl-Ing and Dr-Ing degrees in Electrical Engineering from Gerhard-Mercator-University, Germany in 1991 and 1997, respectively. From 1987 until 1996 he was CEO of MS Steuerungsanlagen GmbH in Kerpen-Horrem, Germany. From 1996 until 2013 he was with University Duisburg as a research scientist. In 1998 and 1999 he joined the National Institute of Information and Communications Technology (NICT) – the former CRL – in Tokyo, Japan where he worked on 60 GHz wireless systems employing radio-over-fiber (RoF) techniques. He also worked on millimeter-wave and THz RoF systems with France Telecom Orange Labs in Lannion, France in 2009 and with Corning Inc. in New York, USA in 2015. Since 2011, he has been Professor and Head of the Optoelectronics Institute within the Center for Semiconductor Technology and Optoelectronics (ZHO) at University Duisburg-Essen (UDE), Germany. His current research interests include III/V integrated microwave photonic device technology and RF photonic integration technologies for millimeter-wave and THz communications, spectroscopy, imaging and test and measurement systems. Professor Stöhr has published more than 200 papers in refereed journals and conferences. He is a senior member of the IEEE Photonics and MTT societies, committee member and chair of a number of international conferences, and IEEE/OSA guest editor. He is a management board member of the national DFG Transregio-Sonderforschungsbereich research program entitled MARIE.

Rattana Chuenchom was born on 29 October 1976. She received a BSc degree in Electronics and Computers and an MSc degrees in Electrical Communication Engineering from King Mongkut's Institute of Technology Ladkrabang, Bangkok, Thailand in 1998 and 2006, respectively. From 2007 to 2013, she worked with the National Institute of Metrology, Thailand in photometry and radiometry. In 2013 she joined optoelectronic research and studied her PhD at the optoelectronic University of Duisburg, Germany. She joined the IPHOBAC-NG in the project of the Integrated

Photonic Broadband Radio Access Units for Next Generation Optical Access Networks. She received the prestigious IEEE Microwave Photonics 2017 best paper award.

Dr Andreas Gerhard Steffan was born on 22 November 1970 in Hilden, Germany. He studied Electrical Engineering and Physics at the Rheinisch-Westfälisch Technische Hochschule (RWTH) Aachen, where he received his Dipl-Ing and Dipl-Phys degrees. His diploma theses were on silicon-on-insulator (SOI) waveguides and the work was carried out at the Institute of Semiconductor Technology (IHT) in cooperation with AMO GmbH, Aachen. In 2002 he received his PhD from University of Cambridge, UK. In 2002 he joined u²t Photonics AG now Finisar. Initially he worked as project manager on the development of a new transmitter, receivers and test and measurement products, including modulators and pulsed laser sources. At Finisar he is now heading optical chip development and the T&M/Analogue Products group. He is a member of the Institute of Physics (IoP), and has received awards from the Engineering and Physical Science Research Council (EPSRC), Renishaw plc. and the Cambridge European Trust.

Stephen Clements is Chief Executive Officer of aXenic. After graduating from Nottingham University, he joined the Caswell Research Centre (then part of Plessey) before moving to the Nortel Harlow Laboratories (UK) where he developed lasers for underwater fiber-optic cables and then led the R&D group in passive optical components. He joined Filtronic in 2001, where he developed their optical semiconductor business beyond the company's traditional wireless applications. Following Filtronic's sale to RFMD, he led the sale of the GaAs Photonics group to u²t Photonics, later acquired by Finisar. He then managed the buy-out of aXenic from Finisar in 2015 to develop it as a specialist supplier of modulators. A senior member of IEEE and Institute of Physics, he has more than 50 publications in semiconductor device design, process and systems and patents to support their exploitation.

Robert Walker is a Chief Technical Designer at aXenic. He studied electronics and electrical engineering at the University of Glasgow, receiving his doctorate in 1981. In 1982 he joined Plessey Research, Caswell (subsequently GEC/Marconi, Bookham and Oclaro), initiating the work on III-V guided-wave device technology with which he has since been closely identified. He joined u²t Photonics (UK) – now aXenic – in 2012 after working as an independent consultant to the company. Dr Walker was awarded the Nelson Gold Medal by GEC in 1994 for his contributions to microwave optoelectronics. His first publication of the GaAs traveling-wave modulator led to an Electronics Letters Premium Award in 1991. He has authored seminal papers on this and related topics, and has contributed a chapter on GaAs modulators to the reference book *Broadband Optical Modulators*, CRC Press, 2012.

Mateusz Lech is Principal Engineer at Orange Polska, Cracow, Poland. He received his MSc in Electrical Engineering from the AGH University of Science and Technology. His areas of expertise include DSL and PON broadband. He has worked with Orange Polska in new transmission systems evaluation, debugging and troubleshooting.

Andrzej Banach is Principal Engineer at Orange Polska, Lublin, Poland. He received his MSc in Electrical Engineering from the Lublin University of Technology. His areas of expertise include optical telecommunication networks, optic metrology. He has worked with Orange Polska, Orange France, Sofercom France in developing, evaluation, debugging and troubleshooting of new transmission systems since 1994. He has participated in EU research projects and in favor of telecom vendors.

Yigal Leiba is a co-founder and CTO of Siklu Communication. In previous positions, Yigal worked for Runcom Technologies, in charge of the design and the development of the world's first mobile-WiMAX (802.16e) chip, and a voting member of the IEEE 802.16 standard committee. Yigal was also a co-founder and CTO of Breezecom's LMDS group, a spin-off of Breezecom (later Alvarion). His career includes project management and chief engineer positions at Breezecom, where he was responsible for the development of wireless access and wireless LAN systems. Before Breezecom, Yigal was an R&D engineer in a technological division of the Israel military intelligence establishment. Yigal holds an MSc Degree in Electrical Engineering from the Technion – Haifa, Israel.

Nelson Muga was born in Mogadouro, Portugal in 1980. He graduated in Physics from University of Porto, Portugal in 2002, and received a master's degree in Applied Physics in 2006 from the University of Aveiro, Portugal. He received his PhD in Physical Engineering at Aveiro University for research on polarization effects in fiber-optic communication systems. After receiving his PhD degree, he joined the Instituto de Telecomunicações as a postdoctoral fellow, where he continued his research in coherent optical fiber communications from 2012 to 2017. He has been a lecturer at the Physics Department of the University of Aveiro since 2016, where he teaches courses related to optoelectronics, optics, and photonics. Currently, he is a senior researcher at the Instituto de Telecomunicações, Aveiro, where, over the years, he has participated as a researcher in 13+ R&D research projects (four directly supported by the industry), developing expertise in the field of high-speed optical communication systems and quantum secure optical communications. His current main research interests include coherent optical systems, all-optical and digital signal processing, advanced modulation formats, and fiber-optic quantum communications. He has a track record of 30+ papers published in international journals and 40+ international conference proceedings. Dr Muga is a Member of the Optical Society-OSA.

Siaka Ajewale Alimi received a BTech(Hons) degree in Electrical and Electronics Engineering (Communication) from Ladoke Akintola University of Technology, Ogbomoso, Nigeria in 2001, and MEng degree in Electrical and Electronics Engineering (Communication) from the Federal University of Technology, Akure, Nigeria in 2010, and a PhD in Telecommunications Engineering from the University of Aveiro, in 2018. He is currently a post-doctoral researcher at the Instituto de Telecomunicações, Department of Electronics, Telecommunications and Informatics, Universidade de Aveiro, Portugal. Presently, he is a lecturer at the Federal University of Technology, Akure, Nigeria. He worked in the Federal Radio Corporation of Nigeria as a Senior Engineer (radio frequency (RF) management and transmission) from June 2004 to December 2012. His research interests include computer networking, network security, radio over fiber (RoF), optical wireless communication (OWC), hybrid RF/FSO systems, and advanced signal processing and their applications for effective resource management in the cloud computing radio access networks (CC-RANs).

Seedahmed S. Mahmoud received a BSc (Hons, 1st class) in Biomedical Engineering from University of Gezira, Sudan (1996) and an MSc in Electronic Systems Engineering from University Putra Malaysia, UPM (1998). He received his PhD degree in Electronic Engineering (Signal Processing/Digital Communication) from RMIT University, Australia in 2004. His PhD was sponsored by the International Postgraduate Research Scholarship (IPRS) for outstanding overseas students from the Australian government. Dr Mahmoud was a former Research Fellow/Assistant Professor at the School of Electrical and Computer Engineering, RMIT University, a senior academic at Applied Engineering College (Riyadh, KSA), and a senior digital signal processing engineer for eleven years at Future Fibre Technologies (FFT) Ltd, where he led the development of artificial intelligent algorithms for FFT's optical fiber intrusion detection systems. His work on machine learning for a perimeter intrusion detection system has led to three international patents. Dr Mahmoud is Associate Professor at College of Engineering, Shantou University, China and a senior member of IEEE. He has written one book in applied signal processing for wireless communications, three book chapters and over 150 technical reports and papers in refereed international journals and conferences in various areas of electrical and electronic engineering. He is also serving on the editorial board of the Journal of the Computer Science (JCS) and is a reviewer for a number of IEEE and Elsevier journals. His research interests are development of wearable systems for stroke-patient rehabilitation, machine and deep learning algorithms for biomedical applications, signal processing for communications, optical fiber sensors, and perimeter intrusion detection and classification.

Bernhard Kozio graduated from Victoria University (Australia) in 2005 with an MSc (Physics) focusing on optical fiber based temperature and strain sensing. He later worked in the optical fiber sensing industry for eleven years and holds several patents in improvements to optical fiber sensing solutions for perimeter intrusion detection techniques. As of 2017 he has been employed as a physics and materials laboratory technician at Swinburne University, Melbourne.

Jusak Jusak received a BSc in Electronic Engineering from Brawijaya University, Malang, Indonesia in 1996. He received a PhD degree in Electrical Engineering (Signal Processing for Wireless Communication) from RMIT University, Australia in 2006. He did post-doctoral research in Massey University, New Zealand between 2009 and 2011. Currently he is a Senior Lecturer in Institute of Business and Informatics Stikom Surabaya, Indonesia. His main research areas include signal processing for wireless communication, wireless sensor networks, and application of Internet of Things (IoT) for public healthcare improvement.

Dr Hassan Hamdoun received his PhD in Advanced Telecommunications Engineering from Swansea University, UK in 2013. Thereafter, he joined the electronics research group and the digital Economy research hub at University of Aberdeen, UK, as an internet engineering and networking post-doctoral research fellow. In February 2018 he joined BT as a senior researcher in video delivery at BT Applied Research Labs, BT Technology, Adastral Park, Ipswich, UK.

His research portfolio is focused on energy efficient 4G/5G cellular mobile networking, cross-layer transmission protocols for 4G and 5G networks, broadband networks service delivery, complex networks modeling applications in ICT and ICT Energy management (architecture, design and optimization) and has published his contributions in highly reputed conferences and journals. He serves as reviewer and TPC member for several IEEE and non-IEEE transactions conferences and other journals. He holds an MSc(Eng) in Electronics Engineering (University of Sheffield, UK) and a BSc(Hons) in Electrical Engineering (University of Khartoum, Sudan). He is a member of the IEEE (USA) and IET (UK), Fellow of the UK higher education academy, a Chartered Engineer at the Sudanese Engineering Council and a certified carbon consultant expert of the European Energy Center and Centro Studi Galileo.

Mohammed Hamid received his PhD in Information and Communication Technology from KTH, The Royal Institute of Technology, Stockholm, Sweden in 2015. Thereafter, he held a post-doctoral researcher position in the Intelligent Signal Processing & Wireless Networks lab (WISENET) at University of Agder, Grimstad, Norway. In October 2017 he joined Ericsson where he is currently working as a system designer and researcher dealing with different radio aspects in 5G networks. Dr Hamid's research areas have been dynamic spectrum access and cognitive radio, signal detection techniques, spectrum cartography, quantitative analysis of spectrum opportunities, cellular traffic modeling, and radio-near algorithms for AAS, and he has published his research findings in highly reputed conferences, journals and as book chapters in the areas of communications and signal processing. Dr Hamid has also been an active reviewer for many IEEE and non-IEEE transactions and a TPC member for many conferences and symposia. Dr Hamid has strong teaching experience in both wireless communications major and non-major courses including radio communication, digital communication, statistical signal processing, and convex optimization.

Shoaib Amin received an MSc degree in Electronics from University of Gävle, Sweden in 2011 and a PhD degree in Electrical Engineering in 2017 from KTH Royal Institute of Technology, Sweden in 2017. Since 2017, he has been working as a senior RF and signal processing engineer at Qamcom Research and Technology and as Radio Design Consultant with Ericsson. His main research interests are in the field of nonlinear high-frequency systems and measurement technique.

Dr Frank Slyne received his BEng in Electrical Engineering from University College Cork in 1988 and MEng in Electronic Engineering from Dublin City University in 1994. He worked as an engineering manager in Eircom (Ireland) for over 20 years, with responsibility for ISP platforms and systems. He joined Professor Marco Ruffini's Optical Network Architecture research group in 2012 and completed his PhD in Software Defined Telecommunications Networks, with particular emphasis on LR-PON architectures. Dr Slyne's research interests include converged optical-wireless (LTE) metro-access architectures, disaggregation of legacy switch functions, and slicing and resource pooling through SDN Control and Network Function Virtualisation, Field Programmable Logic (FPGA).

Bhaskar Prasad Rimal received an MSc degree in Information Systems from the Kookmin University, Seoul, South Korea and a PhD degree in Telecommunications from the Institut National de la Recherche Scientifique (INRS), University of Québec, Montréal, QC, Canada. He is currently a Post-doctoral Fellow with the Electrical and Computer Engineering Department, University of New Mexico (UNM), Albuquerque, NM, USA. Dr Rimal was a Visiting Scholar with the Department of Computer Science, Carnegie Mellon University (CMU), Pittsburgh, PA, USA in 2016. His current research interests include edge/fog computing, cyber-physical systems (CPS), Internet of Things (IoT), and 5G and beyond. He has published over 40 papers in peer-reviewed journals and conference proceedings in the area of cloud and edge computing, FiWi enhanced networks, and IoT. Dr Rimal serves as an Associate Editor of IEEE Access, Associate Technical Editor of IEEE Communications Magazine, Associate Editor of EURASIP Journal on Wireless Communications and Networking (Springer), and an Editor of Internet Technology Letters (Wiley). He was a recipient of the Doctoral Research Scholarship from the Québec Merit Scholarship Program for foreign students of Fonds de Recherche du Québec-Nature et Technologies, the Korean Government Information

Technology Fellowship, the Kookmin University IT Scholarship, and the Kookmin Excellence Award as an Excellent Role Model Fellow. He is a Senior Member of the IEEE, Professional Member of the ACM, and Member of the OSA.

Akeem O. Mufutau holds a MEng Degree in Electrical and Electronics Engineering (Communication) from the Federal University of Technology, Akure (FUTA), Nigeria (2013) and a BEng degree in Electrical Engineering from the University of Ilorin, Nigeria (1994). In 2005, he joined the services of FUTA as a Senior Network Engineer in the Computer Resource Centre and rose to the position of Chief Network Engineer/Head of Systems & Network Engineering Unit before joining the Department of Electrical and Electronics Engineering, FUTA in 2015 as a Lecturer. He is presently a PhD student at the Instituto de Telecomunicações, Universidade de Aveiro, Portugal. His research interests include optical-wireless communications, mobile networks, radio frequency technologies, and computer networking. He is a COREN registered engineer, Chartered Information Technology Practitioner (C.itp), and member of the Computer Professionals [Registration Council] of Nigeria (CPN).

Fernando Guiomar received MSc and PhD degrees in Electronics and Telecommunications Engineering from University of Aveiro, Portugal in 2009 and 2015, respectively. In 2015, he received a Marie Skłodowska-Curie post-doctoral fellowship (two years) to work with the OptCom group of Politecnico di Torino. As a part of this fellowship, he also worked with CISCO Optical GmbH, Nuremberg for a period of five months. In 2017, he joined Instituto de Telecomunicações, Aveiro as a senior researcher responsible for the national research infrastructure ORCIP (www.orcip.pt), aimed at testing and prototyping of future 5G solutions involving novel optical and radio technologies. Fernando Guiomar has co-authored more than 70 scientific publications in leading international journals and conferences. In 2016, he received the Photonics21 Student Innovation Award, distinguishing industrial-oriented research with high impact in Europe.

Cátia Pinho has a degree in Physics Engineering and a Masters in Biomedical Engineering from University of Aveiro. She is currently working toward a PhD degree at the MAP-tele Doctoral Programme in Telecommunications, University of Aveiro, Portugal. Research experience includes optical communications (e.g. photonic integrated circuits, fiber optics and optical systems design and testing), biomedical signal processing, and development of health information technologies. Research/training in the field of optical communications has been undertaken at the Optical Communications and Photonic Integration (OCPI) group of the University of California, Santa Barbara (UCSB), the Centre for Advanced Photonics and Electronics (CAPE) – Electrical Engineering Division of the University of Cambridge, and the Photonic Research Labs – iTEAM Research Institute from the Universitat Politècnica de València (UPV).

Hind Dafallah is a RFIC designer at product development unit at Ericsson AB, Lund, Sweden, developing RF-ASIC for 5G. She received an MSc in Embedded Electronics Engineering from Lund University in 2016, with a focus on integrated radio electronics, and worked as a research assistant in Lund university in the Analog RF group, and a BSc in Electronics and Computer Engineering in 2011 from University of Khartoum, Sudan. Her research interest is the design and validation of mm-wave transceivers for 5G.

Pham Tien Dat received a BEng degree in Electronics and Telecommunication Engineering from the Posts and Telecommunications Institute of Technology, Vietnam in

2003, and MSc and PhD degrees in the Science of Global Information and Telecommunication Studies from Waseda University, Tokyo, Japan in 2008 and 2011, respectively. In 2011, he joined the National Institute of Information and Communications Technology, Japan. His research interests include the field of microwave/millimeter-wave photonics, radio-over-fiber, and optical wireless systems.

Atsushi Kanno received BSc, MSc, and PhD degrees in science from the University of Tsukuba, Tsukuba, Japan in 1999, 2001, and 2005, respectively. In 2005, he was with the Venture Business Laboratory, Institute of Science and Engineering, University of Tsukuba, where he was engaged in research on electron spin dynamics in semiconductor quantum-dot structures using the optical-polarization-sensitive Kerr effect measurement technique. In 2006, he joined the National Institute of Information and Communications Technology, Japan. From 2006 to 2007, he was also a member of the CRESTJST project, "Creation of Novel Functional Devices Using Nanoscale Spatial Structures of the Radiation Field". He is working on microwave/millimeterwave/terahertz photonics, ultrafast optical communication systems, lithium niobate optical modulators, and the study of ultrafast phenomena in semiconductor optical devices. He is a Member of the Institute of Electronics, Information and Communication Engineers, the Japan Society of Applied Physics, and the Laser Society of Japan.

Naokatsu Yamamoto received a PhD degree in electrical engineering from Tokyo Denki University, Tokyo, Japan in 2000. In April 2001, he joined the National Institute of Information and Communications Technology (NICT), Tokyo, Japan. He also joined Tokyo Denki University as a Visiting Professor from May 2008 and the Ministry of Internal Affairs and Communications as a Deputy Director from July 2012 to September 2013. Since 2016, he has been managing and currently the director of the Network Science and Convergence Device Technology Laboratory, NICT, and also has been the Director of Advanced ICT Device Laboratory. His research interests include a heterogeneous quantum dot laser with silicon photonics, and a convergence device technology of photonics and wireless. Recently, he has been interested in the use of a 1.0 μm waveband (thousand-band, T-band) as a new optical frequency band for short-range communications.

Tetsuya Kawanishi received BEng, MEng, and PhD degrees in Electronics from Kyoto University, Kyoto, Japan in 1992, 1994, and 1997, respectively. From 1994 to 1995, he was with the Production Engineering Laboratory, Panasonic. In 1997, he was with the Venture Business Laboratory, Kyoto University, where he was involved in research on electromagnetic scattering and near-field optics. In 1998, he joined the Communications Research Laboratory, Ministry of Posts and Telecommunications (now the National Institute of Information and Communications Technology), Tokyo, Japan, where he is currently a Research Executive Director. In 2004, he was a Visiting Scholar in the Department of Electrical and Computer Engineering, University of California at San Diego. In 2015, he joined the Department of Electronic and Physical Systems, Waseda University, as a Professor. His research interests include high-speed optical modulators and RF photonics. He is the Fellow of the Institute of Electrical and Electronics Engineers.

Amin Ebrahimzadeh received his BSc and MSc degrees in Electrical Engineering from the University of Tabriz in 2009 and 2011, respectively. From 2011 to 2015, he was with the Sahand University of Technology (SUT), Iran. He is currently pursuing his PhD at the Optical Zeitgeist Laboratory at the Institut National de la Recherche Scientifique (INRS), Montreal, QC, Canada. His research interests include fiber-wireless networks,

Tactile Internet, teleoperation, artificial intelligence enhanced mobile edge computing, and multi-robot task allocation. He is a recipient of the Doctoral Research Scholarship from the B2X Program for foreign students of Fonds de Recherche du Quebec-Nature et Technologies (FRQNT).

Mahfuzulhoq Chowdhury received his BSc degree in Computer Science and Engineering from Chittagong University of Engineering and Technology, Bangladesh in 2010. Afterwards, he received his MSc degree in Computer Science and Engineering from Chittagong University of Engineering and Technology in January 2015. He is currently a PhD student at INRS working on the Tactile Internet. His research interests include collaborative computing, fiber-wireless enhanced networks, human-to-robot communications, and the Tactile Internet.

Forough Yaghoubi received her BSc degree in Electrical Engineering and MSc degree in Communication Systems in 2010 and 2013, respectively, both from University of Tehran. She is currently working toward a PhD degree at the School of Electrical Engineering and Computer Science, KTH Royal Institute of Technology, Stockholm, Sweden. Her research is focused on software-defined wireless networking, microwave backhauling, cross-layer design, and reliability of future wireless networks.

Mozhgan Mahloo works as Consultant at ÅF company. She was an Experienced Researcher in the cloud technology area at Ericsson research until April 2018. She received a PhD degree in Communication Systems with specialization in optical networking from KTH Royal Institute of Technology, Sweden, in 2015. Her current research interest covers the cloud technology area with a focus on hardware acceleration, cost optimization, and reliability assessments of the future cloud platforms. Mozhgan has co-authored over 30 scientific papers and 8 patent applications.

Lena Wosinska received her PhD degree in Photonics and a Docent degree in Optical Networking from KTH Royal Institute of Technology, Sweden, where she is currently a Full Professor of Telecommunication in the School of Electrical Engineering and Computer Science (EECS). She is founder and leader of the Optical Networks Lab (ONLab). She has been working on several EU projects and coordinating a number of national and international research projects. Her research interests include fiber access and 5G transport networks, energy and cost efficiency, optical datacenter networks, photonics in switching, as well as optical network control, reliability, security, and survivability. She has been involved in many professional activities including serving on panels evaluating research project proposals for many funding agencies, guest editorship of IEEE, OSA, Elsevier and Springer journals, serving as General Chair and Co-Chair of several IEEE, OSA and SPIE conferences and workshops, serving in TPC of many conferences, as well as being reviewer for scientific journals and project proposals. She has been an Associate Editor of OSA Journal of Optical Networking and IEEE/OSA Journal of Optical Communications and Networking. Currently she is serving on the Editorial Board of Springer Photonic Networks Communication Journal and of Wiley Transactions on Emerging Telecommunications Technologies.

Paolo Monti received a Laurea degree in Electrical Engineering (2001) from the Politecnico di Torino, Italy, and a PhD in Electrical Engineering (2005) from the University of Texas at Dallas (UTD). From 2006 to 2008 he worked as a Research Associate at the Open Networking Advanced Research (OpNeAR) Lab at UTD. He joined KTH Royal Institute of Technology in September 2008, where he is currently an Associate Professor in the School of Electrical Engineering and Computer Science

(EECS) and the deputy director of the Optical Networks Laboratory (ONLab). From 2015 Dr Monti has been the Program Director of the Bachelor of Science in Information and Communication Technology (TCOMK). He has published more than 100 papers in peer-reviewed international journals and conferences. Dr Monti is serving on the editorial boards of the IEEE Transactions on Green Communications and Networking, IEEE Communications Letters, and IEEE Networking Letters journals. He has also served as lead and/or co-lead Guest Editor of a number of special issues focused on optical network design and energy efficiency in IEEE and Springer journals. Dr Monti regularly participates in several TPCs including IEEE Globecom and ICC where he also co-chaired one workshop on network survivability (at ICC 2012) and four workshops on green broadband access (at ICC 2013, Globecom 2014, ICC 2015, ICC 2016). Dr Monti was also the TPC chair of IEEE ONDM 2014 and served as a TPC co-chair for the Symposium on Optical and Grid Computing in ICNC 2014, 2016, and 2017, IEEE OnlineGreenComm 2016, HPSR 2017, and the ONS Symposium at the IEEE Globecom 2017. His main research interests are within the networking aspects of all-optical networks, currently with a focus on the architectural, technological, and sustainability challenges of 5G networks. Dr Monti is a Senior Member of IEEE.

Fabricio Farias received a PhD degree from the Universidade Federal do Pará (UFPA), Brazil, in 2016. He has been professor at the Faculty of Information Systems/UFPA-Cametá since 2014. His research interests include data mining, HetNets, mobile backhaul, digital TV, and DSL networks.

João Crisóstomo Weyl Albuquerque Costa received a PhD degree in Electrical Engineering and Telecommunications from the State University of Campinas (UNICAMP) in 1994. Currently he is professor at the Faculty of Computer and Telecommunication Engineering of the Institute of Technology. He is the author of more than 80 refereed papers in journals and co-author of two international patents. He has supervised 28 master students, 14 PhD thesis, and about 100 undergraduate students.

Jiajia Chen is working as Associate Professor at KTH Royal Institute of Technology, Sweden. She is author/co-author of over 100 publications in international journals and conferences. Her main research interests are optical transport and interconnect technology supporting future broadband access, 5G, and cloud environment. She has been involved in various European research projects and a principle investigator/co-principle investigator of several national research projects funded by Swedish Foundation of Strategic Research (SSF) and Swedish Research Council (VR). Currently, she is heading a research group on data center networks at Sino-Swedish Joint Research Centre of Photonics (JORCEP), which is the largest and most integrated academic collaboration between China and Sweden, founded in 2003.

Preface

Over the last few decades, the mobile market has experienced vast growth in the number of broadband consumers due to consumer trends targeting mobile internet services, being delivered by effective deployment of 3G and 4G legacy networks on a global scale. In essence, 4G mobile networks offer a true broadband solution with data throughputs exceeding gigabit data rates, but their limitations are starting to become apparent in terms of pushing further the boundaries on delivering high-speed connectivity. Although multiple antenna technologies and small cell deployments have been proven to provide significant gains in terms of spectral and area efficiency, they do not go far enough towards targeting the performance indicators being defined by future market trends; it is clear that there is a need to seek new spectral opportunities to accommodate demand. The market is now heading towards 5G communications, which at first was driven by the need for enhanced broadband connectivity to secure gigabit wireless on the move, but also new market applications are emerging such the internet of everything and tactile internet. This has led to the 5G vision that aims to significantly improve on legacy systems to include higher speed (up to 10 Gbps) communication links, to provide almost zero latency networks, and flexible networking to cater for heterogeneous traffic types. There seems to be a consensus among fixed and mobile stakeholders that no single technology or architecture will be able to meet all the requirements mentioned above. In conclusion, it is clear that an integrated wireless communication system will emerge to harness legacy systems and technologies, as well as integrating the future emerging new radio technologies to meet the exceptional targets imposed by the global standardization committees such as the 3rd Generation Partnership Project (3GPP), the 5G Infrastructure Public-Private Partnership (5G-PPP), and the International Telecommunication Union (ITU), which reflect market needs.

In order for 5G to cater for the foreseen demand in data rate, optical communications are widely accepted as a way forward. Today's mobile systems and wide area networks already work in synergy with mobile systems to provide both fronthauling and backhauling capability, and in particular its use in Cloud Radio Access Networks (C-RAN) provides a vehicle for centralized radio resource management providing clear benefits in terms of interference management. Therefore, it is clear that any integrated 5G system will build on the current optical core to route data in a cost-effective manner, raising highly significant challenges in how wireless can coexist with optical to provide a seamless end-to-end communication path.

This book will target enabling technologies for the 5G networks and services, and in particular offer new insight into the so-called ubiquitous and agile 5G-core system, managing a technological ecosystem of mixed and multiple access network communications with a key focus on the optical-wireless convergence. The use-cases to be considered include indoor communications such as fiber, fiber-wireless (FiWi), visible light communication (VLC), and device-to-device communication. In addition, this book will also consider outdoor communications in crowded metropolitan areas (e.g. new 5G RAN). Moreover, state-of-the-art services and technology enablers like tactile internet, fog computing (FC), network functions virtualization (NFV), software-defined networking (SDN), and optical fiber based sensors will be introduced. Finally, the techno-economic framework and associated impact will be presented considering all networking actors.

Acknowledgments

The editors would like to thank not only the collaborators who have contributed chapters to this book but also the Mobile Systems Research Group and Optical Communications Group at the Instituto de Telecomunicações, Aveiro, which have provided valuable comments and contributions towards its compilation. The editors would specifically like to acknowledge the H2020-ITN-2016-722424-SECRET project, which provided the inspiration for Chapter 7, whilst a special mention goes to the OCEAN12(H2020-ECSEL-2017-1-783127) and FCT and the ENIAC JU (THINGS2DO – GA n. 621221) projects that provided contributions towards Chapters 1, 5, 7, and 11, respectively.

Introduction

Over the last few decades, the prolific innovation in telecommunication agnostic systems and their integration with optical technologies in terms of mobile and fixed networking have been driving social and industrial transformations. The evolution of the mobile generation networks from the first-generation (1G) to the more recent fourth-generation (4G) has progressed at an unprecedented rate, to cater for consumer market demands for ever increasing higher speed connectivity, network capacity, reliability, security, low latency, and very high energy efficiency. The successive networking upgrades have been implemented on a piecemeal basis, each time building on legacy technologies; even though new market requirements emerged, it was always important for vendors to ensure backward compatibility for their products in order to minimize the return on investment (RoI) time for mobile operators in introducing new technologies.

In order to understand the motivation for this book, it is necessary to briefly revisit the evolution of mobile communications and where we are today in terms of network connectivity. In brief, the evolution of cellular networks was initiated by the first-generation (1G) that was launched during the 1980s with a maximum speed of about 2.4 kilobits per second (kbps). It was implemented using circuit switching technology that was based on the analog paradigm. The network was engineered to support only voice services. The second-generation (2G) cellular mobile technology, which was introduced in Finland around 1991, was the first significant evolution that offered early internet access. It has three sub-generations, 2G, 2.5G, and 2.75G, and was driven and standardized by the global system for mobile communications (GSM), Code Division Multiple Access (CDMA) IS-95, and other interim standards according to international telecommunication union (ITU)[1]. Specifically, the 2G defined a digital radio signal in conformity with circuit-switched network optimized for full duplex voice telephony, which supported the earliest text messaging service currently known as short message service (SMS); the 2.5 subgeneration entry introduces the first packet data transport in compliance with general packet radio services (GPRS) and coexisted with the circuit switching technique, and finally, the 2.75 presents enhanced data rates for GSM evolution (EDGE). The 2G posed a maximum transfer speed of 50 Kbps with the first phase deployment and 1 megabit per second (Mbps) for GPRS. Services such as SMS, speech encoding, billing

---

1 ITU is one of the United Nations agencies, which is in charge of information and communication technologies (ICTs) by means of assigning the global radio spectrum and satellite orbits, building the technical standards that guarantee networks and technologies ideally interconnect, and compete to enhance access to ICTs to worldwide.

dependent services, and internal roaming innovated by this network generation are still valid today. Moreover, the 2G networks are still running in many countries around the world. Similar to 2G, the subsequent 3G standard evolved toward 3.5G, 3.75G, and 3.9G, each subentry upgrading the data connection speed and introducing new networking service features to provide more VAS (value added services) for the end users. The 3G was the first cellular broadband network to provide high-speed internet access with a minimum speed of 144 kbps. This rate was made possible due to the seamless integration of wireless backhaul over an optical access network, as well as improvements in signal quality, spectral efficiency, and related technical aspects. The 3G was driven by the 3rd Generation Partnership Project (3GPP)[2], which built the Universal Mobile Telecommunications Service (UMTS) system standard of the 3G and was subsequently approved by the ITU. The generation backed up numerous services and applications via smart devices and flexible software. Early examples of 3G improvements include, but are not limited to, the global positioning system (GPS), location-based services, mobile TV, telemedicine, video conferencing, and video on demand, as well as higher security, reliability, and high capacity starting from 384 kbps for Wide CDMA (WCDMA) up to 42 Mbps, which complies with the evolved high-speed packet access (HSPA+) standard from 3GPP. In 2008, the ITU-radio-communication sector (ITU-R) adopted recommendations with other forums such as the aforementioned 3GPP group, Institute of Electrical and Electronics Engineers (IEEE) standards association (IEEE-SA)[3], and the worldwide interoperability for microwave access (WiMAX) Forum[4] to define the International mobile telecommunications-advanced (IMT-Advanced) standard for 4G networks.

The 4G standard has been developed to provide mobile ultra-broadband access and to ensure interoperability within a heterogeneous networking environment. The 4G standard is known as "long-term evolution" (LTE) that was certified by Release 8 and 9 (Rel-8) (Rel-9) standards. The LTE standard is based on orthogonal frequency division multiple access (OFDMA) for the downlink (DL), and single carrier–frequency division multiple access (SC-FDMA) designated to the uplink (UL), together with flexibility in frequency and higher modulation formats. The key benefit of LTE is the use of the OFDM (orthogonal frequency division multiplexing) modulation format that can control and mitigate the effect of inter-symbol interference substantially due to the wideband channel effect, as well as provide additional bandwidth in contrast to 3G. The peak data rate for LTE in the download and upload speed was up to 100 Mbps and 50 Mbps,

---

2 3GPP is an association body of telecommunications standards. The first aim of 3GPP was to build a universally suitable 3G mobile phone system specification dependent upon evolved GSM specifications that are compatible with the International Mobile Telecommunications-2000 project of the ITU. After the first mission was over, 3GPP continue to develop and determine the specification and requirements, and structure them into standards for future mobile wireless networks to support an interoperability amongst heterogeneous technologies.
3 IEEE-SA is an international plethora standardization body that contains numerous organizations and individuals from different technical backgrounds and country leaders worldwide with consensus for collaboration amongst them to build, enhance, and improve universal technologies via IEEE to enable standards growth and standards within the forums.
4 The WiMAX Forum is an international association authorized to give accreditation that accomplishes worldwide interoperability, advance technical requirements dependent upon open standards, work at a good regulatory environment, and support the vision.

respectively. In a bid to increase the data rates even further, the LTE-A (LTE-advanced) standard evolved building on significant upgrades to the 4G baseline.

The LTE-A main specifications were standardized by Rel-10, Rel-11, and Rel-12, and include significant new technologies such as carrier aggregation, multiple inputs multiple outputs (MIMO) or spatial multiplexing, relay nodes, and coordinated multipoint operation (CoMP). Each of these technologies aims to enhance the coverage capacity of the network along the cell profile/diameter by enhancing the peak data rate at the cell center or enhancing coverage at the cell edge. Recently, 3GPP has launched LTE-Advanced Pro (LTE-A Pro) through Release 13, and 14, (Rel-13) and (Rel-14), distinguished as 4.5G, 4.9G, and Pre-5G that is paving the way towards 5G networks. LTE-A Pro has many mature specifications that were included in the previous releases of LTE and LTE-A, as well as integrating new requirement and use cases. These new features include, but are not limited to, interworking with wireless local area networking (WLAN) that complies with the traditional Wi-Fi IEEE 802.11 series (and new Wi-Fi IEEE 802.11ah standards), carrier aggregation enhancements up to 32 component carriers (CCs) (i.e. massive MIMO), machine-type communication (MTC) enhancements, Internet of Things (IoT), public safety features, device to device (D2D) communication and proximity services (ProSe), full-duplex small cell and layout, enhanced LTE-WLAN aggregation (eLWA); licensed assisted access (LAA) (at 5 GHz), MulteFire[5], indoor positioning, 3D space/full dimension (FD) FD-MIMO, cellular vehicle-to-everything (C-V2X), and pursue opportunities for various vertical use cases (e.g. e-industry, e-health, entertainment, automotive). The inspiration behind the LAA, eLAW, MulteFire standards is to allow the LTE-A Pro mobile terminals that traditionally use licensed spectrum to opportunistically exploit the unlicensed spectrum in WLAN to enhance the coverage/capacity of the 4G network.

It is widely accepted that these aforementioned specifications have been achieved because of the synergies of different stakeholders coming together to provide interdisciplinary, integrated solutions to serve emerging vertical markets. Once again, in 2012 the ITU and its partners agreed on the IMT-2020 recommendation that gives the initial vision of the IMT (international mobile telecommunications) for 2020 and beyond. This is in an effort to identify the specifications and requirements that will facilitate development and the adoption of 5G technologies. The smooth migration toward the first phase of the 5G era will enrich the specifications and characteristics of legacy 4G networks and technologies. The significant flexibility improvements for agnostic end-to-end (E2E) architectures will be a quantum evolutionary leap for the IMT-2020/5G networks in contrast to 4G networks. The key enabler for 5G will be softwarization providing a new dimension to the way we manage mobile networks, and facilitating new services to provide more effective end-to-end service delivery. Thanks to methods such as network function virtualization (NFV), and software-defined networking (SDN), this will provide the platform for more flexibility and programmability of the mobile network system that empowers many new capabilities that automatically runs over the next generation networks such as network slicing, among others. This new paradigm will also introduce new business opportunities for third party service providers, such as virtual

---

5 MulteFire[TM] is an LTE-based technology that was built by MulteFire Alliances and runs as a separate entity in an unlicensed and shared spectrum; targetting the global 5 GHz band. According to 3GPP standards stated in Releases 13 and 14, MulteFire technology backs up Listen-Before-Talk for equitable existence with Wi-Fi and other technologies running in a similar spectrum.

mobile network operators (VMNOs), which can acquire networking infrastructure on demand.

The primary applications for IMT-2020/5G include massive machine type connectivity (MTC) to cater for IoT smart verticals by providing access to a vast number of MTCs, enhanced mobile broadband (eMBB) to facilitate and offer excellent viewing experience and high data rate connectivity for interactive media such as 4K or 8K ultra-high-definition (UHD) videos, virtual reality/augmented reality (VR/AR), among others, and ultra-reliable and low-latency communication (URLLC) to support critical applications and services that demand minimum round trip time responses, e.g., fog computing, autonomous driving cars, Tactile Internet, and traffic control. At the time of writing, the global stakeholder bodies like ITU, 5G Infrastructure Public Private Partnership (5G-PPP), 3GPP, the BroadBand Forum (BBF), GSM Association (GSMA), and Next Generation Mobile Networks (NGMN) are still conducting technology field trials and are in the process of defining the standards with a first phase – Release 16 (Rel-16) – scheduled for 2020. However, it can be said with certainty that the 5G standard will include a new 5G Core (5GC) and a 5G New Radio (NR) wireless interface to provide high speed connectivity in the millimeter-wave frequency range. Moreover, the 3GPP group proposes seven scenarios for 5G networks deployments, which include provisioning for a smooth migration phase that has 5G as a standalone deployment and non-standalone in synergy with a legacy evolved packet core (EPC). It is clear that to fulfill the overall network requirements and specifications for next generations networks they should include a converged wireless and optical networking infrastructure. Such convergence will give the MNOs benefits such as reduced capital and operational expenditure (CAPEX and OPEX), as well as high scalability and flexibility.

This book aims to provide the readership with fascinating insights into the 5G era and the role of optical networking as part of a converged wireless-optical networking infrastructure. Mainly, it will focus on a holistic vision of the architectural and technological developments, as well as requirements for the 5G optical wireless convergence. Moreover, we present many intelligent, dynamic, and automatic protocols to enable solutions for real-time IoT applications, and Tactile Internet or internet of skills. In this context, we predict that processing, storage, and interconnecting optical systems will progressively converge towards a unified platform utilizing completely interoperable hardware that will be accomplished via a group of distributed autonomous controllers. This context will allow instantaneous reception of chronological events. This vision will be based on automatic response to dynamically varying services and client requests. The book is underpinned by industrial and academic research contributing to thirteen themed chapters on this topic.

Chapter 1 provides network deployment solutions for 5G and beyond 5G (B5G) networks according to the targeted IMT-2020 applications. It presents 5G transport networks in which cross (X) haul (i.e. integrated fronthaul/backhaul) and routing solutions are considered. Moreover, this chapter elaborates the 5G requirements and challenges that include how to attain flexibility, scalability, and high throughput that facilitate energy, as well as cost-efficient, converged optical and wireless 5G and B5G networks. Moreover, this chapter also presents several advanced technologies like digital/analog radio-over-fiber (D-RoF/A-RoF), optical spatial-division multiplexing (SDM), hybrid radio frequency (RF)/free space optical (FSO), and relay-assisted FSO transmission that are intended to be integrated into 5G transport networks. In Chapter 2, the "hybrid fiber-wireless (HFW) extension for GPON toward 5G" is introduced, an integrated photonics transceiver based on a coherent RoF (C-RoF) approach that is implemented with E2E connections toward 5G mobile networks, and tested in a real-life field trial. This is effort geared towards a seamless extension for attaining gigabit passive optical networks (GPON) systems with downlink and uplink transmission line rates of 2.5 Gbit s^{-1} and 1.25 Gbit s^{-1}, respectively. The presented C-RoF architecture thus represents the physical layer of a converged fiber and optical networks, eventually leading to an integrated hybrid fiber wireless (HFW) network. The developed components have been used to build radio access units (RAUs) for seamless photonic-to-RF and vice versa conversion. The system achieved good performance metrics that includes jitter less than 0.0015 ms and low latency RTT of around 1.1 ms, which is aligned with the stringent specifications of IMT-2020. In the field trial, which took place in Garwolin, Poland, the RAUs were used to extend the coverage of a commercial GPON network operated by Orange Polska through a wireless E-band link. Furthermore, key enabling photonic and electronic technologies developed within the system above for 5G networks have been intensely discussed. Chapter 3, entitled "software defined networking (SDN) and network function virtualization (NFV) for converged access-metro networks", describes how new 5G requirements are shaping the architectural design process, through the convergence of both data and control planes of the access and metro networks. The content of the chapter doesn't only focus on the access-metro network, but also emphasizes the importance of integration with the mobile and cloud edge ecosystems. Moreover, this chapter discusses, from a physical layer perspective, solutions to reduce further the network latency bounds, that includes the cloud-radio access network (C-RAN), and how a wavelength-switching technique in the optical access-metro network dynamically provides the capacity to 5G mobile base stations from different computation centers in the metro area. The services mentioned above have been further supported by network functions virtualization, which allows the functional convergence of multiple and diverse network domains. Finally, the chapter provides an insight into the virtualization of the central office, exploring some of the main software frameworks being developed across the globe.

Multicore fibers (MCFs), a promising solution for future 5G fronthaul, is discussed in Chapter 4. First, an inspirational idea behind these technologies that implement homogeneous and heterogeneous types of the MCFs, which are built by the SDM approach, is reviewed and discussed. The chapter offers a fiber availability solution via the use of the proposed MCFs to support C-RAN architectures. The proposed MCF-based C-RAN solutions are potentially flexible enough to support both digital and analog RoF architectures, capacity upgrades by carrier aggregation, massive MIMO, as well as accurate cloud operation to secure IMT-2020 requirements. More importantly, the spatial diversity inherent to MCF transmission is capable of enabling the abovementioned features by employing electronic spatial switching placed at the shared central office (CO), allowing SDN as well as NFV. In addition, MCF-based C-RAN configurations will be compatible with wavelength-division multiplexing (WDM) and will support the PON overlay. Moreover, the microwave signal processing enabled by multicore fibers over both MCFs mentioned above have been intensely discussed. Chapter 5, which is entitled "enabling VLC and Wi-Fi network technologies and architectures towards 5G", provides an overview of optical wireless communication (OWC) technologies regarding visible light communication (VLC) systems. Moreover, the VLC system principles, technologies, architectures, applications, and challenges have been compared with that of the RF technology counterpart. Furthermore, in addition to purely application scenarios for VLC systems, hybrid VLC and RF solutions are also considered for 5G wireless communications to exploit both the VLC high-speed data transmission and illumination, as well as ubiquitous coverage of the RF.

The key RAN radio-enabling technologies, and their implementations that could be featured as candidate solutions for 5G NR considering fixed wireless access (FWA) convergence for eMBB use cases, are presented in Chapter 6. The authors believe that the main distinction of this chapter is combining the theoretical and hardware aspects of the 5G NR key technologies, which, to the best of authors' knowledge, have mostly been considered separately. Therefore, the chapter concisely explains how hardware implementation and the associated impairments can limit the performance of 5G NR technology. Furthermore, the chapter discusses mitigation techniques for some of these impairments and hardware design challenges. Millimeter-wave (mmW) broadband monopole antennas within the spectrum 30 GHz to 45 GHz have been developed in Chapter 7 for future 5G mobile network applications. The proposed antennas demonstrate promising performances regarding return loss, power gain, current surfaces, and efficiency that ultimately characterizes them as attractive solutions for future 5G wireless systems. Moreover, the chapter describes a family of future antennas for 5G wireless mobile communication.

A promising solution for low-latency and flexible fronthaul for 5G heterogeneous cellular networks is discussed in Chapter 8. In particular, the encapsulation of wireless signals on a seamless fiber-mmW system in the DL and UL directions is experimentally demonstrated and tested. The system has been shown to be scalable to meet future deployment demands concerning capacity, multiple radios, and the number of wireless services. Moreover, advanced technologies, such as wavelength division multiplexing in optical systems, arrayed antenna, beam forming, and multi-hop transmission in wireless communication can be exploited to increase the system capacity and range. Furthermore, different methods and the corresponding transmission impairments on the signal performance are discussed.

Security will become one of the most significant challenges in 5G networks, in particular one that incorporates IoT devices. Therefore, optical fiber based sensors with particular attention to optical fiber network physical layer security is of paramount interest. In this context, Chapter 9 addresses fiber optic sensors used in network security in terms of potential application and services. Different types of fiber optic sensors such as phase-modulated sensors, intensity-modulated sensors and scattering-based sensors are presented with their application to intrusion detection systems.

Chapter 10 addresses how the Tactile Internet can create value for society and human well-being. Unlike the IoT, without any human involvement in its underlying machine-to-machine communications, the Tactile Internet aims to involve the inherent human-in-the-loop (HITL) nature of the haptic interaction, thus allowing for a human-centric design approach towards creating new immersive experiences via the internet. Furthermore, the chapter shows the design of reliable fiber-wireless (FiWi) enhanced LTE-A heterogeneous networks cases (HetNets) with artificial intelligence (AI) embedded multi-access edge computing (MEC) capabilities to meet the quality of experience (QoE) requirements of HITL-centric local and non-local teleoperation regarding low latency performance of about 1–10 ms. Moreover, the chapter introduces the developed human-agent-robot teamwork (HART)-centric multi-robot task allocation strategy, which relies on the proposed unified resource allocation in FiWi based tactile internet infrastructures employing suitable host robot selection and computation task offloading onto collaborative nodes. The proposed collaborative computing based human-to-robot (H2R) task allocation and resource management scheme prove instrumental in enabling future low-latency multi-robot tactile internet applications.

Opportunities and challenges of energy efficiency (EE) in C-RAN for 5G mobile networks are presented in Chapter 11. The chapter focuses on the associated opportunities of MIMO and mmWave implementations in C-RAN, including a comprehensive review and discussion on techniques that may help in enhancing the 5G EE gains. Different economical schemes for supplying the required power to the network entities are discussed, with the aim of providing ubiquitous network connectivity. Moreover, different means of achieving the goal of green wireless communications and networking through the reduction in power consumption by the system are considered. It also presents a comprehensive review and discussion on techniques that not only aid in enhancing the 5G EE gains but also help in reducing the OPEX and carbon dioxide (CO_2) emissions in green wireless cellular networks. The chapter further presents analytical expressions and numerical results that show that the 5G network EE can be improved considerably.

The fog computing fiber-wireless (FC-FiWi) networks based on the integration of ethernet PON (EPON), WLAN, and fog computing for the creation of multi-purpose assets are introduced in Chapter 12. Different protection schemes have been proposed, to evaluate how the FiWi and fog computing segments will be best adapted to their surroundings and continue to survive. An analytical framework to examine the packet delay performance of both FiWi and cloud traffic has been developed. The obtained results reveal the effectiveness of implementing fog computing in FiWi networks. More specifically, the proposed scheme offers low E2E offload delay, and high reliability at the edge of the FC-FiWi network without negatively influencing the network performance of FiWi broadband traffic. The obtained results reveal that for a typical scenario of 32 ONU-APs/MPPs and a FiWi traffic load of 0.6, a fog and cloud packet delay of 33.41 ms and 133.41 ms can be obtained, respectively, without degrading the network performance for FiWi traffic. Furthermore, an experimental FC-FiWi network testbed is presented. The obtained results demonstrate the feasibility and effectiveness of fog computing enhanced FiWi networks to achieve low delay and improved reliability performance.

Finally, to complete the picture of the book, a techno-economic framework for 5G network actors is presented in Chapter 13. The chapter presents a model that provides a complete market analysis of the various business actors, for any type of mobile access network deployment that includes both homogeneous and HetNets. The framework is applied to a case study, where two options for the backhaul have been considered (i.e. microwave and fiber) in a wireless network scenario that caters for both homogeneous (i.e. macro only) and heterogeneous deployments. The case study shows that fiber is the most cost-efficient technology to provide a high capacity backhaul for heterogeneous wireless deployments in dense areas. Regarding profitability, the net present value (NPV) results show that a lower total cost of ownership (TCO) does not necessarily lead to higher profit.

1

Towards a Converged Optical-Wireless Fronthaul/Backhaul Solution for 5G Networks and Beyond

Isiaka Ajewale Alimi[*,1], *Nelson Jesus Muga*[1], *Abdelgader M. Abdalla*[1], *Cátia Pinho*[1], *Jonathan Rodriguez*[1,2], *Paulo Pereira Monteiro*[1,3], *and Antonio Luís Teixeira*[1,3]

[1] *Instituto de Telecomunicações, 3810-193, Aveiro, Portugal*
[2] *University of South Wales, Pontypridd, Wales, UK*
[3] *Department of Electronics, Telecommunications and Informatics (DETI), Universidade de Aveiro, 3810-193, Aveiro, Portugal*

1.1 Introduction

Recently, there has been a high increase in cellular wireless traffic. The traffic surge can be attributed to the exceptional proliferation of smart mobile devices as well as perpetual deployment of bandwidth intensive applications and services. To attend to the network demands, radio access networks (RANs) have been revolutionized from one generation of network to the other. The fifth generation (5G) wireless networks are envisaged to be the viable solutions for meeting the increasing network demands by offering ultra-reliable low-latency communications (URLLC). There are different aspects of the system that need consideration in order to make future wireless communications feasible. One such is the link spectrum efficiency (SE) in which the bandwidth of point-to-point radio frequency (RF) transmissions have been exploited through the implementation of multiple-input and multiple-output (MIMO) techniques, high-level modulation formats, and advanced channel coding schemes such as Turbo and low-density parity-check (LDPC) codes. Nevertheless, the system SE can be further improved by employing schemes such as cognitive radio in order to dynamically exploit the underutilized or "white space" spectrum [Liu et al. - 2014; Alimi et al. - 2017c, b, 2018a, 2017d].

Furthermore, coordinated multipoint (CoMP) transmission can be employed for adjacent cell coordination so that they can jointly transmit signals to cell-edge users. The implementation can considerably improve the system SE by alleviating inter-cell interference and enhancing the data rate of the edge users. In addition, apart from the limited, heavily regulated, and congested RF in the lower-frequency spectrum, there are vast and unexploited spectra at higher RF bands such as the millimeter-wave (mm-wave) band [Liu et al. - 2014; Chang et al. - 2013]. Apart from the fact that the bands are capable of supporting multi-gigabit wireless transmission, the use of

* Corresponding Author:Isiaka Ajewale Alimi iaalimi@ua.pt

Optical and Wireless Convergence for 5G Networks, First Edition.
Edited by Abdelgader M. Abdalla, Jonathan Rodriguez, Issa Elfergani, and Antonio Teixeira.
© 2020 John Wiley & Sons Ltd. Published 2020 by John Wiley & Sons Ltd.

complicated and time-consuming modulation and coding techniques are not necessary [Chang et al. - 2013]. Moreover, small-cell concepts have also been considered for improving the system performance, mainly in terms of capacity. The concept is based on reducing the cell size in order to reuse limited spectral resources among cells. Nonetheless, apart from inter-cell interference management, which is really challenging in small-cell implementation, capital expenditure (CAPEX) and operating expenses (OPEX) are also demanding [Liu et al. - 2014; Chang et al. - 2013].

The chapter is organized as follows. Section 1.2 discusses the cellular network interface and solution. In Section 1.3, we present a comprehensive discussion on the 5G enabling technologies. We introduce the concept of fiber-wireless network convergence in Section 1.4. In Section 1.5, we present a broad explanation of the radio-over-fiber transmission scheme. Optical transport network multiplexing schemes are detailed in Section 1.6. Wireless based transport networks are considered in Section 1.7. Experimental channel measurement and characterization are presented in Section 1.8. The obtained simulation and experimental results as well as comprehensive discussions are presented in Section 1.9 and concluding remarks are given in Section 1.10.

1.2 Cellular Network Interface and Solution

This chapter primarily focuses on centralized/cloud RAN (C-RAN) architectures due to their salient advantages such as cost-effectiveness, efficient centralized processing, better service provisioning, support for dynamic resource allocation and mobile traffic load balancing. So, in this section, we consider C-RAN interfaces such as mobile backhaul (MBH) and mobile fronthaul (MFH). Moreover, more attention is paid to the MFH interface because of the stringent requirements that it presents to current and future networks. To achieve this, we discuss the associated radio interface that can be employed for the in-phase and quadrature (IQ) data transmission between central units (CUs) and distributed units (DUs).

1.2.1 MBH/MFH Architecture

This subsection presents the concepts of MBH and MFH network connectivity segments for a C-RAN.

1.2.1.1 Mobile Backhaul (MBH)

In a C-RAN, the core network is generally connected to the base stations (BSs) through the MBH. The connection can be typically realized by means of an IP/ethernet based network. Consider the long term evolution (LTE) network for instance: the evolved node Bs (eNBs) are connected to one another via X2 interfaces. Moreover, S1 interfaces connect the eNBs to the evolved packet core (EPC). Furthermore, the eNBs are connected to the mobility management entity (MME) and serving gateway (S-GW) through the S1-MME and S1-U interfaces, respectively. It is worth mentioning that the S1 is a logical interface that has the ability to maintain many-to-many relations among MMEs/S-GWs and eNBs. In general, the MBH is the transmission medium for the user data as well as control and management data between the EPC and eNB. In addition, it is the transmission medium for handover and coordination signals between the eNBs [Alimi et al. - 2018a].

1.2.1.2 Mobile Fronthaul (MFH)

The MFH-based network is related to a centralized architecture in which the baseband processing functions are centralized so that not only remote antenna unit (RAU) operation can be simplified but also an enhanced cooperation can be realized among the BSs. In general, the fiber-optic-based MFH can be grouped into analog MFH and digital MFH.

Signal transmission between the network elements such as baseband units (BBUs) and remote radio heads (RRHs) is mainly based on digital radio over fiber (D-RoF) technology. This is in an effort to ensure waveform transparency as well as cost-effectiveness. Furthermore, the connection between the BBUs and RRHs are mainly based on the common public radio interface (CPRI). However, with URLLC, enhanced mobile broadband (eMBB), and massive machine-type communications (mMTCs) that are envisaged by the 5G RANs, there is a considerable need for a diverse spectrum and carrier aggregation (CA) of radio carriers to enhance the system capacity, throughputs, and efficiency. In addition, mm-wave and massive MIMO (M-MIMO) antenna are promising technologies that will be integrated into both standalone and non-standalone versions of 5G new radio for a significant network capacity improvement. However, implementation of these technologies give rise to enormous capacity requirements in the MFH network.

The exigent bandwidth required by the MFH network, as discussed in subsection 1.3.2, is really difficult to be addressed by the CPRI-based fronthaul solution. Therefore, alternative solutions are required for an effective and viable MFH. One such solution is analog radio-over-fiber (A-RoF) transmission technology [Alimi et al. - 2018a]. The associated advantages of A-RoF implementation are discussed in Section 1.5. In the following subsections, we discuss the ideas of MBH/MFH transport network integration.

1.2.2 Integrated MBH/MFH Transport Network

As stated earlier, due to the high-cost of a limited radio spectrum and the growing demand for mobile broadband capacity, radio features require disruptive infrastructural change and redesign. So, for effective network redesign, consolidated BS schemes such as C-RAN can be adopted. Moreover, adoption of C-RAN systems aids in the implementation of schemes such as CoMP and CA, which can be of great help in radio coverage expansion and optimization. Furthermore, since C-RAN is based on decoupling the BBUs and remote radio units (RRUs), the link between them is no longer a mere one-to-one connection because of the increase in distance as well as consequential transmission latency that may significantly impact radio performance [Eri - 2018a]. The subsequent network link is known as a fronthaul transport network. The fronthaul networks are mainly based on CPRI for the baseband samples distribution. It is remarkable that, besides the Open Base Station Architecture Initiative (OBSAI), enhanced CPRI (eCPRI) can as well be used for connecting eCPRI radio equipment and eCPRI radio equipment control [Parties - 2017]. The major function of the fronthaul is to ensure that the BBUs connect seamlessly to the RRUs without impacting radio performance. To achieve this, certain stringent radio requirements have to be met in cost-efficient ways. Moreover, in the LTE C-RAN architectures, a backhaul transport network signifies the internet protocol (IP) network from the centralized BBUs to the EPC [Eri - 2018a].

It is envisaged that the 5G transport networks will employ heterogeneous data plane technologies for supporting both fronthaul and backhaul traffic. So, there have been different innovative integrated fronthaul/backhaul transport networks such as 5G-Crosshaul [Dei et al. - 2016; Xhaul et al. - 2018; Anyhaul Nok - 2017]. The transport networks are majorly software-controlled for flexible and efficient management of the fluctuating nature of the new generation network capacity demand. A way of achieving this is by employing a common frame format and forwarding abstraction for information exchange in the network [Dei et al. - 2016].

1.3 5G Enabling Technologies

The International Mobile Telecommunication 2020 (IMT-2020) envisioned 5G application scenarios/use cases towards unprecedented mobile broadband communications considering high data rate, ultra-low latency as well as ubiquitous access are URLLC, eMBB, and mMTC. There are a number of enabling technologies that can be employed for effective realization of the 5G concepts in order to support the use cases. In general, there will be tremendous need for ultra-dense deployment of small cell and M-MIMO with the integration of high-data-rate MFH networks based on mm-wave frequencies. Apart from being license-free or light licensed, mm-wave enables higher frequency reuse. Also, C-RAN architecture can also be employed for better radio resource management and coordination so as to mitigate inter-cell interference. So, in this section, fundamental concepts and technologies such as ultra-densification, mm-wave small cells, M-MIMO, advanced radio coordination, C-RAN and RAN virtualization, as well as optical-wireless convergence are presented.

1.3.1 Ultra-Densification

In 5G and B5G networks, there will be huge demand for ultra-dense deployment of small cells. The major concept of ultra-densification is to enhance the RAN capacity by means of dense deployment of low-power and low-cost small cells for both indoor and outdoor scenarios. This can be achieved by overlaying small cells on the conventional macro cells that are primarily deployed for coverage purposes. Consequently, the transport networks are facing huge increase in the number of connected network elements. The growing connections not only increase the transport network complexity concerning cost but also regarding power consumption. Therefore, there is a need for further efforts to achieve more energy-efficient and cost-effective transport solutions for the 5G and B5G networks [Fiorani et al. - 2015].

1.3.2 C-RAN and RAN Virtualization

Conceptually, C-RAN is based on separation of conventional cell sites digital BBUs from the mostly analog RAUs/RRHs. Moreover, for centralized signal processing and management, the BBUs are redeployed to the "cloud" (BBU pool). This decoupling simplifies the usually complicated conventional cell sites, consequently enabling the implementation of power-efficient and cost-effective RRHs. This also helps in reducing the environmental effects and CAPEX for deploying a massive small-cell system by helping in reducing the footprint and cooling/power requirements at the cell sites. Furthermore, with centralized processing, network management can be simplified considerably. Also, advanced and efficient coordination among cells can be achieved to enhance the system performance and reduce the OPEX [Liu et al. - 2014].

There are different measures that can be employed in achieving the cloud-RAN objectives. BBU centralization or BBU hoteling can be implemented. In this scheme, several independent BBUs running on specialized hardware (HW) are co-located at the central office (CO). Moreover, BBU pooling is another scheme that can be employed. In this scheme, dedicated HW resources are shared between the co-located BBUs. In addition, a "cloud" RAN platform can also be employed. In this scheme, BBU functions are supported and running on commercial off-the-shelf HW that is located in the cloud. It should be noted that BBU pooling and cloud RAN platforms are associated with the existing development in network function virtualization (NFV). The NFV idea is based on sharing of resources by isolation and abstraction of network functionalities. Centralization of baseband processing functions is highly essential for HW resources sharing and virtualization. This helps HW resources to run on general purpose processors (GPPs) in the RAN virtualization schemes. There are various advantages to C-RAN and RAN virtualization. A notable one is the reduction of CAPEX and OPEX through HW sharing that can enable the use of low-cost GPPs. Moreover, it supports centralization of baseband processing, control and management functions. This facilitates CoMP by ensuring tighter and dynamic coordination between cells [Liu et al. - 2014].

Furthermore, it should be noted that, within the 5G and B5G context, ultra-dense deployment of RRHs is envisaged. This results in a deployment challenge considering the availability and cost of fiber-optic for the MFH/MBH networks. It is noteworthy that stringent requirements are also imposed on the fronthaul links concerning jitter, latency, and bandwidth for transporting multiple duplex radio transmissions. This is even more challenging with the application of CoMP since the number of RRHs that can access the same BBU pool at the same time will be limited.

D-RoF based CPRI is the most extensive means of conveying digital baseband oversampled I/Q streams in the C-RAN fronthaul. Nevertheless, optical links based on CPRI require huge throughput and capacity of the backhaul/fronthaul networks [Kuwano et al. - 2014; Chang et al. - 2013]. In addition, due to the separation of the media access control (MAC) and PHY layer functionalities at the BBU and RRH, jitter and latency become considerable issues in the digital-sampling-based C-RAN MFH links [Liu et al. - 2014; Chang et al. - 2013]. Meanwhile, the requirements for transmitting multiple streams, multi-RRH joint processing, high number of radio channels, and channel monitoring and estimation demand a high bit-rate in the order of Tbit s^{-1} [Monteiro and Gameiro - 2016, 2014].

Practically, multiple antennas as well as multiple radio access technologies (RATs) are usually supported by the mobile network operators (MNOs) MFH network. Consequently, it is desirable to aggregate the bit-rates into multiple tens of Gbps with several RRHs that are connected to a common BBU pool by the CPRIs. Moreover, the achievable data rate is contingent on features like the carrier bandwidth, the RAT being employed, as well as the number of employed multiple antennas. Considering these, the CPRI bandwidth for multi-sector and multi-antenna configurations can be defined as [Alimi et al. - 2018a]

$$B_{\text{CPRI}} = N_s M f_s \upsilon N C_w C, \tag{1.1}$$

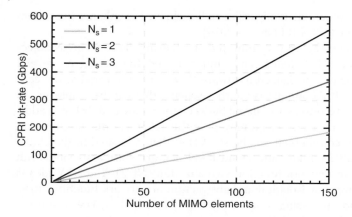

Figure 1.1 CPRI data rate for multi-sector and multi-antenna configurations.

where N_s represents the number of sectors per RRH, M signifies the number of antennas per sector, f_s is the sampling rate (frequency) that is employed for digitization (sample/s/carrier), N denotes the sampling bit-width (bits/sample) for I/Q samples, the number $v = 2$ is a multiplication factor to account for the (IQ) data, C_w denotes the factor of CPRI control word, and C is a coding factor (either 10/8 for 8B/10B code or 66/64 for 64B/66B code).

Figure 1.1 illustrates the required bandwidth of a CPRI for multi-sector and multi-antenna configurations. It can be inferred that the required bandwidth increases with an increase in the number of sectors N_s and/or the number antennas M per sector. For instance, when $N_s = 3$, the required bandwidth for 4×4 antenna configurations is 14.75 Gbps, this eventually increases to 29.49 Gbps with 8×8 antenna configurations.

In 5G and B5G networks, the required bandwidths are expected to be even more challenging. For instance, with CA of five 20 MHz LTE, a mobile signal with 8×8 MIMO antennas and $N_s = 3$, about 147.5 Gbps fronthaul data rate will be required by the CPRI. Consequently, to attain the high-bandwidth requirement of the current and future mobile fronthaul technologies, installation of new optical fibers or a substantial amount of high-speed optical ON–OFF-keying transceivers will be needed to support high data rate MFH implementation [Liu et al. - 2016]. Consequently, employment of advanced measures is essential to realize bandwidth-efficient MFH that is capable of addressing the bandwidth and flexibility limitations [Liu et al. - 2014]. Furthermore, the 5G RANs are envisaged to have as a feature, 100 MHz channels with M-MIMO; this brings about enormous capacity demand in the MFH. Therefore, the CPRI-based over-sampled approach might be challenging considering effective scalability. Subsequently, this can result in the implementation of the conventional analog A-RoF transmission technology, as discussed in Section 1.5.

1.3.3 Advanced Radio Coordination

It is envisaged that future networks will adopt different innovative radio coordination schemes like enhanced inter-cell interference coordination (eICIC) as well as CoMP in order to improve the system performance. The schemes can be implemented for adjacent cell coordination. This will ensure that the signal is jointly and effectively transmitted to the cell-edge users. The implementation can considerably improve the radio network spectral efficiency specifically by alleviating inter-cell interference and enhancing the data rate of the cell-edge users [Fiorani et al. - 2015; Alimi et al. - 2018a]. In addition, there are a number of coordination levels, such as moderate, tight, and very tight, that present distinct limitations on the transport network. For instance, a moderate coordination technique like eICIC introduces no precise requirement regarding transport performance. However, tight coordination schemes present stringent latency constraints (i.e. between 1 and 10 ms), although very strict capacity constraints (i.e. < 20 Mbps) are not presented. Furthermore, unlike the moderate and tight coordination schemes, a very tight coordination scheme poses very challenging constraints regarding both capacity (i.e. several Gbps using CPRI) and latency (< 0.5 ms) [Fiorani et al. - 2015].

1.3.4 Millimeter-Wave Small Cells

It has been observed that the lower-frequency bands of RF are limited, heavily regulated and highly congested. Consequently, the recent research trend for 5G wireless communication is towards the employment of higher-frequency bands that are under utilized or unexploited. The higher-frequency band that is under consideration is the mm-wave band (30–300 GHz). The interest in this band is as a result of its sufficient spectral resource and ability to offer 1000× throughput demanded by the 5G networks. It is remarkable that there are vast unexploited spectra in the 70–80 GHz E-band in which the 60 GHz band, with 7 GHz license-free bandwidth that is more appropriate for the 5G mobile communications, is the major band of interest [Liu et al. - 2014; Kalfas et al. - 2016; Stephen and Zhang - 2017; Zhang et al. - 2016; Alimi et al. - 2018a].

The band has been attracting significant attention due to its support for compact and high-dimensional transmission and reception antenna array equipment [Kalfas et al. - 2016; Stephen and Zhang - 2017; Alimi et al. - 2018a]. In addition, the band has been taken into consideration because of the issued 60 GHz standards like WiHD (WirelessHD Consortium), 802.11ad (Wireless Gigabit Alliance), and 802.15.3c (IEEE 802.15.3 Task Group 3c) for multi-gigabit (high rate) wireless communication systems [Alimi et al. - 2018a]. Furthermore, like wired-based optical fiber solutions, implementation of mm-wave in RAN systems for MFH/MBH has been considered. This is due to certain offered benefits such as easy of deployment. Unlike wire-based solutions, the benefit makes it promising for the envisaged ultra-dense small-cell RRHs to be deployed in 5G networks. However, implementation of mm-wave for mobile cellular communications possesses a number of challenges. The requirements for line-of-sight (LOS) communication as well as high-speed but low-cost MFH/MBH links for supporting several small-cell RRHs is really challenging. In addition, inter-cell interference and mobility management deserve further consideration. Similarly, due to the characteristic as well as high propagation losses, mm-wave communications offer a shorter network range, although this is advantageous for the 5G small-cell C-RAN implementations [Liu et al. - 2014; Chang et al. - 2013; Kalfas et al. - 2016; Stephen and Zhang - 2017; Zhang et al. - 2016; Alimi et al. - 2018a]. This is due to the fact that implementation of small-cell C-RAN architecture ensures effective exploitation of the mm-wave bandwidth as well as minimization of the associated inter-cell interference and mobility management cost [Chang et al. - 2013].

1.3.5 Massive MIMO

The conventional MIMO concepts have been employed in standards such as 802.11ac (Wi-Fi), 802.11n (Wi-FI), WiMAX, HSPA+, and LTE. M-MIMO, an enhancement of conventional MIMO, is expected to play an important role in 5G and B5G networks by being one of the key enablers. As 5G and B5G networks are envisaged as supporting a huge amount of traffic, devices, applications and services, the capability of a large antenna array-based M-MIMO for spatially multiplexing multiple autonomous terminals concurrently at the physical layer enables it to be a promising and effective technology for attending to the envisioned next generation network requirements. In general, M-MIMO is a wireless technology that employs multiple transmitters (Txs) and receivers (Rxs) with a minimum of a 16×16 array for more data transfer. M-MIMO technology presents viable features that are really promising for 4G network evolution toward 5G and B5G [Eri - 2018b].

M-MIMO normally exploits the terminals' prolific as well as unique propagation signatures by using smart processing at the corresponding array in order to realize a superior performance. Consequently, it is able to present an outstanding throughput and better spectral efficiency desired by MNOs and subscribers by employing the same time-frequency resource for multiple terminals spatial multiplexing [Eri - 2018b]. Moreover, M-MIMO offers improved energy efficiency through array gain exploitation that enables a reduction in the radiated power. Thus, large antennas aid in more precise spatial energy focus in the miniature elements so as to achieve substantial throughput improvements as well as enhanced energy efficiency. These features enable M-MIMO base stations to offer better performance and at the same time help in reducing wireless network interference and eventually enhancing the end-user experience. However, for a transport network, M-MIMO present an enormous transmission capacity that might be really challenging for the CPRI-based fronthaul networks, as discussed in Section 1.3.2 [Fiorani et al. - 2015].

1.3.6 New Multicarrier Modulations for 5G

In an effort to meet the 5G requirements, innovative modulation and multiple access schemes have been investigated. It should be noted that the existing wireless networks have been leveraging on orthogonal frequency division multiplexing (OFDM) based transmission schemes due to their effectiveness and simplicity. However, OFDM modulation exhibits some drawbacks when employed in a system. A notable one is due to the reduction in power and spectral efficiency that is a result of cyclic prefix redundancy. Also, narrow band interference and low tolerance to frequency offset/synchronization errors are other implementation disadvantages. These limitations might be really challenging in an un-coordinated multi-user environment. These drawbacks necessitate innovative modulation formats for the 5G and B5G networks. A number of modulation formats that are based on precoding, pulse shaping and filtering have been presented in an attempt to reduce the out-of-band (OOB) power radiation or OFDM side-lobe leakage. However, a simple method for reducing the OOB leakage is filtering. Consequently, multi-carrier schemes such filter bank multi-carrier, generalized frequency division multiplexing, and universal filtered multi-carrier schemes have been presented. Furthermore, each waveform exhibits a tradeoff depending on the implementation scenario [Cai et al. - 2018; Maliatsos et al. - 2016].

1.4 Fiber-Wireless Network Convergence

With the inception of C-RAN, the network architecture is classified as a centralized unit (BBU pool), transport network, and remote unit (RU), which is typically the RRHs. As aforementioned, there have been different evolutions in RAN systems in order to cope with continuous growth in the network traffic. Subsequently, it is imperative for the transport network to also evolve accordingly. The MFH transport network is an essential segment of the 5G C-RAN due to the stringent requirement imposed on it [Liu et al. - 2014].

The transport networks are typically implemented with T1 lines, microwave point-to-point links, and optical fiber links. However, with the need for high-capacity transport network development in accordance with evolving RAN systems, optical fiber solutions are promising and attractive transport links with inherent huge and scalable capacity for 5G and B5G fronthaul and midhaul architecture implementations [Liu et al. - 2014]. As a result of the high data rate required by MFH, network centralization demands a huge number of fiber cores that are not only scarce but also expensive. Moreover, transport technologies like an optical transport network and wavelength-division multiplexing (WDM) can be employed to provide protection and conserve fiber consumption; however, the deployment costs for additional transport equipment presents economic concerns [Chih-Lin et al. - 2015]. Therefore, for fixed-mobile convergence, there has been active research effort on an effective means of leveraging different passive optical network (PON) architectures like 10 gigabit-class PONs and the next generation PON stage 2 (NG-PON2) defined by IEEE, Full Service Access Network (FSAN) and International Telecommunication Union, Telecommunication Standardization Sector (ITU-T) for fixed access and backhaul networks [Liu et al. - 2014].

Furthermore, for microwave-photonics technologies, optical fiber links can offer ultra-high-speed internet access for fiber to the x applications and backhaul networks while high-speed wireless access provisioning can be achieved by employing small cell based on C-RAN architectures as well as mm-wave MFH/MBH. It is remarkable that network convergence can be realized both for indoor and outdoor conditions. For instance, apart from indoor local area network visible light communication with the capability to offer energy-efficient lighting and high-speed, short-range wireless communication concurrently [Gupta and Chockalingam - 2018], WiFi systems and mm-wave small cells that are based on high-rate in-building optical fiber networks can be employed in the indoor scenario for enhancing the link throughput. Furthermore, outdoor networks can be achieved by employing optical fiber systems for mobile MFH/MBH of macro-cell and C-RAN systems [Liu et al. - 2014].

In general, the transport network should be made of high-capacity switches as well as heterogeneous transmission links such as high-capacity wired (e.g. copper, optical fiber) and wireless (e.g. mm-wave, optical wireless communication) links for connecting different network elements like the RRHs and BBUs. Also, they should have provisions for different cell types such as macro and small cells to be supported by the 5G and B5G networks. Furthermore, another important reason for heterogeneous transmission links is to achieve reliable communications. For instance, since outdoor small cells are normally mounted on lampposts instead of rooftops, it might be challenging to have a reliable connection. This is as a result of several obstacles in the rural–urban settings that can block the LOS between the transmitting and receiving network entities. Consequently, this shows the need for non-LOS technologies that can serve as alternative/complementary solutions [Oliva et al. - 2015].

1.5 Radio-Over-Fiber Transmission Scheme

The inception of C-RAN has brought about major architectural and functionality changes to cellular systems. A notable instance is in the BBU centralization that facilitates simplification of the RRH design in a power-efficient and cost-effective way [Monteiro et al. - 2015; Maier and Rimal - 2015]. This assists in dense deployment of the RRHs as close as possible to the network subscribers in order to enhance the enjoyed quality of experience.

In the C-RAN based architecture, the baseband signal processing and MAC layer functions have been moved to the CO BBU [Liu et al. - 2014; Chang et al. - 2013]. Additionally, with employment of smart antenna systems, the C-RAN architecture has the capability of increasing the system capacity considerably with low-energy consumption [Zakrzewska et al. - 2014].

1.5.1 Digital Radio-Over-Fiber (D-RoF) Transmission

As discussed in Section 1.2.1.2, signal transmission between the BBU pool and RRHs in the C-RAN is primarily based on D-RoF technology. This is normally achieved by a serial constant bit rate interface, CPRI, due to its efficient mapping methods. The PHY layer of the CPRI is generally optical fiber and is small form pluggable connectivity based. Moreover, the baseband processing centralization in the C-RAN architecture exploits the processing power multiplexing gain and helps in achieving a better energy-efficient cooling. These benefits comparatively enable the C-RAN architecture to be a more energy efficient system. Furthermore, it remarkable that the C-RAN architecture initiates innovative low-latency and high-speed transmission connectivity fronthaul links between the RRHs and the BBUs. However, as discussed in Section 1.3.2, stringent requirements are imposed on the fronthaul links due to D-RoF technology based CPRI being employed. Moreover, another digitized wireless transport technology is digitized intermediate frequency over fiber. The limitations of digitized wireless transport technologies demand effective MBH/MFH networks. Furthermore, efficient transport links can be realized by exploiting analog wireless transport technologies such as A-RoF and intermediate IFoF. It is remarkable that each technology has its benefits and drawbacks, depending on the application [Urban et al. - 2016; Alimi et al. - 2018a].

1.5.2 Analog Radio-Over-Fiber (A-RoF) Transmission

Additionally, the RAN system architecture can further be simplified by employing the A-RoF (RoF) transmission technologies. The A-RoF implementation helps in shifting expensive analog-to-digital converters (ADCs) and digital-to-analog converters (DACs) to the BBU pool [Urban et al. - 2016; Alimi et al. - 2018a]. This results into another technology that is known as a cloud-RoF (C-RoF) access system. It is remarkable that, compared to the C-RAN architecture, the C-RoF access system simplifies the RU functionality by moving not only the ADC/DAC but also the RF frontend functions to the BBU pool. Subsequently, RF antennas with O/E and E/O converters are the major components that are left at the RU [Urban et al. - 2016; Liu et al. - 2014; Chang et al. - 2013].

Moreover, the C-RoF access scheme offers a number of advantages that makes it attractive for signal transmission. Due to the fact that the RF frontend has been moved to the BBU pool, then the generated RF signals at the BBU are transmitted to RUs over optical fiber fronthaul. With this, extensive RF ranges (i.e. ~1–100 GHz) can be transmitted through the shared C-RoF infrastructure. In addition, compared to the typical digital-baseband-transmission method that usually allows one band or service at a time, the RoF schemes support multi-service, multi-band and multi-operator coexistence in a shared infrastructure. The protocol-agnostic and infrastructure-sharing attributes of the RoF scheme assist in reducing the associated cost of small-cell deployment. With an IFoF system, multiple wireless signals can be assigned with multi-IFs and then multiplexed in the frequency domain in order to support MIMO services [Cho et al. - 2014]. Also, compatibility issues between the existing services and the future ones can be addressed by the RoF scheme [Urban et al. - 2016; Liu et al. - 2014; Chang et al. - 2013].

1.6 Optical MBH/MFH Transport Network Multiplexing Schemes

MBH/MFH transport network provisioning is largely based on optical fiber solutions. These can be achieved with single mode fiber (SMF) and/or multi-core fiber (MCF) [Alimi et al. - 2018a]. In the subsequent subsections, we expatiate on schemes such as WDM and spatial-division multiplexing (SDM) that can be employed for performance enhancement of MBH/MFH transport networks. Also, we put emphasis on SDM due to its salient features that can help in attending to the capacity crunch in the network.

1.6.1 Wavelength-Division Multiplexing (WDM) Based Schemes

Furthermore, for MFH network flexibility, optical wavelength division multiplexing PON (WDM-PON), coarse-WDM (CWDM) PON, as well as dense-WDM (DWDM) PON schemes can be implemented [Alimi et al. - 2018a]. In [Alimi et al. - 2018a], we presented a comprehensive discussion on the use cases of the PON schemes. The schemes' implementations will further enhance the C-RoF system performance by enabling the coexistence of multiple MNOs in a shared infrastructure in which different WDM wavelengths are employed. Likewise, different wireless services on lower- and higher-RF bands can be easily propagated concurrently in the RoF MFH/MBH of each MNO. Also, multiple sub-bands as well as multiple MIMO data streams can coexist in the RoF link for each wireless service without incurring detrimental interference, which may hinder high-rate services. Similarly, with centralized management, multiple services, multiple operators, as well as multiple wireless schemes can coexist in a shared small-cell infrastructure while still maintaining autonomous configurability. Consequently, the centralized scheme not only facilitates swift implementation of innovative radio technologies such as mm-wave radio and CoMP but also aids in more effective network management techniques. Also, with the implementation of schemes such as software-defined networking and NFV, it can offer further CAPEX and OPEX reductions [Liu et al. - 2014; Chang et al. - 2013].

It should be noted that there are certain associated design challenges for the C-RoF networks. The design of optoelectronics (i.e. O/E and E/O) interfaces at both RUs and BBU sides differ considerably from the typical digital baseband-over-fiber system. For instance, C-RoF application demands high linearity for generating, detecting, and transmitting optical signals in the network. Moreover, it is noteworthy that severe performance impairments may result from the RoF system nonlinearity-induced interference. This is even more intricate when multi-band multi-carrier RoF signals are under consideration [Liu et al. - 2014; Chang et al. - 2013].

1.6.2 Spatial-Division Multiplexing (SDM) Based Schemes

We are now in an era in which traffic growth, dominated by high-definition video streaming, multimedia file sharing, and other information technologies, is increasing faster than system capacity [Cisco - 2017]. By 2020 it is estimated that 30 billion devices will be connected to a 5G-enabled Internet of Things [Cisco - 2017]. This combination of broadband services and large scale of network-connected end devices will shortly lead to a capacity crunch, making the increment capacity of optical communication networks mandatory. Such high-demand for capacity is shared by core, metro, and access networks, as well as by optical networks directly supporting 5G infrastructure, i.e. optical fronthaul and backhaul.

In addition, SDM is a new approach to avoid capacity crunch. SDM has been evolving worldwide as key technology capable of breaking the limitation of conventional single-core single-mode optical fiber based communications systems, where the transmission capacity is considered limited to at most about 0.1 petabits s^{-1} [Qian et al. - 2011] due to the limitation on the launched input power or interference between signals [Mizuno et al. - 2016].

1.6.2.1 State-of-the-Art of SDM in 5G Infrastructure

Due to its natural potential to increment the capacity of optical communications, SDM is being proposed in the context of short-reach application networks, e.g. data-center connectivity or 5G infrastructure transport. Notice that, besides its potential role to avoid a future capacity crunch, the SDM concept can also be used for providing key solutions to such issues as achieving effective infrastructure operation/management and lower power consumption [Nakajima et al. - 2016]. In [Llorente et al. - 2015], a next generation RAN was proposed, employing MCFs in optical wireless fronthaul and mode-division multiplexing in the backhaul to overcome the capacity limitation of single-mode fibre systems. Such proposed optical fronthaul was based on RoF transmission on MCF. The combination of RoF and MCF permits cost- and energy-efficient support of MIMO wireless present in actual 4G cellular systems, which is a key requirement for next generation 5G radio technology. Recently, MIMO SDM on MCF was experimentally proposed and evaluated in multi-antenna LTE-advanced systems [Morant and Llorente - 2018].

MCF was also proposed as a compact medium to implement fiber-distributed signal processing, which may provide both radio access signal distribution (including MIMO antenna connectivity) and RF signal processing [García and Gasulla - 2016]. Applications of such a solution may cover a variety of application scenarios that will be especially demanded in 5G and B5G communications [Gasulla et al. - 2017].

1.6.2.2 Spatial Division Multiplexing Enabling Tools

The SDM techniques can be concretized by mode-multiplexing, using for instance few-mode fibers (FMF), or by different optical paths in different fiber optical cores, i.e. using MCFs [Richardson et al. - 2013]. Actually, MCFs can also be designed in order to support multiple modes in each fiber core [Mizuno et al. - 2016]. MCFs can be roughly grouped in coupled-core MCFs and in uncoupled MCFs. In the first case, the spacing between the cores, i.e. the pitch, is smaller than the one used in uncoupled MCFs, which results in a higher spatial density of cores. In this case, all the spatial channels are strongly coupled and the spatial channel can be described as a super-mode, or spatial superchannel, for a given wavelength. On the other hand, in uncoupled MCFs, each core is used as an individual waveguide and one can neglect any kind of interaction between the different cores supported by the MCF. This approach has received the most attention among all of the SDM technologies due to the possibility of avoiding MIMO processing [Fernandes et al. - 2017b]. In general, the overall bit rate of an SDM system is increased in proportion to the number of spatial modes/cores, with diverse SDM transmission systems being explored, demonstrating its potential by setting milestones in transmission capacity and capacity-distance product per fiber [Mizuno et al. - 2016]. For instance, a 2.15 petabit s^{-1} transmission over 31 km of a homogeneous 22-core single-mode multi-core fiber using 399×25 GHz spaced, 6.468 terabits s^{-1} spatial-super-channels comprising 24.5 GBaud PDM-64QAM modulation in each core was obtained in [Puttnam et al. - 2015]. Recently, the record-long WDM/SDM transmission experiments over FMFs with transmission reach exceeding 2500 km for a 12-core 3-mode MC-FMF and 6300 km for a single-core FMF have been demonstrated [Shibahara et al. - 2018]. Moreover, 159 terabit s^{-1} data transmission over 1045 km distance using a three-mode optical fiber was reached in [Rademacher et al. - 2018]. Such a high-capacity value was reached as mode multiplexing was used in combination with 16-QAM (quadrature-amplitude modulation), which is a practical high-density multi-level modulation optical signal, for all 348 wavelengths. It is worth mentioning that the FMF used in that experiment had a standard 125 µm outer diameter, which represents a key advantage as it can be cabled using existing equipment.

Figure 1.2 shows a schematic of an SDM transmission system that is based on optical coherent detection and advanced digital signal processing (DSP). On the Tx side, dual-polarization IQ modulators (DP-IQM) are driven by four signals generating a set of DP-quadrature amplitude modulation (DP-QAM) signals that, after optical space multiplexing, are propagated through the SDM transmission link. After optical SDM de-multiplexing, each tributary is detected in a digital optical coherent receiver (coherent Rx), comprised of an optical front end, analog-to-digital stage and advanced DSP subsystems.

Analogously to the case of single-mode fiber based systems, the real implementation of fiber-optic SDM systems requires assistance at both optical and digital domains. Such supporting tools can be employed as key optical subsystems, e.g.

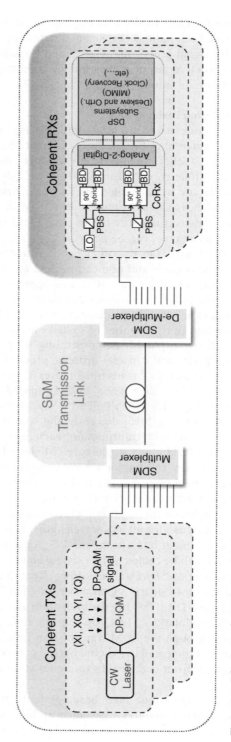

Figure 1.2 Schematic of an SDM transmission system.

optical switchers/couplers or amplifiers, or as compensation/mitigation techniques to deal with the impairments suffered during the signal propagation. As shown in [Fernandes et al. - 2015], an optical signal propagating in a particular core can be switched to any other core or distributed over all the cores exploring the acousto-optic effect. It was also shown that by tuning the acoustic wave amplitude, one can adjust the amount of optical power transferred between the cores [Fernandes et al. - 2015]. The light transfer/switching can also be attained inline by using long-period gratings inscribed in heterogeneous multi-core fibers [Rocha et al. - 2016]. The two previously mentioned techniques have the advantage of being inline techniques, thus avoiding coupling/decoupling processes and respective insertion losses. Regarding optical amplification and also employing an inline approach, a 6-mode and 580-wavelength multiplexed transmission with an dual C+L-band cladding-pumped few-mode erbium doped fiber amplifier (EDFA) was reported in [Wakayama et al. - 2018]. Such systems allow a total capacity of 266.1 terabit s^{-1} over a 90.4 km transmission line at a spectral efficiency of 36.7 bit s^{-1} Hz^{-1}. In real scenarios, mechanical perturbations are expected to cause mode coupling variations on time scales as short as tens of microseconds, requiring fast adaptive digital MIMO techniques with tolerable computational complexity [Arik et al. - 2014]. In that way, SDMs must use different signal processing approaches than wireless and SMF systems to ensure near optimum DSP with tolerable complexity.

Analogously to the case of single-mode fiber based systems, SDM techniques can also benefit from the combination of coherent detection and advanced DSP techniques [Fernandes et al. - 2017a]. A particular example is the adaptive MIMO equalization to compensate for modal crosstalk and modal dispersion. Such techniques can be implemented both in time and frequency domains, with different tradeoffs between performance and computational complexity. The least mean squares and recursive least squares frequency-domain equalization algorithms have been deeply analyzed in [Arik et al. - 2014]. Recently, a new approach based on Stokes space analysis has also been proposed for signal equalization both in classical single mode fiber systems [Muga and Pinto - 2015] and in SDM systems. Again, this approach can also be implemented both in time [Fernandes et al. - 2017a] and frequency [Caballero et al. - 2016] domains. The space-demultiplexing algorithm based on higher-order Poincaré spheres, presents important advantages when compared with its counterparts, e.g. modulation format agnostic, free of training sequences and robust to the local oscillator phase fluctuations and frequency offsets [Fernandes et al. - 2017a]. Figure 1.3(a) illustrates a block diagram of a higher-order Poincaré sphere based digital space-demultiplexing subsystem in a coherent transceiver with polarization, phase, and space diversity. Moreover, Figure 1.3(b) shows an error vector magnitude (left y-axis) and remaining penalty (right y-axis) as a function of the number of samples for the space-demultiplexed signal. Also, the left- and right-hand side insets show the quadrature phase shift keying (QPSK) constellations for one tributary before and after demultiplexing, respectively. Results show a fast convergence speed as the technique reaches negligible penalty values for a number of samples smaller than one thousand.

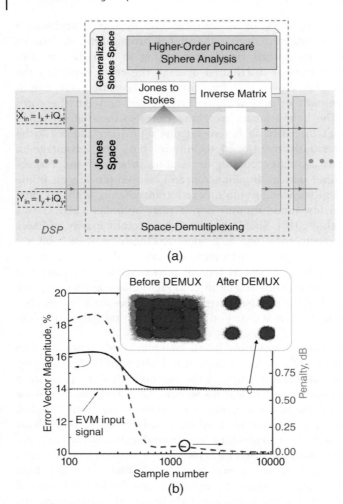

Figure 1.3 Block diagram of (a) a higher-order Poincaré sphere system, (b) an error vector magnitude (left *y*-axis) and remaining penalty (right *y*-axis) with insets that show the quadrature phase shift keying (QPSK) constellations.

1.7 Wireless based MFH/MBH

As aforementioned, C-RAN MFH can be achieved by a number of solutions, nevertheless, the low-latency and high-capacity requirements of the fronthaul render optical fiber-based solutions attractive. Furthermore, it is noteworthy that the flexibility and cost-effectiveness advantages of the C-RAN may be hindered by the implementation of optical fiber-based solutions for ultra-dense RRH deployment. In general, fiber deployment is time consuming and cost-intensive in an ultra-dense network, especially when trenching is required. In addition, wireless-based MFH solutions are highly attractive ways of exchanging information between the CUs and the DUs. The interest can be ascribed to the associated higher flexibility, lower cost, and ease of deployment compared with the fixed wired fronthaul solutions. Therefore, scalable and flexible solutions such as mm-wave (discussed in Section 1.3.4) and free-space optical (FSO) communication systems are means of realizing efficient and realistic wireless MFHs.

1.7.1 FSO Communication Systems

The FSO system is an optical wireless alternative technology of fiber-based fronthaul networks. In contrast to the RoF scheme discussed in Section 1.5, implementation of the FSO system does not depend on optical cables. Thus, without a fiber medium, FSO can be used for RF signals transmission via the free space between the transmit and receive apertures. Moreover, being an optical wireless technology, its deployment is not contingent on trenching. This feature is highly beneficial regarding cost and time compared with the usual optical fiber technology.

Furthermore, the fact that FSO can be employed in areas where physical connections by optical fiber cables are unrealistic is one of its major benefits. Also, as a result of the inherent FSO advantages like high-bit rates, full duplex transmission, ease of deployment, protocol transparency, and high transmission security, it has been acknowledged as an effective broadband access scheme that is capable of attending to the bandwidth requirements using MFH and MBH networks [Alimi et al. - 2017a, 2016, 2017e].

Figure 1.4 illustrates an FSO system as well as different typical components that are needed at both sides of the link. However, seamless connection FSO system does not demand signal conversion from optical to electrical and vice versa before transmission or reception via the free space. Therefore, optical-electrical as well as electrical-optical converters are not needed. Thus, this facilitates a data rate and protocol transparent FSO link. In addition, the system can be used along with other cutting-edge optical schemes like an EDFA and WDM in order to improve the system capacity considerably [Alimi et al. - 2018a; Kazaura et al. - 2010, 2009; Dat et al. - 2010].

In addition, the notion of transmitting radio signals over FSO (RoFSO) leverages the high-transmission capacity of optical systems and ease of deployment of wireless technologies. Consequently, DWDM RoFSO systems are capable of supporting multiple wireless signals concurrently. Nevertheless, FSO systems are considerably vulnerable to factors such as local weather conditions and atmospheric turbulence effects. It should be noted that, atmospheric turbulence is caused by the air refractive index variations through the transmission path [Alimi et al. - 2017a, 2016, 2017e,f; Sousa et al. - 2018; Alimi et al. - 2018b].

There are a number of statistical models for defining FSO intensity fluctuation in the literature for different turbulence regimes. For instance, log-normal (LN) distribution has been widely used due to the experimental measurement fits. Other models that have been extensively used are gamma–gamma (ΓΓ), negative exponential, K and I–K distributions [Alimi et al. - 2017c,b]. This work focuses on the LN and ΓΓ distribution models.

1.7.1.1 Log-Normal Distribution (LN)

The LN model is just appropriate for weak turbulence situations with a link range of about 100 m. Consequently, the weak turbulence intensity fluctuation probability

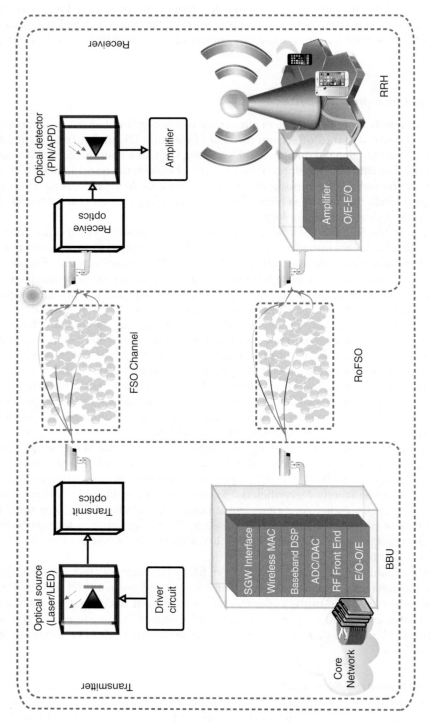

Figure 1.4 Block diagram of an RoFSO system.

density function (pdf) for the LN distribution is defined as [Alimi et al. - 2016, 2017c,b]

$$f_{h_a}(h_a) = \frac{1}{2h_a\sqrt{2\pi\sigma_x^2}} \exp\left(-\frac{(\ln(h_a) + 2\sigma_x)^2}{8\sigma_x^2}\right), \tag{1.2}$$

where h_a denotes atmospheric turbulence-induced fading, $\sigma_x^2 = \sigma_l^2/4$ denotes the log-amplitude variance defined for plane waves and spherical waves, respectively, as [Alimi et al. - 2016, 2017c,b]

$$\sigma_x^2|_{\text{plane}} = 0.307C_n^2 k^{7/6} L^{11/6}, \tag{1.3a}$$

$$\sigma_x^2|_{\text{spherical}} = 0.124C_n^2 k^{7/6} L^{11/6}, \tag{1.3b}$$

$$\sigma_l^2|_{\text{plane}} = 1.23C_n^2 k^{7/6} L^{11/6}, \tag{1.3c}$$

$$\sigma_l^2|_{\text{spherical}} = 0.50C_n^2 k^{7/6} L^{11/6}, \tag{1.3d}$$

where σ_l^2 represents the log-irradiance variance, $k = 2\pi/\lambda$ denotes the optical wave number, L represents the distance, and C_n^2 signifies the altitude-dependent index of refraction structure parameter.

1.7.1.2 Gamma-Gamma (ΓΓ) Distribution

The ΓΓ distribution is typically used in modeling the scintillation effects in the strong turbulence regimes where the LN distribution characterization is not valid. Furthermore, the ΓΓ model is also suitable for modeling and characterizing the fading gains from the weak to strong turbulence regimes. The pdf of h_a for the ΓΓ distribution can be expressed as [Alimi et al. - 2017c,b]

$$f_{h_a}(h_a) = \frac{2(\alpha\beta)^{(\alpha+\beta)/2}}{\Gamma(\alpha)\Gamma(\beta)}(h_a)^{\frac{(\alpha+\beta)}{2}-1}K_{\alpha-\beta}(2\sqrt{\alpha\beta h_a}), \tag{1.4}$$

where $\Gamma(\cdot)$ denotes the gamma function, $K_\nu(\cdot)$ represents the modified Bessel function of the second kind of order ν, and α and β represent the effective number of large-scale and small-scale eddies of the scattering process, respectively. The parameters α and β are expressed respectively for the plane wave as [Alimi et al. - 2016, 2017c,b]

$$\alpha = \left[\exp\left(\frac{0.49\sigma_R^2}{(1 + 1.11\sigma_R^{12/5})^{7/6}}\right) - 1\right]^{-1}, \tag{1.5a}$$

$$\beta = \left[\exp\left(\frac{0.51\sigma_R^2}{(1 + 0.69\sigma_R^{12/5})^{5/6}}\right) - 1\right]^{-1}, \tag{1.5b}$$

also for the spherical wave, they can be defined as [Alimi et al. - 2017c,b]

$$\alpha = \left[\exp\left(\frac{0.49\sigma_R^2}{(1 + 0.18d^2 + 0.56\sigma_R^{12/5})^{7/6}}\right) - 1\right]^{-1}, \tag{1.6a}$$

$$\beta = \left[\exp\left(\frac{0.51\sigma_R^2(1 + 0.69\sigma_R^{12/5})^{-5/6}}{(1 + 0.9d^2 + 0.62d^2\sigma_R^{12/5})^{5/6}}\right) - 1\right]^{-1}, \tag{1.6b}$$

where $d \triangleq (kD^2/4L)^{1/2}$, D denotes the diameter of the receiver aperture, σ_R^2 represents the Rytov variance that is a metric for the strength of turbulence fluctuations. The σ_R^2 is expressed for the plane and spherical waves, respectively, as [Alimi et al. - 2017c,b]

$$\sigma_R^2|_{\text{plane}} = 1.23 \quad C_n^2 \ k^{7/6} \ L^{11/6}, \tag{1.7a}$$

$$\sigma_R^2|_{\text{spherical}} = 0.492 \quad C_n^2 \ k^{7/6} \ L^{11/6}. \tag{1.7b}$$

In addition, the normalized variance of the irradiance that is also known as the scintillation index (σ_N^2) can be defined with respect to σ_x^2 as well as eddies of the scattering process (α and β), respectively, as [Alimi et al. - 2017c,b]

$$\sigma_N^2 \triangleq \frac{\langle h_a^2 \rangle - \langle h_a \rangle^2}{\langle h_a \rangle^2} \tag{1.8a}$$

$$= \frac{\langle h_a^2 \rangle}{\langle h_a \rangle^2} - 1 \tag{1.8b}$$

$$= \exp(4\sigma_x^2) - 1 \tag{1.8c}$$

$$= 1/\alpha + 1/\beta + 1/(\alpha\beta). \tag{1.8d}$$

As aforementioned, an atmospheric turbulence-induced fading and local weather conditions bring about the received optical intensity fluctuation. Therefore, the resulting impairments hinder the optical wireless scheme from being a reliable technology like a typical optical fiber solution. This does not only limit RoFSO system performance but also hinders FSO from being an effective standalone MFH solution. Several PHY layer schemes like maximum likelihood sequence detection, diversity schemes and adaptive optics are normally used for mitigating turbulence-induced fading [Alimi et al. - 2018a]. In the following subsection, we present advanced technologies that can be employed to improve FSO-based MFH system performance.

1.7.2 Hybrid RF/FSO Technology

RF/FSO systems have been presented in an effort to make FSO a better technology. The RF/FSO system is a hybrid scheme that exploits and integrates high-transmission capacity exhibited by optical schemes as well as ease of deployment of wireless systems while addressing the associated weaknesses of both technologies. This enables a reliable and concurrent transmission of heterogeneous wireless services (multiple analog and digital signals). Likewise, the idea is also to integrate the RF solution's scalability and cost-effectiveness with the FSO solution's low-latency and high-data rate. Consequently, the hybrid scheme helps in realizing the low-latency and high-throughput requirements of the future networks in a cost-effective way. In a hybrid RF/FSO system, based on the application and deployment scenario, there are two parallel links between the Tx and Rx that have the ability for data transmission. Nevertheless, either of the links can be used to transmit data based on the electromagnetic interference levels and weather conditions.

1.7.3 Relay-Assisted FSO Transmission

The spatial diversity technique is one of the effective means of mitigating turbulence-induced fading. This entails deployment of multiple transmit/receive apertures with the intention of exploiting and establishing additional spatial degrees of freedom. The technique has been extensively employed for mitigating fading because of the inherent redundancy. However, multiple aperture deployment presents challenges not only in terms of an increase in the system complexity but also regarding cost. Moreover, stringent requirements like adequate aperture spacing have to be met to prevent spatial correlation.

In RF communication, dual-hop relaying has been widely used not only to simplify spatial diversity implementation but also to enhance the quality of the received signal and extend the coverage area considerably. Furthermore, the benefits of MIMO schemes can be achieved by employing a relay-assisted transmission in which *virtual* multiple-aperture systems are logically created. A relay-assisted scheme utilizes both the FSO and RF features in order to provide an efficient system. In addition, it is a mixed RF/FSO dual-hop technology where the source-relay links are RF links whereas relay-destination links are FSO links.

It is remarkable that the mixed RF/FSO dual-hop relay system is comparatively different from the hybrid RF/FSO approach since parallel RF and FSO links exist between the source (Tx) and the destination (Rx). Moreover, the main function of the FSO link in the mixed RF/FSO dual-hop approach is to enable the RF users to connect with the backbone network so as to bridge the connectivity gab between the last-mile and the backbone networks. This scheme efficiently addresses the system last-mile transmission bottleneck by multiplexing and aggregating multiple RF users into a single shared high-speed FSO infrastructure with the intention of harnessing the intrinsic optical bandwidth. In addition, any kind of interference can be prevented due to the fact that FSO and RF operate on dissimilar frequency bands. Thus, this configuration provides improved performance compared with the conventional RF/RF transmission system. In Figure 1.5, we present an implementation scenario for some of the discussed schemes for a converged fiber-wireless access network.

1.8 Experimental Channel Measurement and Characterization

The performance of an FSO link is studied experimentally under a real atmospheric turbulence condition. We measure σ_N^2 from the channel samples acquired in an attempt to determine the extent of atmospheric turbulence as well as consequent effects on the FSO link performance. To achieve this, we use the setup shown in Figure 1.4. The setup comprises a point-to-point link that is based on intensity modulation/direct detection scheme. We generate a 10 Gb s^{-1} non-return-to-zero signal using a $2^{23} - 1$ pseudorandom binary sequence. Furthermore, the generated electrical signal is injected into a laser that operates at 1548.51 nm. The optical output signal is then launched from the laser to a 3 mm diameter collimator using a standard single-mode fiber (SSMF). In addition, the input optical power of the collimator is set to 0 dBm. Consequently, the collimated beam is conveyed through a 54 m length FSO channel. At the receiver, the converged optical signal focuses on the collimator that is SSMF coupled to the photodetector. The subsequent optical signal is converted to electrical signal by means of a 10 Gb s^{-1} photodiode. Then, C_n^2 is estimated as discussed earlier.

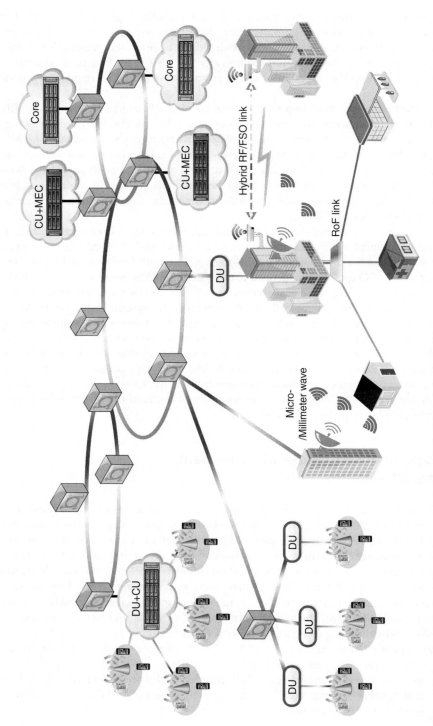

Figure 1.5 Converged fiber-wireless access networks scenario for 5G implementation.

1.9 Results and Discussions

Figure 1.6 presents both theoretical and experimental results of the weak turbulence condition for a link range of 54 m. Figure 1.6(a) shows the LN pdf for different values of log-irradiance variance using the logarithmic scale. It is evident that, as σ_l^2 increases, the distribution is further tilted. This behavior indicates the irradiance fluctuation magnitude of the system. Furthermore, the refractive-index structure parameter C_n^2 characterization is realized by fitting the nearest LN as well as the $\Gamma\Gamma$ pdf curves to that of the received data. Figure 1.6(b) shows the resulting fittings of the measurement taken around 9:30 pm on 18 November 2015. The scintillation index σ_N^2 is evaluated and the obtained value is 0.0052. Under this condition, i.e. $\sigma_N^2 = 0.0052$, the LN as well as the $\Gamma\Gamma$

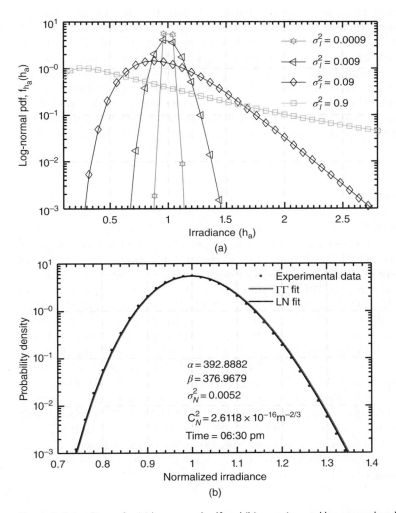

Figure 1.6 Irradiance for (a) log-normal pdf and (b) experimental log-normal and gamma-gamma fittings.

are perfectly fitted to the measured channel samples σ_N^2. This result indicates that both models are suitable for weak atmospheric fading characterization.

1.10 Conclusion

In this chapter, we have presented a number of fundamental features and techniques that can facilitate effective deployment of a 5G wireless network. Being one of the main supporting architectures for the 5G and B5G networks, we have focused on the small-cell C-RAN and the related RAN virtualization. We have also emphasized the need for high-capacity and low-latency architecture as well as energy- and cost-efficient mobile fronthaul links. Furthermore, implementation of mm-wave and optics-based schemes that support multi-service, multi-band and multi-operator coexistence in a shared network infrastructure have been comprehensively discussed. Moreover, the main requirements of the MFH/MBH networks for supporting 5G and B5G networks have been presented considering optical and wireless access network convergence. In addition, we have presented not only analytical expressions but also numerical and experimental results to demonstrate the effects of atmospheric turbulence-induced fading that can hinder the number of RRHs that can access the same BBU pool at the same time. We also discuss research challenges and open-ended issues on means of achieving the requirements of 5G network in terms of jitter, capacity, and latency in an economical manner.

Acknowledgments

This work is supported by the Fundação para a Ciência e a Tecnologia (FCT) under the PhD grant PD/BD/52590/2014. Also, it is supported by the European Regional Development Fund (FEDER), through the Regional Operational Programme of Lisbon (POR LISBOA 2020) and the Competitiveness and Internationalization Operational Programme (COMPETE 2020) of the Portugal 2020 framework, Project 5G (POCI-01-0247-FEDER-024539), ORCIP (CENTRO-01-0145-FEDER-022141) and SOCA (CENTRO-01-0145-FEDER-000010). It is also funded by FCT through national funds under the project COMPRESS - PTDC/EEI-TEL/7163/2014 and by the European Regional Development Fund (FEDER), through the Regional Operational Program of Centre (CENTRO 2020) of the Portugal 2020 framework [Project HeatIT with Nr. 017942 (CENTRO-01-0247-FEDER-017942)]. It is also supported in part by Fundação para a Ciência e a Tecnologia (FCT) through national funds, and when applicable co-funded by FEDER-PT2020 partnership agreement, under the project UID/EEA/50008/2013 (actions COHERENTINUOUS and OPTICAL-5G). Our gratitude are also extended to the following funding bodies: Ocean12-H2020-ECSEL-2017-1-783127, and FCT and the ENIAC JU (THINGS2DO–GA n. 621221) projects.

Bibliography

I. Alimi, A. Shahpari, V. Ribeiro, N. Kumar, P. Monteiro, and A. Teixeira. Optical wireless communication for future broadband access networks. In *2016 21st European Conference on Networks and Optical Communications (NOC)*, pages 124–128, June 2016. doi: 10.1109/NOC.2016.7506998.

I. A. Alimi, A. M. Abdalla, J. Rodriguez, P. P. Monteiro, and A. L. Teixeira. Spatial Interpolated Lookup Tables (LUTs) Models for Ergodic Capacity of MIMO FSO Systems. *IEEE Photonics Technology Letters*, 29(7):583–586, April 2017a. ISSN 1041-1135. doi: 10.1109/LPT.2017.2669337.

I. A. Alimi, A. L. Teixeira, and P. P. Monteiro. Toward an Efficient C-RAN Optical Fronthaul for the Future Networks: A Tutorial on Technologies, Requirements, Challenges, and Solutions. *IEEE Communications Surveys Tutorials*, 20(1):708–769, Firstquarter 2018a. doi: 10.1109/COMST.2017.2773462.

Isiaka Alimi, Ali Shahpari, Vítor Ribeiro, Artur Sousa, Paulo Monteiro, and António Teixeira. Channel characterization and empirical model for ergodic capacity of free-space optical communication link. *Optics Communications*, 390:123 –129, 2017b. ISSN 0030-4018. doi: https://doi.org/10.1016/j.optcom.2017.01.001. URL http://www .sciencedirect.com/science/article/pii/S0030401817300019.

Isiaka Alimi, Ali Shahpari, Artur Sousa, Ricardo Ferreira, Paulo Monteiro, and António Teixeira. Challenges and opportunities of optical wireless communication technologies. In Pedro Pinho, editor, *Optical Communication Technology*, chapter 02, pages 5–44. InTech, Rijeka, 2017c. ISBN 978-953-51-3418-3. doi: 10.5772/intechopen.69113. URL https://cdn.intechopen.com/pdfs-wm/55559.pdf.

Isiaka A. Alimi, Paulo P. Monteiro, and António L. Teixeira. Analysis of multiuser mixed RF/FSO relay networks for performance improvements in cloud computing-based radio access networks (CC-RANs). *Optics Communications*, 402:653 –661, 2017d. ISSN 0030-4018. doi: http://dx.doi.org/10.1016/j.optcom.2017.06.097. URL http://www .sciencedirect.com/science/article/pii/S0030401817305734.

Isiaka A. Alimi, Paulo P. Monteiro, and António L. Teixeira. Outage Probability of Multiuser Mixed RF/FSO Relay Schemes for Heterogeneous Cloud Radio Access Networks (H-CRANs). *Wireless Personal Communications*, 95(1): 27–41, Jul 2017e. ISSN 1572-834X. doi: 10.1007/s11277-017-4413-y. URL https://doi.org/10.1007/ s11277-017-4413-y.

Isiaka A. Alimi, Ali Shahpari, Paulo P. Monteiro, and António L. Teixeira. Effects of diversity schemes and correlated channels on OWC systems performance. *Journal of Modern Optics*, 64(21):2298–2305, 2017f. doi: 10.1080/09500340.2017.1357851. URL https://doi.org/10.1080/09500340.2017.1357851.

Isiaka A. Alimi, Akeem O. Mufutau, António L. Teixeira, and Paulo P. Monteiro. Performance Analysis of Space-Air-Ground Integrated Network (SAGIN) Over an Arbitrarily Correlated Multivariate FSO Channel. *Wireless Personal Communications*, 100 (1):47–66, May 2018b. ISSN 1572-834X. doi: 10.1007/s11277-018-5620-x. URL https://doi.org/10.1007/s11277-018-5620-x.

Sercan Ö. Arik, Daulet Askarov, and Joseph M. Kahn. Adaptive frequency-domain equalization in mode-division multiplexing systems. *J. Lightwave Technol.*, 32(10):1841–1852, May 2014.

Common Public Radio Interface (CPRI): Requirements for the eCPRI Transport Network. C. Parties, August 2017. eCPRI Transport Network D0.1, [Online]. Available: http://www .cpri.info/downloads/Requirements_for_the_eCPRI_Transport_Network_d_0_1_2017_08_30.pdf.

F. J. V. Caballero, A. Zanaty, F. Pittala, G. Goeger, Y. Ye, I. Tafur Monroy, and W. Rosenkranz. Efficient SDM-MIMO Stokes-Space Equalization. In *ECOC 2016; 42nd European Conference on Optical Communication*, pages 1–3, 2016.

Y. Cai, Z. Qin, F. Cui, G. Y. Li, and J. A. McCann. Modulation and Multiple Access for 5G Networks. *IEEE Communications Surveys Tutorials*, 20 (1): 629–646, Firstquarter 2018. doi: 10.1109/COMST.2017.2766698.

Gee-Kung Chang, C. Liu, and Liang Zhang. Architecture and applications of a versatile small-cell, multi-service cloud radio access network using radio-over-fiber technologies. In *2013 IEEE International Conference on Communications Workshops (ICC)*, pages 879–883, June 2013. doi: 10.1109/ICCW.2013.6649358.

Seung-Hyun Cho, Heuk Park, Hwan Seok Chung, Kyeong Hwan Doo, Sangsoo Lee, and Jong Hyun Lee. Cost-effective next generation mobile fronthaul architecture with multi-IF carrier transmission scheme. In *OFC 2014*, pages 1–3, March 2014. doi: 10.1364/OFC.2014.Tu2B.6.

Cisco. Cisco Visual Networking Index: Global Mobile Data Traffic Forecast Update, 2016–2021 White Paper, 2017.

P. T. Dat, A. Bekkali, K. Kazaura, K. Wakamori, and M. Matsumoto. A universal platform for ubiquitous wireless communications using radio over FSO system. *Journal of Lightwave Technology*, 28(16):2258–2267, Aug 2010. ISSN 0733-8724. doi: 10.1109/JLT.2010.2049641.

T. Deiß, L. Cominardi, A. Garcia-Saavedra, P. Iovanna, G. Landi, X. Li, J. Mangues-Bafalluy, J. Núñez-Martínez, and A. de la Oliva. Packet forwarding for heterogeneous technologies for integrated fronthaul/backhaul. In *2016 European Conference on Networks and Communications (EuCNC)*, pages 133–137, June 2016. doi: 10.1109/EuCNC.2016.7561019.

Ericsson Fronthaul. Ericsson, Mar 2018a. Available: https://www.ericsson.com/ ourportfolio/networks-products/fronthaul?nav=fgb_101_0561%7Cfgb_101_0516.

Going Massive with MIMO. Ericsson, January 2018b. URL https://www.ericsson .com/en/news/2018/1/massive-mimo-highlights.

G. Fernandes, N. J. Muga, Ana M. Rocha, and A. N. Pinto. Switching in multicore fibers using flexural acoustic waves. *Optics Express*, 23(20):26313–26325, October 2015. ISSN 1094-4087.

G. Fernandes, N. J. Muga, and A. N. Pinto. Space-demultiplexing based on higher-order Poincaré spheres. *Optics Express*, 25(4):3899–3899, February 2017a. doi: 10.1364/OE.25.003899.

G. Fernandes, N. J. Muga, and A. N. Pinto. *Optical Fibers: Technology, Communications and Recent Advances*, chapter Space-Division Multiplexing in Fiber-Optic Transmission Systems. Nova Publisher, New York, 2017b.

Matteo Fiorani, Björn Skubic, Jonas Mårtensson, Luca Valcarenghi, Piero Castoldi, Lena Wosinska, and Paolo Monti. On the design of 5G transport networks. *Photonic Network*

Communications, 30(3):403–415, Dec 2015. ISSN 1572-8188. doi: 10.1007/s11107-015-0553-8. URL https://doi.org/10.1007/s11107-015-0553-8.

Sergi García and Ivana Gasulla. Dispersion-engineered multicore fibers for distributed radiofrequency signal processing. *Opt. Express*, 24(18): 20641–20654, Sep 2016.

Ivana Gasulla, Sergi García, David Barrera, Javier Hervás, and Salvador Sales. Fiber-distributed signal processing: Where the space dimension comes into play. In *Advanced Photonics 2017 (IPR, NOMA, Sensors, Networks, SPPCom, PS)*, page PW1D.1. Optical Society of America, 2017.

A. K. Gupta and A. Chockalingam. Performance of MIMO Modulation Schemes With Imaging Receivers in Visible Light Communication. *Journal of Lightwave Technology*, 36(10):1912–1927, May 2018. ISSN 0733-8724. doi: 10.1109/JLT.2018.2795698.

C. L. I, H. Li, J. Korhonen, J. Huang, and L. Han. RAN Revolution With NGFI (xhaul) for 5G. *Journal of Lightwave Technology*, 36(2):541–550, Jan 2018. ISSN 0733-8724. doi: 10.1109/JLT.2017.2764924.

G. Kalfas, N. Pleros, L. Alonso, and C. Verikoukis. Network planning for 802.11ad and MT-MAC 60 GHz fiber-wireless gigabit wireless local area networks over passive optical networks. *IEEE/OSA Journal of Optical Communications and Networking*, 8(4):206–220, April 2016. ISSN 1943-0620. doi: 10.1364/JOCN.8.000206.

K. Kazaura, K. Wakamori, M. Matsumoto, T. Higashino, K. Tsukamoto, and S. Komaki. RoFSO: A universal platform for convergence of fiber and free-space optical communication networks. In *Innovations for Digital Inclusions, 2009. K-IDI 2009. ITU-T Kaleidoscope:*, pages 1–8, Aug 2009.

K. Kazaura, K. Wakamori, M. Matsumoto, T. Higashino, K. Tsukamoto, and S. Komaki. RoFSO: A universal platform for convergence of fiber and free-space optical communication networks. *IEEE Communications Magazine*, 48(2):130–137, February 2010. ISSN 0163-6804. doi: 10.1109/MCOM.2010.5402676.

S. Kuwano, J. Terada, and N. Yoshimoto. Operator perspective on next-generation optical access for future radio access. In *2014 IEEE International Conference on Communications Workshops (ICC)*, pages 376–381, June 2014. doi: 10.1109/ICCW.2014.6881226.

I. Chih-Lin, Y. Yuan, J. Huang, S. Ma, C. Cui, and R. Duan. Rethink fronthaul for soft RAN. *IEEE Communications Magazine*, 53 (9):82–88, September 2015. ISSN 0163-6804. doi: 10.1109/MCOM.2015.7263350.

C. Liu, J. Wang, L. Cheng, M. Zhu, and G. K. Chang. Key Microwave-Photonics Technologies for Next-Generation Cloud-Based Radio Access Networks. *Journal of Lightwave Technology*, 32(20):3452–3460, Oct 2014. ISSN 0733-8724. doi: 10.1109/JLT.2014.2338854.

X. Liu, H. Zeng, N. Chand, and F. Effenberger. Efficient mobile fronthaul via DSP-based channel aggregation. *Journal of Lightwave Technology*, 34(6): 1556–1564, March 2016. ISSN 0733-8724. doi: 10.1109/JLT.2015.2508451.

Roberto Llorente, Maria Morant, Andrés Macho, David Garcia-Rodriguez, and Juan Luis Corral. Demonstration of a Spatially Multiplexed Multicore Fibre-Based Next-Generation Radio-Access Cellular Network. In *International Conf. on Transparent Optical Networks - ICTON*, volume 1, page Th.A1.4, 2015.

M. Maier and B. P. Rimal. Invited paper: The audacity of fiber-wireless (FiWi) networks: revisited for clouds and cloudlets. *China Communications*, 12(8): 33–45, August 2015. ISSN 1673-5447. doi: 10.1109/CC.2015.7224704.

Konstantinos N. Maliatsos, Eleftherios Kofidis, and Athanasios G. Kanatas. A Unified Multicarrier Modulation Framework. *CoRR*, abs/1607.03737, 2016. URL http://arxiv.org/abs/1607.03737.

T. Mizuno, H. Takara, A. Sano, and Y. Miyamoto. Dense space-division multiplexed transmission systems using multi-core and multi-mode fiber. *Journal of Lightwave Technology*, 34(2):582–592, 2016.

P. P. Monteiro and A. Gameiro. Hybrid fibre infrastructures for cloud radio access networks. In *2014 16th International Conference on Transparent Optical Networks (ICTON)*, pages 1–4, July 2014. doi: 10.1109/ICTON.2014.6876549.

P. P Monteiro and A. Gameiro. Convergence of optical and wireless technologies for 5G. In F. Hu, editor, *Opportunities in 5G Networks: A Research and Development Perspective*, chapter 9, page 179–215. CRC Press, CRC Press, 2016.

P. P. Monteiro, D. Viana, J. da Silva, D. Riscado, M. Drummond, A. S. R. Oliveira, N. Silva, and P. Jesus. Mobile fronthaul RoF transceivers for C-RAN applications. In *2015 17th International Conference on Transparent Optical Networks (ICTON)*, pages 1–4, July 2015. doi: 10.1109/ICTON.2015.7193452.

M. Morant and R. Llorente. Performance Analysis of Carrier-Aggregated Multiantenna 4 x 4 MIMO LTE-A Fronthaul by Spatial Multiplexing on Multicore Fiber. *Journal of Lightwave Technology*, 36(2):594–600, 2018.

N. J. Muga and A. N. Pinto. Extended Kalman Filter vs. Geometrical Approach for Stokes Space Based Polarization Demultiplexing. *J. Lightw. Technol.*, 33 (23):4826 –4833, 2015.

K. Nakajima, T. Matsui, K. Saito, T. Sakamoto, and N. Araki. Space division multiplexing technology: Next generation optical communication strategy. In *2016 ITU Kaleidoscope: ICTs for a Sustainable World (ITU WT)*, pages 1–7, Nov 2016.

Mobile Anyhaul. Nokia, Dec. 2017. URL https://pages.nokia.com/14265.Mobile.Anyhaul.html.

A. D. La Oliva, X. C. Perez, A. Azcorra, A. D. Giglio, F. Cavaliere, D. Tiegelbekkers, J. Lessmann, T. Haustein, A. Mourad, and P. Iovanna. Xhaul: toward an integrated fronthaul/backhaul architecture in 5G networks. *IEEE Wireless Communications*, 22(5):32–40, October 2015. ISSN 1536-1284. doi: 10.1109/MWC.2015.7306535.

B. J. Puttnam, R. S. Luís, W. Klaus, J. Sakaguchi, J. M. Delgado Mendinueta, Y. Awaji, N. Wada, Y. Tamura, T. Hayashi, M. Hirano, and J. Marciante. 2.15 pb/s transmission using a 22 core homogeneous single-mode multi-core fiber and wideband optical comb. In *2015 European Conference on Optical Communication (ECOC)*, pages 1–3, Sept 2015.

Dayou Qian, Ming-Fang Huang, E. Ip, Yue-Kai Huang, Yin Shao, Junqiang Hu, and Ting Wang. 101.7-tb/s (370×294-gb/s) pdm-128qam-ofdm transmission over 3×55-km ssmf using pilot-based phase noise mitigation. In *2011 Optical Fiber Communication Conference and Exposition and the National Fiber Optic Engineers Conference*, pages 1–3, 2011.

Georg Rademacher, Ruben S. Luís, Benjamin J. Puttnam, Tobias A. Eriksson, Erik Agrell, Ryo Maruyama, Kazuhiko Aikawa, Hideaki Furukawa, Yoshinari Awaji, and Naoya Wada. 159 tbit/s c+l band transmission over 1045 km 3-mode graded-index few-mode fiber. In *Optical Fiber Communication Conference Postdeadline Papers*, page Th4C.4. Optical Society of America, 2018.

D. J. Richardson, J. M. Fini, and L. E. Nelson. Space-division multiplexing in optical fibres. *Nature Photonics*, 7:354–362, 2013.

Ana M. Rocha, R.N. Nogueira, and M. Facão. Core/wavelength selective switching based on heterogeneous mcfs with lpgs. *IEEE Photonics Technology Letters*, 28(18):1992–1995, 2016.

K. Shibahara, T. Mizuno, L. Doowhan, Y. Miyamoto, H. Ono, K. Nakajima, S. Saitoh, K. Takenaga, and K. Saitoh. Dmd-unmanaged long-haul sdm transmission over 2500-km 12-core x 3-mode mc-fmf and 6300-km 3-mode fmf employing intermodal interference cancelling technique. In *Optical Fiber Communication Conference Postdeadline Papers*, page Th4C.6. Optical Society of America, 2018.

Artur N. Sousa, Isiaka A. Alimi, Ricardo M. Ferreira, Ali Shahpari, Mário Lima, Paulo P. Monteiro, and António L. Teixeira. Real-time dual-polarization transmission based on hybrid optical wireless communications. *Optical Fiber Technology*, 40:114 –117, 2018. ISSN 1068-5200. doi: https://doi.org/10.1016/j.yofte.2017.11.011. URL http://www.sciencedirect.com/science/article/pii/S1068520017302900.

R. G. Stephen and R. Zhang. Joint millimeter-wave fronthaul and OFDMA resource allocation in ultra-dense CRAN. *IEEE Transactions on Communications*, 65 (3):1411–1423, March 2017. ISSN 0090-6778. doi: 10.1109/TCOMM.2017.2649519.

P. J. Urban, G. C. Amaral, and J. P. von der Weid. Fiber Monitoring Using a Sub-Carrier Band in a Sub-Carrier Multiplexed Radio-over-Fiber Transmission System for Applications in Analog Mobile Fronthaul. *Journal of Lightwave Technology*, 34(13):3118–3125, July 2016. ISSN 0733-8724. doi: 10.1109/JLT.2016.2559480.

Yuta Wakayama, Daiki Soma, Shohei Beppu, Seiya Sumita, Koji Igarashi, and Takehiro Tsuritani. 266.1-Tbit/s Repeatered Transmission over 90.4-km 6-Mode Fiber Using Dual C+L-Band 6-Mode EDFA. In *Optical Fiber Communication Conference*, page W4C.1. Optical Society of America, 2018.

A. Zakrzewska, S. Ruepp, and M. S. Berger. Towards converged 5G mobile networks-challenges and current trends. In *ITU Kaleidoscope Academic Conference: Living in a converged world - Impossible without standards?, Proceedings of the 2014*, pages 39–45, June 2014. doi: 10.1109/Kaleidoscope.2014.6858478.

H. Zhang, Y. Dong, J. Cheng, M. J. Hossain, and V. C. M. Leung. Fronthauling for 5G LTE-U ultra dense cloud small cell networks. *IEEE Wireless Communications*, 23(6):48–53, December 2016. ISSN 1536-1284. doi: 10.1109/MWC.2016.1600066WC.

2

Hybrid Fiber Wireless (HFW) Extension for GPON Toward 5G

Rattana Chuenchom[1], Andreas Steffan[2], Robert G. Walker[3], Stephen J. Clements[3], Yigal Leiba[4], Andrzej Banach[5], Mateusz Lech[5], and Andreas Stöhr[*,1]

[1]*Center for Semiconductor Technology and Optoelectronics (ZHO), University of Duisburg-Essen, 47057, Lotharstr. 55, Germany*
[2]*Finisar Germany GmbH, 10553, Berlin, Reuchlinstraße 10 - 11, Germany*
[3]*Axenic Limited, TS21 3FD, Sedgefield, Durham, Thomas Wright Way, United Kingdom*
[4]*Siklu Communication Ltd., 49517, Petah Tikva, Ha-Sivim St 43, Israel*
[5]*Orange Polska, Wholesale Orange Labs, 21040, Świdnik, Czereśniowa 8, Poland*

2.1 Introduction

A numerous number of communication services and applications have been developed in the current decade and it can be stated that the bandwidth demand for data services is constantly increasing. This is because services like voice over internet protocol (VoIP), video on demand (VoD), internet protocol TV (IPTV), and direct web applications requiring high speed internet access are widely used [Osseiran et al. - 2014; Cisco - 2016]. This increasing need for data communications makes broadband internet access essential for citizens in the metropolis, but of course also for people living in rural areas. Providing broadband internet to all customers, i.e. also to those living in rural or underdeployed areas, requires the development of new internet access technologies.

A seamless hybrid fiber-wireless (HFW) architecture represents an alternative access technology that is especially beneficial for customers living in rural and underdeployed regions. HFW systems combine fiber-like speed and the flexibility of a wireless network. It opens up the possibility for operators to deploy wireless broadband internet access in an almost ad hoc manner without the necessity to deploy a new fiber infrastructure beforehand. Not only does this avoid delays associated with the fiber deployment, but it also enables alternative economic business models for operators [Chuenchom et al. - 2017].

HFW systems are therefore of particular interest in connecting new commercial customers, e.g. working in new business areas or private customers living in rural areas or underdeployed areas where an optical fiber infrastructure is either inaccessible or economically not viable.

As for the optical infrastructure, gigabit passive optical network (GPON) technology is widely used nowadays in Europe and around the globe.

---

* Corresponding Author: Andreas Stöhr; andreas.stoehr@uni-due.de

Optical and Wireless Convergence for 5G Networks, First Edition.
Edited by Abdelgader M. Abdalla, Jonathan Rodriguez, Issa Elfergani, and Antonio Teixeira.
© 2020 John Wiley & Sons Ltd. Published 2020 by John Wiley & Sons Ltd.

GPON is adapting coarse wavelength division multiplexing (CWDM) for bidirectional communications and a point-to-multipoint topology. Today, many key operators worldwide use GPON technology in their fiber infrastructure. Normally, GPON provides fiber to the end consumers, both for domestic and commercial end users. Currently, it is a leading technology for passive optical networks (PONs). GPON offers a splitting ratio up to 1:64 on a single fiber, in contrast to standard copper wires [Syambas and Farizi - 2015]. Obviously, it is therefore of commercial interest for operators around the world to provide a low-cost solution for connecting new urban environments, newly built enterprises or multi-dwelling houses to the existing GPON infrastructure.

2.2   Passive Optical Network

Communication technology has been continuously upgraded and developed by improving capacities and efficiency in optical fiber and equipment in the communication network, in which the optical fiber medium is used because it can support the demands of users such as speed and bandwidth. The advantages of optical fiber are less noise and it is less susceptible to electromagnetic interference than copper wire. However, there are also restrictions on access to the fiber optic network, such as the high cost and the area for installation network of connected consumer and business applications.

Over the past decade, PONs have become the most attractive technological alternative for network providers and today several generations of PON standards for high speed bidirectional communication have been deployed [Srivastava - 2013].

In general, PONs can be mainly divided into two main categories: GPONs and ethernet-based passive optical network (EPONs). The evolution of the GPON standard has paved the way for GPON generations supporting ever increasing capacities for downstream and upstream traffic. Basically, the PON standard consists of an optical line termination (OLT) and an optical network unit (ONU). The communication from the OLT to the ONU is named downstream communication and consequently, the other communication direction from the ONU to the OLT is called upstream communication. GPON supports bidirectional communication, but uses a different wavelength for downstream and upstream traffic, i.e. a CWDM is applied [ITU-T - 2008; Effenberger et al. - 2007].

The EPON standard merges low-cost ethernet technology and low-cost fiber-optic network architecture targeting next generation network technologies. The first EPON standard approved by the IEEE in the year 2002 was IEEE 802.3ah supporting 1.25 Gbit s^{-1} data speed for downstream and upstream communication [Srivastava - 2013].

2.2.1   GPON and EPON Standards

The first standard that eventually led to GPON was a PON utilizing the asynchronous transfer mode (ATM-PON or APON). The APON standard was drafted by the FSAN and subsequently ratified by ITU-T. APON uses a single mode fiber with 1310 nm and 1550 nm wavelengths for upstream and downstream communication, respectively. The next passive optical network standard after APON was BPON. BPON used a similar topology as APON. The major difference was that the wavelengths for downstream and upstream communications were changed with respect to APON. In BPON, 1490 nm was used for upstream and 1310 nm for downstream communications, while the 1550 nm wavelength was reserved for the transmission of additional RF video signals. For both standards, i.e. APON and BPON, the line speed for upstream and downstream traffic is 155 Mbit s^{-1} and 622 Mbit s^{-1}, respectively.

Because of the huge demand from consumers, the line rates had to be increased. This eventually led to the GPON standard, which supports a line speed of 2.488 Gbit s^{-1} for downstream and 1.24 Gbit s^{-1} for upstream traffic. GPON utilizes a time division multiple access scheme to allocate a specific time slot to each ONU for upstream communication to the OLT. The maximum upstream data rate is 1.24 Gbit s^{-1}, but it must be shared with all other ONUs connected to the OLT. In a typical GPON network, usually a 1:32 or 1:64 optical splitter is used, i.e. one OLT serves up to 32 to 64 ONUs [ITU-T - 2008].

An EPON is based on the IEEE standard. GPON uses an IP-based protocol and ATM and GEM (GPON encapsulation method) end coding; EPON is a similar PON system, which can be reached with optical fiber length up to 20 km distance. EPON is completely compatible with all ethernet standards, which means no encapsulation or conversion is necessary when connecting to the ethernet based networks. In EPON, both upstream and downstream line rates are 1.25 Gbit s^{-1}. It provides bidirectional with 1.25 Gbit s^{-1} link using 1490 nm wavelength for downstream and 1310 nm wavelength for upstream, while 1550 nm wavelength preserved for the added services of providers such as RF video, optional overlay service [Kramer and Pesavento - 2002; Jiang and Zhang - 2010].

2.3 Transparent Wireless Extension of Optical Links

Radio-over-fiber (RoF) is a technique to transport information over optical fiber by modulating the light with an analog RF signal, e.g. to transmit a wireless communications waveform from an OLT to a remote antenna [Jiang - 2012]. Figure 2.1 shown the system approach of the optical wireless access architecture. Based on this concept, the technology RoF is an attractive option for emerging access wireless network access architecture to reduce infrastructure costs and complexity of antennas, and generate high microwave carrier frequencies optically.

2.3.1 Transparent Wireless Extension of Optical Links Using Coherent RoF (CRoF)

Figure 2.2 shows the general architecture of a bidirectional wireless link in which both remote antennas are interconnected by fiber using RoF technology [Pooja et al. - 2015]. Such an architecture could serve as a wireless bridge, e.g. for a GPON network. Assuming the OLT on the left hand side then, for downstream communications, the data signal is applied to an optical-to-electrical (O/E) converter for generating the radio frequency signal to be transmitted wirelessly. The output signal from the OLT is always an optical baseband signal, i.e. the O/E converter must not only photodetect the optical baseband signal, but also upconvert the electrical baseband signal to the desired RF carrier frequency. At the receiving antenna, the RF signal must be first downconverted to baseband and then remodulated onto the standardized optical carrier so it can be correctly received by a standard ONU.

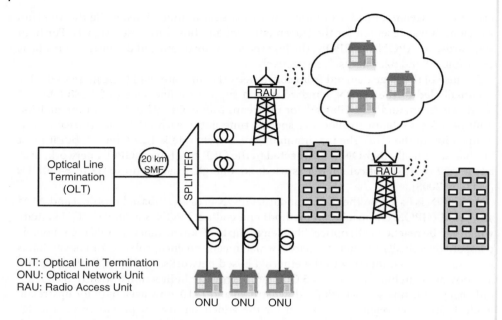

Figure 2.1 Optical-wireless access network architecture using RoF in the radio access units.

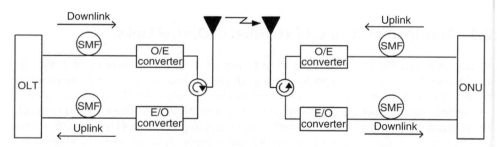

Figure 2.2 Diagram of a bidirectional wireless link of RoF technology.

There exist different options for realizing the O/E converter. The baseband downlink signal from the OLT can be transformed directly and transparently (no re-modulation) into the RF wireless downlink signal by optical means. In this case, the baseband optical signal is multiplexed with another optical cw signal from a local oscillator (LO) laser before it is coherently detected using a photomixer. In this CRoF approach, the wavelength difference between the OLT baseband signal and the LO laser sets the desired RF carrier frequency [Stöhr et al. - 2014]. Figure 2.3 shows the full-duplex system approach for a wireless bridge in a GPON network using CRoF. Depending on the implementation, the required LO laser can either be located close to the antenna or it can be located remotely at the OLT. For receiving the signal using a standard ONU, the received wireless signal is first downconverted to baseband using a Shottky barrier diode (SBD) and then remodulated transparently onto an optical carrier. Remodulation could be done using a direct or externally modulated laser. The CRoF approach has an inherent advantage concerning chromatic dispersion as compared to analog transmission of the RoF (ARoF)

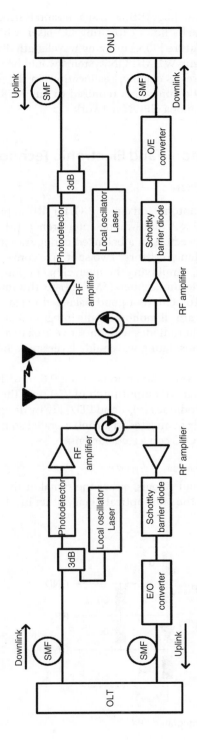

Figure 2.3 A full-duplex CRoF approach.

because in CRoF baseband-over-fiber (BBoF) transmission is used [Babiel et al. - 2014]. In addition, the CRoF approach allows exploiting LO laser, which is uncorrelated to the OLT laser. Especially for future PONs utilizing wavelength division multiplexing to increase capacity, such coherent-WDM-PON systems would be very beneficial since it allows flexible optical and wireless channel reallocation. In the following sections, we will describe some of the key components required to build CRoF radio access units (RAUs) for extending the reach of a GPON network.

2.4 Key Enabling Photonic and Electronic Technologies

2.4.1 Coherent Photonic Mixer

The design of a coherent photonic mixer (CPX) for RF applications is shown in Figure 2.4. It consists of a 2×2 multi-mode interference (MMI) coupler and two waveguide integrated photodetectors in a balanced configuration together with a bias network and two $100 \, \Omega$ matching resistors. Two spot-size converters (SSCs) are added to the MMI input waveguides, simplifying the fiber/chip coupling process by matching the mode of a standard single-mode fiber (SMF-28) to the input mode of the SSC [Chuenchom et al. - 2015]. To achieve a good balanced performance between all of these function elements, they are all monolithically integrated onto a single InP die.

The evanescently coupled photodiodes with an active area of $4 \times 15 \, \mu m^2$ were grown by MOVPE on top of a semi-insulating waveguide layer stack and biased via the integrated bias circuitry.

The produced CPX chip has the dimensions of $3 \times 0.6 \, mm^2$. The chip has an input waveguide pitch and a coplanar RF output pitch of 250 µm. The waveguides between the MMI and each of the photodiodes (PD1 and PD2) are by design of equal length and the photodiodes are electrically connected in an anti-parallel configuration, resulting in an on-chip subtraction of the O/E converted currents.

$$P_{RF,out} \alpha (I_{PD1} - I_{PD2})^2 \alpha P_{opt,signal} * P_{opt,LO}. \tag{2.1}$$

If only one of the optical input signals, either the signal or the LO is supplied to the CPX, the two photodiodes see the same optical signal, either in-phase or with 180° phase

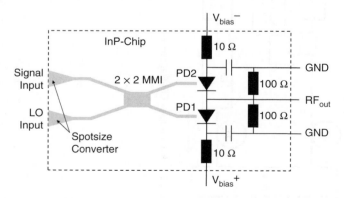

Figure 2.4 Schematic view of the integrated CPX.

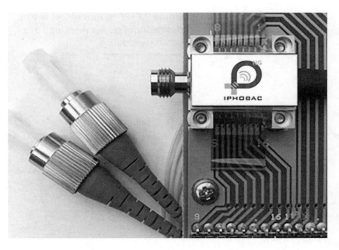

Figure 2.5 Photograph of the CPX module with a V-connector mounted on a printed circuit board (PCB) for easier DC connection.

difference, and in the case of good symmetry, i.e. high common mode rejection ratio, the subtracted current is zero, resulting in a vanishing RF output signal, independent of the input signal.

Alternatively, if the optical LO power is kept constant, one can directly see that the electrical RF output power scales linearly with the optical signal power.

The chip was then mounted in a standard gold-box package with a coaxial V-connector and fiber/chip coupled with two SMF-28s. Both PDs showed a responsivity of 0.3 A W^{-1} with a PDL of only 0.5 dB at 1.55 μm wavelength. The electrical output signal is provided by a short coplanar waveguide (CPW) which is connected by multiple short bonding wires to a following low-loss CPW on a quartz substrate leading to the output connector. A picture of the final module is shown in Figure 2.5.

Unlike commercial photodiodes, the CPX is comprised of a 3 dB coupler and a balanced millimeter-wave photodiode, as can be seen in Figure 2.5. Due to the balanced approach, the total coherent power of the optical signal is used for optical-to-RF conversion and the large DC currents due to the optical LO are subtracted on the chip. This is in contrast to single PDs where already half of the optical power is simply lost due the required optical 3 dB coupler. This way, the conversion efficiency or the responsivity can be improved by at least 3 dB as compared to commercial photodiodes, if one assumes that otherwise the performance of the individual photodiode chips is the same [Chuenchom et al. - 2016].

The RF output power versus optical input power measured in a heterodyne setup where one signal is applied to the signal and the other to LO fiber input is shown in Figure 2.6. The maximal output power slightly drops from nearly 0 dBm at 70 GHz to −6 dBm at 90 GHz due to the roll-off of the photodiode S21 response.

In a second step the chip was packaged together with a commercially available mid-power amplifier (HMC-AUH320 amplifiers, 71–86 GHz, GaAs HEMT, Analog Devices, specification: gain = 16 dB at 74 GHz, P_{1dB} = +15 dBm) and CPW-to-WR12 transition in a newly designed hermetic package suitable for direct connection to a rectangular waveguide.

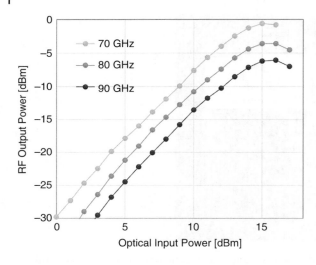

Figure 2.6 RF output power versus optical input power measured in a heterodyne setup.

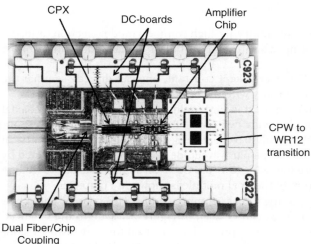

Figure 2.7 Overview of the CPX module assembly with a medium power amplifier and CPW-to-WR12 transition.

The CPW-to-WR12 transition was designed to match the balanced photodiode chips CPW RF output port and to launch the RF power in the rectangular waveguide port (WR12) on the package, as shown in Figure 2.7.

The final packaged WR12 CPX module is shown in Figure 2.8, here without a lid and an additional WR12-to-W1 adapter to ease the characterization. Due to the band pass behavior of the antenna and the WR12 waveguide, the RF output power is maximal at around 78 GHz with a clear roll-off towards the lower and higher pass band edges at 60 GHz and 90 GHz (see Figure 2.9). A similar behavior can be seen for the RF output power versus optical input power shown in Figure 2.10. The saturated RF output power at the center of the band at 77.5 GHz is just short of 15 dBm and at the frequencies of 73 GHz and 83 GHz still around 10 dBm, respectively.

Figure 2.8 Photograph of the CPX module with WR12 output and attached WR12-to-W1 adapter.

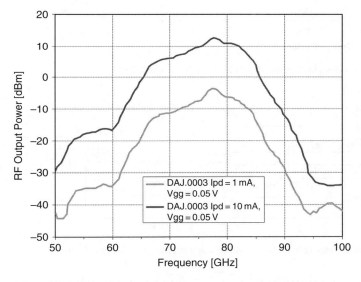

Figure 2.9 RF ouput power versus frequency at 1 and 10 mA.

2.4.2   Single Side Band Mach–Zehnder Modulator

Single-sideband (SSB) modulation discards half of the conventional double-sideband (DSB) spectrum and is therefore spectrally efficient and dispersion tolerant. The remaining sideband represents an up-shift of the baseband spectrum, and this is conveniently recovered by a heterodyne or homodyne mix with an LO laser. GaAs SSB modulators designed here used the I-Q (in-phase–quadrature) configuration – a dual-parallel combination of two integrated Mach–Zehnder (MZ) interferometers. With two identical, but 90° phased RF inputs, SSB modulation results from the mixing of two sets of carrier-suppressed DSB spectra (Figure 2.11].

The traveling-wave MZ modulators and 2x2 recombiner are monolithically integrated into MBE-grown GaAs/AlGaAs epitaxial layers on a 6-inch SI GaAs substrate, using technology based on industry-standard foundry processes common to a monolithic

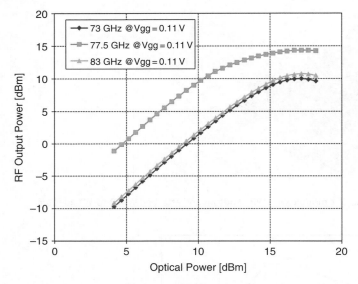

Figure 2.10 RF output power versus optical input power at 73, 77.5 and 83 GHz.

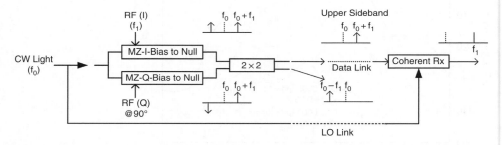

Figure 2.11 The GaAs SSB modulator chip is shown within the dotted outline SSB modulation.

microwave integrated circuit (MMIC) and pseudomorphic high electron mobility transistor (p-HEMT) fabrication.

The E-band emphasis (70–80 GHz) of this program posed unique challenges for modulator technology, which had previously only been demonstrated up to 40 GHz. Design models indicated that lowest drive-voltage over the E-band would be achieved using a reduced active length to reduce the effects of RF loss. By prioritizing the RF/optical velocity match within the E-band, the response flatness of low frequency could be sacrificed for best performance where most required. The traveling-wave electrode structures are inherently dispersive, so to achieve this requires excellent RF models and control of the electrical properties.

The SSB chip designs used, for the first time, a compact folded configuration, whereby a short straight RF feed from one end of the chip is aligned with the traveling-wave propagation direction. This is combined with a folded optical path, with all optical input/output (I/O) via the opposing end of the chip (Figure 2.12). Advantages accrue in terms of RF input simplification and loss (there are no RF 90° bends) and optical

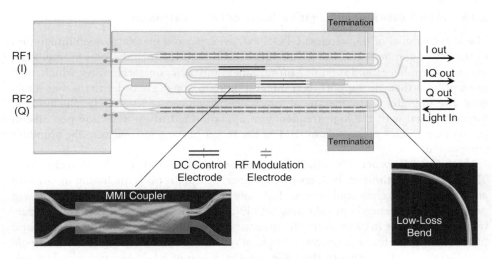

Figure 2.12 GaAs SSB modulator with folded optics to enable straightened RF.

fiber management. This configuration was enabled by a significant investment in the development of low-loss optical 90° corner bends and other guided-wave elements.

Although SSB modules of this type achieved a useful response up to 70 GHz, the desired extension into E-band proved elusive. Analysis highlighted that the effect of RF loss (both on and off chip) had initially been underestimated.

However, the main difficulty of achieving exactly the right velocity match at such high frequencies proved critical, with the best response seen in variants aimed at a 50 GHz DP-IQM application (Figure 2.13, green). Based on this learning, and now with improved models, a new shorter single MZ modulator (also designed for a 50+ GHz) has shown improved response up to 70 GHz (Figure 2.13, blue). This incorporates package and RF interface improvements as well as a new chip design; however, chip-on-carrier measurements using RF probes reveal there is still significant RF loss overhead above 50 GHz due to packaging.

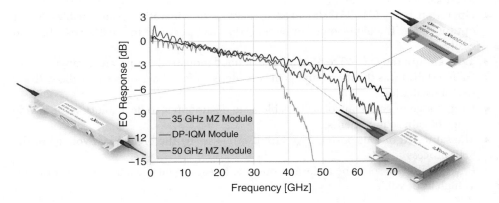

Figure 2.13 Measured electro-optic modulation responses of straightened RF GaAs MZ modulators compared with a 35GHz RF side-entry module of older design. The dual polarization I-Q modulator (DP-IQM) module shown, used a variant of the SSB chip and the same package style.

2.4.3 High Power Amplifier in the E-band for GPON Extension

The low-cost goal of the wireless GPON extension system mandates deep integration of the active components. The RF amplifier chain required includes power amplifiers, low-noise amplifiers, and variable gain amplifiers and their control circuitry, which must use a process that can support their integration onto one die in order to meet cost targets. The amplifier chain operates at mm-wave frequencies of 70/80 GHz. The low cost target coupled with the high frequency of operation led to the selection of a silicon germanium (SiGe) process from IBM as the target process for realizing the amplifier component.

The SiGe 8HP process is a BiCMOS process based on 130 nm CMOS technology. The process is optimized for high-performance RF applications, including microwave and mm-wave wireless applications, last mile communications and dense metro-area networks and advanced optical transport (40G to 100G). The process is therefore particularly promising in its capability to integrate many of integration photonic broadband radio access units for next generation optical access network components onto a single die. The major alternative to the SiGe process is use of III–V technologies like gallium arsenide (GaAs), but these are more expensive to manufacture and do not possess the integration potential offered by the SiGe process. The SiGe process supports SiGe heterojunction bipolar transistors (HBTs), which possess superior linearity and higher breakdown and operating voltages than CMOS-based options, hence the SiGe process can support higher power amplifiers and better LNAs. The SiGe 8HP process was developed by IBM and has been available since 2005. Since then the process has matured and stabilized and is available for large scale commercial manufacturing.

The amplifier building block targeted for the system is primarily a high power amplifier, but may also serve as general purpose gain block. The amplifier specifications are listed below.

- Form factor used a bonded form factor the complicated development of a package can be avoided.
- I/O type used single-ended configuration eliminates the design of a complex balanced to unbalanced on the PCB.
- Operation band (71–76 GHz) used to narrow the operation band enables better optimization. The high-power amplifier (HPA) can be tuned to the 81–86 GHz band easily.
- Temperature range (−40 to 85 °C) has extended industrial temperature range.
- Gain (40 dBi) was the maximum gain considered safe to have on a single die without risk of oscillation.
- Large range of gain control will enable a single amplifier to be a solution for a wide range of gain requirements.
- The saturated output power, P_{SAT} (15 dBm), is set by the breakdown voltage of the 8HP process.
- P_{1dB} (12 dBm) is set to be a close as possible to P_{SAT} to enable a linear power output as high as possible.

The amplifier consists of three gain stages, as shown in block diagram of Figure 2.14. Power amplifiers (PAs) in SiGe technology targeted at mm-wave frequencies are similar in concept to PAs operating at lower frequency bands. The challenges are associated

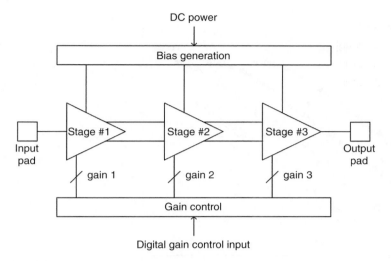

Figure 2.14 Block diagram of high power amplifier.

with the characteristics of the SiGe process. Power amplifiers require a high operating voltage of the output transistors in order to pump significant power into the load, but the SiGe process HBT transistor BVCEO is about 2 V, while the BVCBO is about 6 V. Device performance is highly dependent at bias current stability over process and temperature spread, so dynamic bias control is typically required in order to stabilize bias currents. Power amplifiers might also be subject to bad VSWR conditions when not properly matched. Such conditions might degrade or even destroy the output stage if not protected against. In addition to these issues, realizing high gain is made harder by the fact that the operation frequency of the power amplifier is not far from the cutoff frequency of the transistors it consists of, which is about 200 GHz.

The design deployed in the system consists of three sequential stages interconnected internally using differential signals, but interfacing the outside of the chip using a single-ended interface. The output and input are designed to work with a 50 Ω load, although some additional matching is required to account for the fact that bonding wires should be used to connect the die to a PCB, and these bonding wire characteristics depend upon their dimensions and upon the exact method with which the die is mounted onto a PCB.

The supply voltage is 2.5 V for all stages except the last stage, which may have higher supply voltage ranging from 2.5 V and up to about 3.5 V. The final selection of output voltage for the last stage determines the exact power output capability and the linearity of the PA and has been designed to be determined based on testing of the die. The desired output compression performance was designed to be attained with a nominal supply voltage of 3 V. Under this condition the expected total DC current is about 350 mA without a signal, and might become slightly higher when the amplifier is loaded and driven into saturation.

The design procedure of the baseline power amplifier was layout oriented in the sense that initial schematics were taken into the layout. After that the layout was done so that design parasitics were extracted, and both design and subsequent layout were improved in order to reduce parasitics and account for their presence. This process was repeated

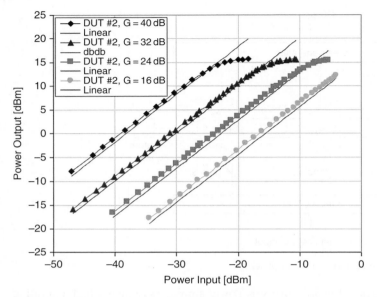

Figure 2.15 Gain and compression at various gains.

iteratively until the desired performance was achieved. Whenever possible, the symmetry associated with differential amplification stages was reflected in the layout in order to ensure the optimal differential performance of the amplification stage.

The bias arrangement of the power amplifier was one of the places in which enhancements were focused, as stable and consistent bias is a key factor for ensuring high reliability and consistent performance of the die over process and temperature variations. The on-die band-gap voltage reference source drive and stabilized local current sources were dedicated to each power amplifier stage. Finer bias control is also available via a digital control interface and enables fine tuning of the device after it is installed in the target system. The same digital control interface is also used to control other local current sources that determine the current through each stage gain transistor and thus enable control of the gain at each stage.

Some measurement results for the amplifier are shown in Figures 2.15 and 2.16.

2.4.4 Integrated Radio Access Units

In this section, the development of RAUs for transparently bridging a GPON network using an E-band (71–76 GHz) wireless point-to-point link are discussed. Figure 2.17 shows the architecture of an RAU employing the optical single-sideband modulator described above for uplink communication to the OLT. For downlink transmission, coherent heterodyne detection in conjunction with a frequency-agile low-linewidth laser serving as a photonic LO is used. The downlink optical baseband (BBoF) signal transmitted from the OLT is transparently converted to the desired RF carrier frequency using the CPX described in Section 2.3.1. The RF signal is amplified and then transmitted wirelessly to the RAU located at the ONU site. Note that the architecture of the RAU at the ONU site is analogous to the one shown in Figure 2.17. Therefore,

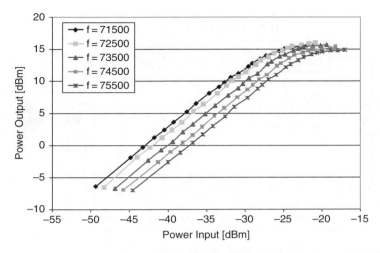

Figure 2.16 Gain and compression at various frequencies.

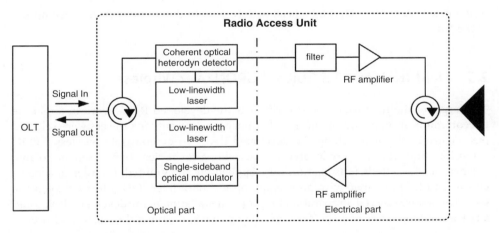

Figure 2.17 Schematic of the RAU.

after wireless transmission, the received RF signal is RF-to-optic converted using an SSB before it is sent to the ONU.

In Figure 2.18, a slightly different RAU concept that is especially suited for a wireless extension in a GPON network is shown. For downlink transmission, the proposed RAU also uses the CRoF approach. Note that the LO laser is not phase locked to the OLT Tx laser. This enables low complexity and thus low cost, but it also leads to non-frequency-stable RF signals. Consequently, we propose using an envelope detector as a wireless receiver instead of a heterodyne receiver. Whereas the RAU architecture shown in Figure 2.18 is more flexible and could in principle also be implemented in a more advanced coherent-WDM-PON network (NG-PON2), the RAU presented in Figure 2.17 provides advantages with respect to power consumption and cost. It allows the use of low-cost standard GPON laser diodes instead of RF single-sideband modulators. Furthermore, the high-power RF amplifier required to drive the SSB-MZM

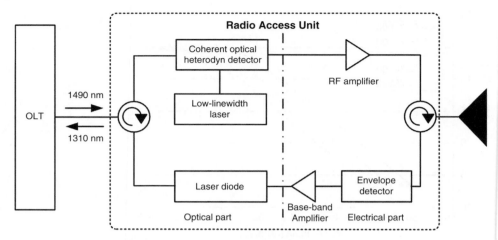

Figure 2.18 Hybrid RAU architecture for the GPON extension approach.

is no longer needed, which is beneficial with respect to the overall power consumption of the RAU.

2.5 Field Trial for a 2.5 Gbit s⁻¹ GPON over Wireless

The target of the field trial was to verify that the developed RAUs work in a real life telecommunication environment. For that purpose, two antenna towers were selected (see Figure 2.19] for installing the developed RAUs to demonstrate wireless GPON extension. Both towers were located in Garwolin, a town close to Warsaw in Poland. Permissions for antenna installation were applied for and granted. In order to achieve end-to-end (E2E) connectivity, an existing optical dark fiber (ODF) link between the two towers was reserved. Preparation for proper installation of the developed RAUs and a risk assessment were done beforehand during an inspection of the Garwolin towers.

Figure 2.19 The Tx tower was installed for the transmitter antenna (right) and the Rx tower was installed for the receiver antenna (right).

The Tx tower, which was connected to the OLT, has a height of 40 m. Its base is 127 m above sea level (127 MASL). The height of the installed RAU antenna was 30 m above ground. The height of the Rx tower is 30 m above ground level and the Rx antenna was implemented on the top of the tower, i.e. also at a height of 30 m. The base of the Tx tower is 128 MASL and the distance between the two towers is 445 m.

Besides the two towers, an office (central office, CO) was also reserved close to the Tx tower for implementing the commercial GPON and test equipments.

For carrying out the field trial, it was necessary to apply for radio transmission acceptance at the Polish regulator (UKE – Office of Electronic Communication https://www.uke.gov.pl/). For that reason, radio permission was requested and granted by UKM for the test period. Since the requested frequency allocation (71–76 GHz) covered not only commercial but also a part of the military spectrum allocation the Polish army also had to accept the application beforehand.

The fundamental properties of the two antenna towers are summarized below.

Tx tower	Rx Tower
Base level: ~128 MASL	Base level: ~127 MASL
Height (total): ~40 m	Height (total): ~30 m
Facility: close to the CO building	Facility: equipment cabinet
ODF (SC/APC connectors in the CO)	ODF (SC/APC connectors)
Fiber length to the CO: 60 m	Fiber length to the CO: 658 m

As mentioned before, the main part of the GPON network and test equipment was localized in the CO building close to the Tx tower. This included the GPON OLT devices. From the GPON optical Tx interface, the 1490 nm downlink wavelength signal was extracted and connected to the RAU transmission module located at the Tx tower via SMF. The transmission antenna was installed on the Tx tower on a platform 30 m above ground with an azimuth angle of 141,85°. The receiver antenna was installed on the roof platform of the elevator engine room at the Rx tower. This platform was also used for 3G/4G mobile network equipment and, therefore, a 644 m long ODF to the CO was already installed. To achieve E2E connectivity during the field trial, this ODF was exploited to connect the RAU at the Rx tower with the ONUs, which were also located in the CO. The reverse direction from subscriber devices to the OLT was purely optical.

Figure 2.20 schematically shows the system architecture, including the two towers and a map of the site.

For testing the system performance, an IP tester was used for generating and monitoring the traffic in the system during the field trial, as shown in Figure 2.20. In addition to the traffic generated by the IP tester, the traffic from a video streaming source was also multiplexed to reach the full-blown capacity of the GPON architecture. In detail, for downlink transmission, two times the 1 Gbit s⁻¹ traffic generated by the IP tester was multiplexed together with the traffic from the video stream source, which was basically a computer. This led to total downlink traffic of up to 2.5 Gbit s⁻¹. For uplink transmission, two times 0.6 Gbit s⁻¹ traffic was generated by the IP tester, thus also reaching approximately the maximum line rate for uplink transmission in a GPON network.

Note that commercial SFP+ modules were used at all ports of the OLT and the IP tester. Two optical wavelength splitters were used to separate the wavelength the optical

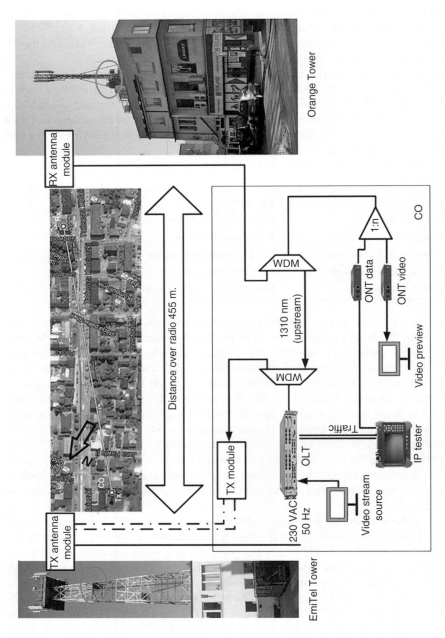

Figure 2.20 The system setup for the field trial for a 2.5 Gbit s^{-1} GPON extension with 445 m wireless distance.

downlink (1490 nm) and uplink (1310 nm) carriers after the OLT and before the ONU. For uplink transmission, the optical splitter between the OLT and the ONU were directly connected via optical fiber. For downlink transmission, the optical carrier at 1490 nm wavelength was connected to the RoF unit for wireless transmission. In the experimental setup, a wavelength converter was used to convert the wavelength from 1490 nm to 1550 nm. This was necessary because the the developed CPX had been optimized for 1550 nm wavelength. It still operates also at 1490 nm and 1310 nm but with a power penalty of 12.73 dB between 1490 nm and 1550 nm. To avoid this penalty, a transparent wavelength converter was developed and implemented during the field trial. For commercial applications, the layer structure and fiber optic packaging CPX must be optimized for supporting 1310 and 1490 nm wavelengths. After the wavelength converter, the 2.5 Gbit s⁻¹ downlink signal was transparently converted to an E-band (71–76 GHz) RF signal using the developed CPX and a local LO in the Tx RAU.

In order to estimate the maximum wireless link the developed RAUs could support, the system was first tested in a lab using an attenuator to represent the wireless path loss. The set-up of the lab system is shown in Figure 2.22. As can be seen from Figure 2.22, the electrical output from the HPA after the CPX was directly connected to an RF attenuator and then to the SBD in the Rx RAU. After downconverting the received E-band signal to baseband using the SBD, it was re-modulated onto the 1490 nm optical downlink carrier using the laser of a commercial SFP B+ module with an optical output power of +3 dBm. Then, the re-modulated 1490 nm signal was connected to three ONUs, as shown in Figure 2.21. For performing bit error and frame loss tests, the two ONUs were re-connected to the IP tester whereas the ONU streaming the video signal was connected to another computer playing the video on its screen. This also allowed a direct inspection of the transmission quality as a high number of lost packets would result in poor video quality.

The maximum wireless distance that can be supported by the system depends on the available gain in the system and the SNR. With respect to power, bottlenecks are also given by the power levels that can be accommodated by the commercial SBD and SFP modules. If the optical power into the SFPs is lower than the predefined limits, the ONUs cannot synchronize with the OLT. On the other hand, the maximum power level is defined by the maximum safe input power for the SBD in front of the SFP module. Figure 2.23 shows the operational range of the baseband signal power to the SFP module. It extends from −19.72 dBm to −12.72 dBm. As can be seen from Figure 2.23, the baseband input power level to the SFP module changes quadratically with the received RF input power to the SBD because the SBD is a square-law detector.

For the field trial, an HPA was used for the wireless transmission with a maximum gain and saturation output power of 70 dB and 17.33 dBm, respectively. In addition, an LNA was installed at the receiver site with maximum gain and saturation output power of 20 dB and 0 dBm, respectively. Two 43 dBi cassegrain antennas were installed at the Tx and Rx RAUs. Based upon the measurement of power level variation the maximum wireless distance can be estimated to be above 2 km.

The experimental results of the field trial using the antennas described above are summarized in Table 2.1 and Table 2.2 showing specifically the results for ONU1. From Table 2.1, it can be observed that the packed loss for the 1 Gbit s⁻¹ downlink transmission to ONU1 is only about one lost packet per second, which basically has no impact on the link quality. The measured maximum latency and the jitter are less than 0.064 ms

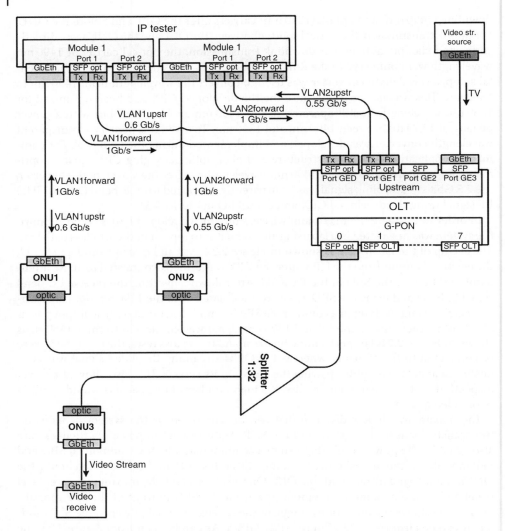

Figure 2.21 Experimental setup diagram for the wireless GPON extension by using an IP tester.

and 0.015 ms, respectively. In uplink transmission there were no lost packets, as can be seen from Table 2.2. The latency is 1.047 ms and the jitter is 0.293 ms.

2.5.1 RX Throughput and Packet Loss

As can be seen from Tables 2.1 and Table 2.2, the developed RAUs support a 2.5 Gbit s2.1 and Table 2.2 wireless GPON extension using an E-band PtP wireless link in a real life GPON network employing all commercial OLT, ONU and SFP modules. The ITU-T Rec. G.984.2 (03/2003) requirement is that the bit error rate is not worse than 1×10^{-10} in the extreme case. To observe a frame loss downstream, more than 8 bytes (64 bits) in the forward error correction frame with the given payload must have been broken.

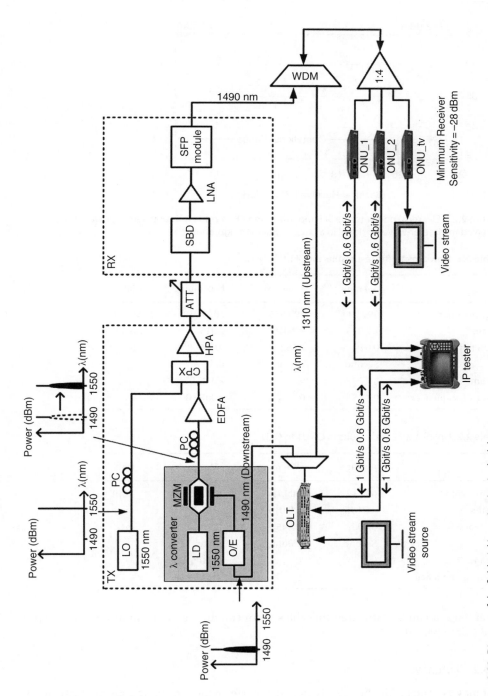

Figure 2.22 Diagram of the field trial investigation in the laboratory.

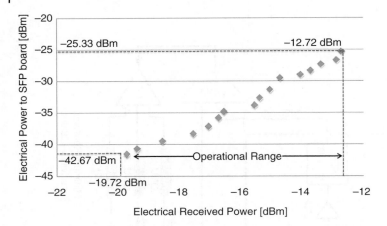

Figure 2.23 Electrical power to the SFP module in the RF-to-optic RAU unit versus power of the received millimeter-wave RF signal in a back-to-back configuration.

Table 2.1 Downlink traffic test results for OLT to ONU1.

	Average	Minimum	Maximum	Unit
Throughput	999.999	999.975	1000.012	Mbit s^{-1}
Jitter	<0.015	<0.015	0.015	ms
Latency	0.06	0.052	0.064	ms
	Second	Count	Rate	
Frame loss	8	9	7.35E-07	
Out-of-sequence	0	0	0.00E+00	

Table 2.2 Uplink traffic test results in ONU1 to OLT.

	Average	Minimum	Maximum	Unit
Throughput	600.12	600.12	600.139	Mbit/s
Jitter	<0.015	<0.015	0.293	ms
Latency	0.095	0.045	1.047	ms
	Second	Count	Rate	
Frame loss	0	0	0.00E+00	
Out-of-sequence	0	0	0.00E+00	

Therefore, it can be estimated that the system can drop at a maximum of 1 frame per 258 s time period.

2.5.2 Latency

Latency is crucial for the GPON network performance in many aspects. First of all, Rec. ITU-T G.984.1 (03/2008) defines a maximum mean signal transfer delay of 1.5 ms between the ONU downlink and the OLT uplink reference points. Therefore, any transmission link in a GPON network must ensure at least this level of delay. In new services such as very high broadband (VHBB) the TCP protocol transceiver must wait for confirmation. For that, a certain level of round trip time (RTT) is required to achieve proper TCP performance. The following simple formula allows calculating the maximum RTT latency for a giver performance.

$$\frac{\text{TCP window size bits}}{\text{Desired throughput in bits per second}} = \text{Maximum RTT latency} \qquad (2.2)$$

In order to support this, a 1000 Mbit s^{-1} download line rate with a standard 64 kbit s^{-1} window, an E2E RTT better than 0.52 ms is required. For 100 Mbit s^{-1}, the RTT is 5.2 ms. Of course, there are TCP protocol enhancements that would allow such high throughputs even with a somewhat higher RTT, but in general, the formula represents the delay requirements for high speed transmissions. In reality, the total average delay introduced by a traditional GPON system is about 0.2 ms and also the developed wireless extension would allow for VHBB.

2.5.3 Jitter

The other problem is jitter in the E2E transmission, which can affect real time services like VoIP or IPTV. With high jitter (above 40 ms) we can expect problems with voice quality or see some artifacts/pixels on the IPTV service (for jitter above 100 ms).

Another thing is jitter between the R/S (ONU uplink) and the S/R (OLT downlink). It is very low compared with limits for services, but both downstream and upstream it is a critical parameter for lossless transmission. Propagation delay in a GPON system is considered a constant. Timing calculations are done with an accuracy of 1 µs. If an upstream transmission window drifts by more than 7 ns the alarm will be raised and the ONU deactivated. Another aspect is that transmission of a single bit in GPON systems takes about 0.42 ns in downstream and 0.84 ns in upstream. If we have some timing problems within the transceivers/receivers, we can expect transmission errors and frame loss. To estimate this jitter an additional eye diagram measurement at the receiver side must be performed.

2.6 Conclusions

In this chapter, a CRoF system is presented, developed, and tested in a real world field trial for seamlessly extending the reach of standard GPON networks with downlink and uplink transmission line rates of 2.5 Gbit s^{-1} and 1.25 Gbit s^{-1}, respectively. The presented CRoF architecture thus represents the physical layer of a converged fiber and optical network, eventually leading to integrated HFW networks.

Some novel key photonic and electronic components required for seamlessly converting the optical baseband signals to E-band (71–76 GHz) wireless signals and vice versa are developed. This includes a CPX for an optical baseband to RF conversion, single-sideband MZM modulators for RF to optical baseband conversion and E-band HPAs. For the developed CPX, the maximum output power slightly drops from 0 dBm at 70 GHz to −6 dBm at 90 GHz. The single-sideband MZM modulators can be achieved frequency response up to 70 GHz. For the HPA, the operation band is the E-band (71–76 GHz), which can be tuned to a 81–86 GHz band. The maximum gain of the HPA is 40 dBi, which is considered safe to have on a single die without risk of oscillation with P_{sat} (15 dBm) and P_{1dB} (12 dBm).

The developed components were used to further develop RAUs for seamless photonic-to-RF and vice versa conversion. In the field trial, which took place in Garwolin, Poland, the RAUs were used to extend the reach of a commercial GPON network operated by Orange Polska using a wireless E-band link. During the field trial, an IP tester was used for generating and monitoring the traffic in the system. Experimentally, 2.5 Gbit s^{-1} downlink transmission of GPON data over a 455 m long wireless 71–76 GHz bridge with a low frame loss of about 1 packet per second was successfully demonstrated.

Bibliography

A. Stöhr, O. Cojucari, F. van Dijk, G. Carpintero, T. Tekin, S. Formont, I. Flammia, V. Rymanov, B. Khani, and R. Chuenchom. Robust 71-76 ghz radio-over-fiber wireless link with high-dynamic range photonic assisted transmitter and laser phase-noise insensitive sbd receiver. In *OFC 2014*, pages 1–3, March 2014. doi: 10.1364/OFC.2014.M2D.4.

S. Babiel, R. Chuenchom, A. Stöhr, J. E. Mitchell, and Y. Leiba. Coherent radio-over-fiber (crof) approach for heterogeneous wireless-optical networks. In *Microwave Photonics (MWP) and the 2014 9th Asia-Pacific Microwave Photonics Conference (APMP) 2014 International Topical Meeting on*, pages 25–27, Oct 2014. doi: 10.1109/MWP.2014.6994480.

R. Chuenchom, X. Zou, V. Rymanov, B. Khani, M. Steeg, S. Dülme, S. Babiel, A. Stöhr, J. Honecker, and A. G. Steffan. Integrated 110 ghz coherent photonic mixer for crof mobile backhaul links. In *2015 International Topical Meeting on Microwave Photonics (MWP)*, pages 1–4, 2015.

R. Chuenchom, X. Zou, N. Schrinski, S. Babiel, M. Freire Hermelo, M. Steeg, A. Steffan, J. Honecker, Y. Leiba, and A. Stöhr. E-band 76-ghz coherent rof backhaul link using an integrated photonic mixer. *Journal of Lightwave Technology*, 34(20):4744–4750, Oct 2016. ISSN 0733-8724. doi: 10.1109/JLT.2016.2573047.

R. Chuenchom, A. Banach, Y. Leiba, M. Lech, N. Schrinski, M. Yaghoubiannia, A. Steffan, J. Honecker, and A. Stöhr. Field trial of a hybrid fiber wireless (hfw) bridge for 2.5 gbit/s gpon. In *2017 19th International Conference on Transparent Optical Networks (ICTON)*, pages 1–4, 2017.

Cisco. Global mobile data traffic forecast update. In *2015-2020 White paper*, 2016.

F. Effenberger, D. Cleary, O. Haran, G. Kramer, R. D. Li, M. Oron, and T. Pfeiffer. An introduction to pon technologies [topics in optical communications]. *IEEE Communications Magazine*, 45(3):S17–S25, March 2007. ISSN 0163-6804. doi: 10.1109/MCOM.2007.344582.

ITU-T. Gigabit-capable passive optical networks. Technical report, International Telecommunication Union, 2008.

H. Jiang. Based on the rf rof technology in the property of the wireless network. In *2012 2nd International Conference on Consumer Electronics, Communications and Networks (CECNet)*, pages 197–199, April 2012. doi: 10.1109/CECNet.2012.6201630.

J. Jiang and X. Zhang. Research on epon of broadband access technology and broadband network deployment. In *2010 3rd International Conference on Advanced Computer Theory and Engineering(ICACTE)*, volume 3, pages V3–148–V3–152, Aug 2010. doi: 10.1109/ICACTE.2010.5579682.

G. Kramer and G. Pesavento. Ethernet passive optical network (epon): building a next-generation optical access network. *IEEE Communications Magazine*, 40(2):66–73, Feb 2002. ISSN 0163-6804. doi: 10.1109/35.983910.

A. Osseiran, F. Boccardi, V. Braun, K. Kusume, P. Marsch, M. Maternia, O. Queseth, M. Schellmann, H. Schotten, H. Taoka, H. Tullberg, M. A. Uusitalo, B. Timus, and M. Fallgren. Scenarios for 5g mobile and wireless communications: the vision of the metis project. *IEEE Communications Magazine*, 52(5):26–35, May 2014. ISSN 0163-6804. doi: 10.1109/MCOM.2014.6815890.

Pooja, Saroj, and Manisha. Advantages and limitation of radio over fiber system. In *International Journal of Computer Science and Mobile Computing Computer Science and Mobile Computing*, 2015.

Anand Srivastava. Next generation pon evolution. In *Proceedings of SPIE - The International Society for Optical Engineering*, 2013.

N. R. Syambas and R. Farizi. Performance analysis of gigabit passive optical network with splitting ratio of 1:64. In *2015 1st International Conference on Wireless and Telematics (ICWT)*, pages 1–5, Nov 2015. doi: 10.1109/ICWT.2015.7449249.

3

Software Defined Networking and Network Function Virtualization for Converged Access-Metro Networks

Marco Ruffini* and Frank Slyne

Department of Computer Science and Statistics, University of Dublin, Trinity College, CTVR, Dunlop House, Ireland

3.1 Introduction

The development of a network capable of supporting 5G services will require efforts going far beyond the development of a next generation, high-capacity, radio access network. As many upcoming services will require assured end-to-end delivery, the architectural design process will need to extend from the mobile access network to the fixed access and metro area network, offering coordinated network operations. A strong convergence of the network technologies and domains is also necessary to provide economically viable solutions to the massive increase in capacity and improvement of other network performance indicators (e.g., latency, availability, etc.). This will be achieved through the design of cost effective architectures that replace massive capacity over provisioning with automatic and intelligent flexible resource allocation.

This chapter describes how new 5G requirements are shaping the architectural design process, through the convergence of both data and control planes of the access and metro networks. The first section provides a brief introduction of the techno-economic challenges posed by 5G's key performance indicators (KPIs) of large increase in capacity, number of devices, network reliability, and decrease in network latency. The second section delves into the topic of convergence of fixed access and metro networks. Here, we first describe how this trend initiated, through the physical process of central office (CO) consolidation that exploited the long transmission reach of fiber access network technology to bridge the gap between access and core networks. Initial architectural solutions, such as long-reach passive optical networks (LR-PONs), achieved optical access reach of over 100 km, enabling CO consolidation of up to two orders of magnitude. Then we describe how the expectation for new latency bound services, such as cloud-RAN and others under the ultra-reliable low latency communication umbrella, have driven the network design principles away from the original LR-PON objectives of massive CO consolidation, leading instead to a more distributed approach, with several data processing facilities spread across the network edge. From a physical layer perspective, we discuss solutions that use wavelength switching in the optical access-metro network to dynamically provide capacity (i.e. making use of statistical multiplexing) to 5G mobile

* Corresponding Author:Marco Ruffini marco.ruffini@scss.tcd.ie

Optical and Wireless Convergence for 5G Networks, First Edition.
Edited by Abdelgader M. Abdalla, Jonathan Rodriguez, Issa Elfergani, and Antonio Teixeira.
© 2020 John Wiley & Sons Ltd. Published 2020 by John Wiley & Sons Ltd.

base stations from different computation centers in the metro area. This is complemented by further development of the virtualization of network functions, which allows the functional convergence of multiple and diverse network domains. Finally, Section 4 provides an insight into the virtualization of the CO, exploring some of the main software frameworks being developed across the globe.

3.2 The 5G Requirements Driving Network Convergence and Virtualization

5G is the next generation of networking that has been driving future network requirements for the past few years. The International Mobile Telecommunications (IMT) and the Next Generation Mobile Networks (NGMN) forums released in 2015 a recommendation [ITU-R M - 2015] and a white paper [Hattachi and Erfanian - 2015], respectively, that summarized the main requirements and possible application areas for 5G. Typical expected latency values are in the order of 10 ms, with an expectation of 1 ms or below for ultra-low latency services. In terms of capacity, the user experience is expected to be between 100 and 300 Mbit s^{-1} outdoor in dense areas, and 1 Gbit s^{-1} indoor, while peak cell capacity is expected at 20 Gbit s^{-1}. If we compare these with the 4G values, we see a projected 20-fold increase in cell capacity, and a decrease in latency between 5 and 50 times (e.g. for ultra-low latency applications). As work on 5G technology and architectures has progressed, these targets have been further investigated and contextualized. An analysis of possible latency performance was recently published in the NGMN white paper on 5G RAN Functional Decomposition [MacKenzie - 2018], showing the achievable end-to-end latency for different architectures, depending on the location of the different functions in the network. The results are reported in Figure 3.1 [MacKenzie - 2018]. The figure shows the possible location of different virtual functions in a RAN decomposition use case. The distributed unit (DU), which carries out the MAC and radio link control and in some cases part of the physical layer operations (high-PHY) can be co-located with the remote unit (RU), which comprises the antenna, RF, and all or part (low-PHY) of the physical layer processing, or it can be partially centralized at an aggregation site serving more than one RU, or else located at the tier 1 node, potentially serving a larger number of RUs. More options are available for the central unit (CU), which carries out the packet data convergence protocol and the radio resource control operations and can be co-located with the DU and RU, or centralized at a tier 1 or tier 2 node, thus serving several more RUs. Finally, user plane function (UPF) and multi-access edge computing (MEC) operations can be either co-located with the RU or else at tier 1 and tier 2 nodes.

We can see that as we move services and network functions from the edge (tier 1) deeper into the network towards the core (tier 3), the number of nodes decreases (these numbers are to be considered only in relative terms, e.g., the number of tier 3 nodes is 10 times less than the tier 2 nodes). Thus, running an instance of a service deep in the network can save data processing resources: for example, if it is hosted on a tier 3 node it does not need to be replicated across the 100 tier 1 nodes covered by that node. However, as seen in the figure, the transport latency increases as the service needs to reach the network core, generating a trade-off between the amount of computer resources required and the achievable end-to-end latency. It is also interesting to see that ultra-low latency

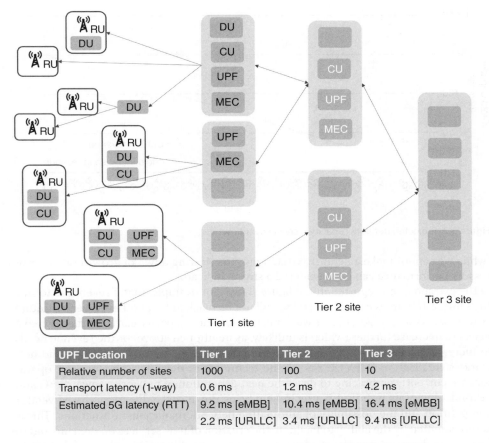

UPF Location	Tier 1	Tier 2	Tier 3
Relative number of sites	1000	100	10
Transport latency (1-way)	0.6 ms	1.2 ms	4.2 ms
Estimated 5G latency (RTT)	9.2 ms [eMBB]	10.4 ms [eMBB]	16.4 ms [eMBB]
	2.2 ms [URLLC]	3.4 ms [URLLC]	9.4 ms [URLLC]

Figure 3.1 Different tier levels for 5G services and associated latency [MacKenzie - 2018].

services cannot be achieved beyond tier 1, meaning they will necessarily be localized functions, i.e. with end points only locally interconnected.

One of the issues that is often mentioned in regards to 5G networks, which are expected to provide an unprecedented leap in network performance, is its profitability, e.g., the ability to generate enough revenue to support its cost while also generating profits for the stakeholders. Figure 3.2 shows the potential increase in network capacity that might occur in the future (for the same cost), if the growth rate remains the same as that experienced over the past 20 years. The curves were generated using data from [Hagel et al. - 2013], showing that from 1999 to 2012 the average yearly increase of capacity (for the same cost) was around 36%. The blue (lower) curve shows the projections when the revenue is maintained the same as today's, while the orange (upper) curve is the projection assuming a (exceedingly optimistic) four-fold increase in revenue for the network.

It should be noticed that the reason behind the decrease in cost per bit over time shown in the curves is the continuity in investments and innovations in networking technologies across the globe. While we would hope that such innovation will continue, it is unlikely that the growth rate will accelerate further in the near to medium term

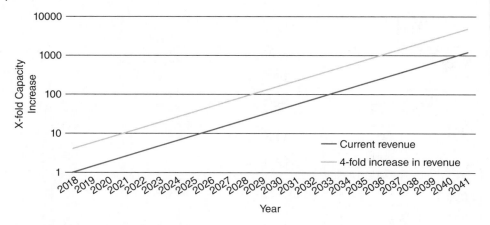

Figure 3.2 Projected network capacity increase over time.

(while we cannot make a definitive statement, the slowing down of growth rate in processors performance can be considered a strong indicator).

The extrapolation reported in the figure shows, for example, that if revenue is to stay the same, a 100 times capacity increase will not be reached for another 15 years, around year 2033. As a consequence, if we expect to see such capacity earlier, either the cost needs to decrease further (which is unlikely, as mentioned above) or the revenue needs to increase (as shown by the orange curve in the figure). Network function (and network) virtualization (NFV) is one of the technologies that most vendors and operators are currently embracing to drive the next generation of networks, towards 5G and beyond. One of its promises is precisely to bring down costs, due to increased vendor competition and the possibility of automating network management functions. This is complemented by the ability to facilitate the creation of new services on demand and to increase end-to-end network performance. However, while NFV will certainly help in decreasing costs and improving performance, one could argue that such a cost decrease might be already factored into the curve of Figure 3.2 (e.g., to offset the fact that the growth rate of processor performance has slowed down).

Rather than protracting a pessimistic prophecy of the future of 5G, this analysis aims at emphasizing that cost decrease of capacity cannot provide alone the necessary financial means to support the anticipated 5G growth, and new business models generating new revenue streams are required. Revenue growth linked to GPD is too small (a few percentage points in average for the developed world) to have an effect, and in recent years operators have not been able to monetize the increase in network capacity. For example most mobile operators, after trying to sell 4G connectivity for a higher price, had finally to revert to offering it as a no-cost upgrade. Similarly, fixed broadband operators find it difficult to sell higher rates in location where rates of about 100 Mbit s^{-1} are easily available. One of the main reasons is that there is limited benefit today in increasing rates above such values, especially as these do not assure that any of the applications will work seamlessly, especially those requiring real time two-way interaction, which need low and stable latency.

New revenue streams thus need to be provided, in addition to the legacy mobile and fixed broadband market, by attracting new types of services into the network.

5G architectures, by unifying mobile and fixed networks and computational domains (e.g., edge cloud and cloud CO) can provide new performance levels that could attract such new services. For example, end-to-end resource orchestration carried out by an automated control plane can provide quality of service (QoS) with bounded latency and high availability, and more dynamic and granular control of QoS (e.g., at application flow level). This in turn can enable applications such as AR and VR, potentially even for mission-critical services, and provide support for autonomous driving, etc.

The goal is, however, not to replace current best effort internet with an extensive end-to-end QoS guarantee, which would become largely ineffective, especially as most of today's applications have adapted to work acceptably without dedicated QoS. Rather, the vision is that of increasing revenue by building a network that, in addition to future broadband services (higher definition streaming for virtual reality, multi-stream, etc), can also support new high-value applications that cannot run over today's networks.

3.3 Access and Metro Convergence

One of the main features of network convergence is enabling resource sharing (either infrastructure or human expertise) at multiple levels, e.g., data plane, control plane, and management plane. For example, in the past two decades network convergence has driven the reduction in cost of network ownership by moving voice services from the synchronous TDM transmission systems (i.e. Sonet and SDH) to the packet switched networks used to transport internet data, through the adoption of voice over IP (VoIP) technology [Ruffini - 2017]. This also created the opportunity for operators to offer bundled services, such as voice, internet, TV, and mobile, enabling economy of scale benefits.

More recently, the mass deployment of access fiber technology has provided a new opportunity for network convergence: by removing the distance-bandwidth limit imposed by legacy copper access technology, optical access fiber enabled the consolidation of several of the network COs [Ruffini - 2016]. Increasing the access distance by over an order of magnitude triggered the investigation of new architectural solutions across the entire network, as longer access distance means that a smaller number of COs can be used to serve the same number of end users. In addition, while the number of users served by each CO becomes larger, the use of shared fiber technologies such as passive optical networks allow the multiplexing of the number of network-side ports among several users (typically 32 to 64), thus reducing the equipment footprint associated with a larger user base at the consolidated COs. If we consider that the access tends to be the most expensive part of a network (i.e. in terms of cost per user), due to the lower possibility of infrastructure sharing (i.e. in comparison with metro and core), we can understand that intuitively it is the access technology of choice that can shape the rest of the network architecture. For example, architectural choices for most of the current network have been dictated by the limit of copper transmission in the access (i.e. an average about 1 mile), which defined the maximum distance between end users and central offices (which in turn defined the overall number of COs and their interconnection). Similarly, today's access technology will be shaped by the (revenue-generating) services and applications that will run in the future, and will shape the way the metro and core network will develop accordingly.

3.3.1 Long-Reach Passive Optical Network

Since optical transmission technology can cover distances of several thousand kilome-
tres without electronic regeneration, one question that arises is what distance should the
optical access cover? The LR-PON architecture was proposed in [Payne et al. - 2002]
in order to provide an answer to this complex question, taking into consideration the
network from an end-to-end perspective [Ruffini et al. - 2017]. Because, as mentioned
above, the choice of the optical access technology determines the architecture in the
rest of the network, the LR-PON optimizes the access reach distance by minimizing the
cost across access, metro, and core networks. A typical LR-PON architecture is shown
in Figure 3.3

Since by reducing the access distance the number of required COs decreases pro-
portionally, the LR-PON architecture proposes to increase the access distance until the
number of central offices becomes small enough to allow a full mesh of wavelength
interconnections between them. For typical European country size, for example, it was
determined that a maximum reach in the order of 100 km would enable reducing the
number of COs to about 100 or less. While extending the PON reach to 100 km and
above required an optical amplifier to be placed in the optical distribution network
(ODN) before the first stage power splitter (typically three stages were considered to
enable a split of 512 ways), the LR-PON enabled the bypass of the metro transport
network, bringing the access fiber directly into the core CO nodes (called metro-core
nodes). Combined with a flat core architecture (e.g., a core interconnected by a full mesh
of wavelength channels) this enabled the reduction of the optical–electrical–optical
(OEO) interfaces, as every end point could be linked to any other by going through only
two OEO conversions, at the source and destination COs.

Studies on the possibility of moving from current hierarchical core networks to
flat cores showed [Raack et al. - 2016] that a smooth transition was indeed possible,
and identified the crossing point where the economic benefits of a flat core started to
prevail over hierarchical core to an average sustained end user capacity of 5 Mb s^{-1} [1].

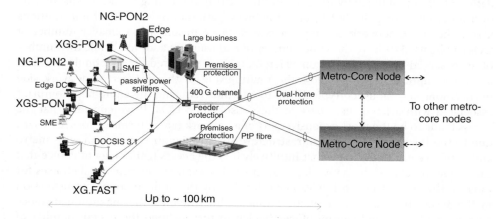

Figure 3.3 Typical LR-PON architecture enabling massive node consolidation.

---

1 Notice that sustained rate refers to the overall core network capacity divided by the number of subscribers

These results proved insightful, as recently large national networks, for example BT, have adopted fully meshed, flat network designs in their core [Lord - 2018].

One important issue that LR-PON brings up is that of protection. Access networks serving residential users are typically unprotected, (PONs typically fit in this description), and service level agreements tend to have loose availability agreements. However, the LR-PON replaces not only the access, but also the metro transmission network, which is typically protected. Thus LR-PONs require some level of protection, at least above the first stage split, over the feeder fiber part of the ODN. Work carried out on this topic spanned from the study of cost-effective load transfer techniques for minimizing protection equipment across the network [Ruffini et al. - 2010; Nag et al. - 2016] to SDN-based techniques to speed up the PON protection [McGettrick et al. - 2013a,b], even across large-scale transmission networks, using an SDN control plane to orchestrate access and core protection [McGettrick et al. - 2015; 2016].

Overall, while the LR-PON architecture has shown some interesting potential economical advantages, by minimizing the use of electronics in the network, it struggles to meet the low latency requirements of future 5G networks due to the long-reach of the transport fiber that can generate round trip time propagation latencies in excess of 1 ms. The next section provides insight into novel architectures that combine the convergence of access/metro with the ability to meet 5G's strict latency requirements.

3.3.2 New Architectures in Support of 5G Networks, Network Virtualization and Mobile Functional Split

5G networks have significantly changed the KPIs from previous generation networks. While until now the latency experienced by networked applications was put in perspective of human reaction times, e.g., in the order of several tens millisecond for most applications, the use of the network to link multiple virtual network functions (VNFs) or machine-to-machine type of applications (e.g., related to autonomous driving and other mission critical services) requires support for much lower latency. In addition, 5G networks are expected to support novel end-user applications that include advanced sensory functions, such as haptic feedback and augmented/virtual reality, which require maximum end-to-end application-level latency below 15 ms [Elbamby et al. - 2018].

Cloud-RAN, i.e. the disaggregation of hardware and software in mobile base stations, which enables centralization of their digital processing functions, is being considered as one of the main architectural choices to reduce overall cost of network ownership. This is especially important for 5G deployments aiming towards high cell densification, potentially 100-fold compared to current deployment in urban areas, which is considered the main mechanism to increase mobile network capacity (as was the case for the past 50 years of mobile network evolution – a trend known as the Cooper's law of spectral efficiency). The initial approach to C-RAN was the separation of the antenna unit on one location and all baseband processing on another, with I/Q samples being transmitted across the two by some standard interface (typically [CPRI Specification - 2013]). Today this concept has evolved to provide several other split options of the mobile protocol stack [MacKenzie - 2018]. Different splits offer different trade-offs between complexity of the remote radio unit, latency requirements, capacity requirements of the fronthaul link and different ability to carry out advanced physical layer processing functions, such as MIMO and other coordinated operations across base stations.

Some of these configurations involve the use of more than two processing points along the functional chain, with a fronthaul link (the F2 interface) connecting the antenna unit with a first stage distributed unit (DU) processing point, and a second link (the F1 interface) linking the DU to the centralized unit (CU) location where the rest of the protocol stack is processed. Even within this dual-split configuration, several options are available, depending on how the protocol stack functions are allocated to the DU and CU.

New optical interconnection architectures are required to support these emerging requirements in a cost-effective manner, i.e. with technologies that favor resource sharing and statistical multiplexing of capacity. Among the work carried out in this area, we present here two interesting directions, both shown in Figure 3.4.

1. Upgrading traditional passive optical network (PON) configurations with mechanisms for inter-optical network unit (ONU) communication that bypass the optical line termination (OLT) [Pfeiffer - 2010]. Thus, when the PON is used to link remote RUs, these can exchange information directly, without messages being relayed by the OLTs or other centralized units, thus minimizing transmission latency. This is useful for the implementation of some of the coordinated multi-point (CoMP) functions that require low latency information exchange across base stations, in cases where

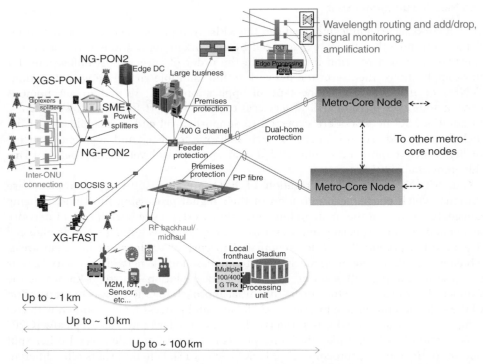

Figure 3.4 5G-enabled converged access/metro architecture with inter-ONU communication and the ability to drop signals locally (e.g. edge cloud).

the RAN is not fully centralized [Dötsch et al. - 2013] (i.e. when CoMP functions need processing at the distributed RUs). The inter-ONU communication configuration is shown on the left-hand side of Figure 3.4 and operates by interconnecting neighbouring ONUs over a local broadcast network overlaid on top of the PON, using multiple wavelength channels. Collision-free interconnection can be used between $N + 1$ neighbours by using N wavelength channels.

2. Co-locating edge processing equipment with PON split stages [Ruffini - 2017], providing the flexibility to terminate locally the services requiring low-latency, while permitting aggregation of other services into more centralized parts of the network. This is important because the location of services and their level of aggregation plays a primary role in determining the cost of network ownership. As shown in Figure 3.1, there is a trade-off between the ability to aggregate and locate functions in a node (tier 1 to tier 3) and the minimum achievable latency. A function located at the edge (tier 1) will provide, for example, lowest latency; however, it will only provide local connectivity (i.e. within a small coverage area of tier 1) and will need to be replicated in every other tier 1 where that particular service is required to run. A function implemented in a tier 3 node instead, although it will experience higher latency, will cover a much larger area (e.g., 100 tier 1 nodes, according to the example in the figure). This configuration is shown in the top part of Figure 3.4, where the splitter node includes one (or more) OLTs that can terminate a wavelength channel locally to carry out edge processing (e.g., potentially both for NFV and application level functionality).

While we have so far described C-RAN as the main driver behind innovation of the optical access/metro network, we should notice that this is only the tip of the iceberg. The C-RAN can be indeed considered a special case of the much wider network virtualization framework which, with the CO re-architected as a data center (CORD) [Peterson et al. - 2016] was extended to include more CO functionalities. The CORD, which exemplifies the concept of virtual or cloud CO, (described in more detail in the next section), virtualizes most of the CO functions, opening up the possibility of unprecedented flexibility in dynamic service allocation and ownership models. NFV indeed opens up the possibility of dynamic generation of new services via creation of new software instances, whose ownership can be transferred to a given virtual network operator in a multi-tenant environment [Cornaglia et al. - 2015; Elrasad et al. - 2017]. "Softwarization" and slicing of network functions are also instrumental in enabling network convergence, an essential aspect of 5G networks, making it possible for a higher layer orchestrator to concatenate (or "chain") shared resources across multiple domains to set up an end-to-end service from end users to data processing centers. This is important because, as mentioned earlier in this chapter, new revenue streams can be brought to the network once end-to-end quality assurance is implemented. Considering that the majority of end users will connect wirelessly to the network, e.g., using dense 5G RANs, the assurance of end-to-end quality of service will require the orchestration of at least the following three domains: the wireless edge, the computational facilities, and the fixed optical network linking the two.

In addition, in order to satisfy the requirements of new revenue-generating applications that need lower latency, some of the network functions and applications will need to be moved towards the edge (e.g., towards the tier 1 shown in Figure 3.1]. This trend goes under the name of multi-access edge computing [Taleb et al. - 2017] (previously known as mobile edge computing), which aims at bringing together computational facilities distributed over a metro area into a unified ecosystem, so that resources can be dynamically assigned to satisfy the latency, bandwidth, storage, and processing requirements of the applications.

This flexibility in the service creation and resource allocation needs however to be paired with flexibility in physical capacity allocation. While network virtualization approaches can assign dedicated slices of already provisioned resources and assure isolation between them, physical capacity can only be increased by operating at the physical layer, for example to alleviate network congestion or create new low-latency paths. A typical use case is that of redirecting optical network capacity towards the mobile cells, as their utilization varies throughout the day. If this can be achieved dynamically (e.g., in the order of a few tens to hundreds of milliseconds), then the architecture can become highly effective in reducing resource overprovisioning, as both the optical transport and the computing resources can be statistically multiplexed across several mobile base stations.

This requires the development of an agile optical access/metro network, and optical network disaggregation. By opening up the control layer of optical systems, it will enable its integration with the mobile and computing domains, allowing the network orchestrator to control the entire end-to-end path, from mobile user to cloud. While a thorough investigation of optical layer disaggregation is outside the scope of this chapter, we would like to remark that there is currently much debate on this topic. On the one hand, optical disaggregation both opens up the sourcing of the optical layer components, thus increasing cost savings and scalability [Ruffini and Kilper - 2018], and it enables its integration into the converged architecture described above. On the other hand, it requires higher margins due to the higher uncertainty of the effect of the physical impairments on the channel transmission, although this can be offset by an increase in the physical layer monitoring activity. For this reason, some believe that optical disaggregation is more suitable in the access and metro part of the network, where optical margins are looser [Belanger et al. - 2018].

In parallel, interesting work is being carried out on the use of machine learning in optical networks [Musumeci et al. - 2018], for example to better predict quality of transmission, using data made available by increased monitoring of the optical channel. This for example could help reduce the gap in performance between open (i.e., disaggregated) and closed systems.

After providing requirements for network virtualization and showing some architectural solution, the next section will provide details of some of the main NFV and SDN frameworks being developed.

3.4 Functional Convergence and Virtualization of the COs

Telecoms operators' COs have traditionally been centers of critical infrastructure located at the edge of the operator network. COs have been constrained physically and functionally due to lack of space and due to the dominance of incumbent operators and vendors, respectively. With the advent of the paradigms of SDN and NFV, the COs have become an arena for innovation (and similar patterns are evident within the head-end of cable operators). This section delves into the details of some of the latest technological innovation in the NFV and SDN areas, attempting to provide some clarity over the many different functional elements and frameworks that have recently appeared.

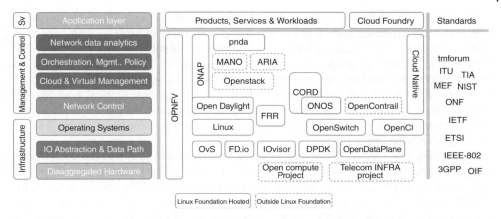

Figure 3.5 Classification of NFV-related development frameworks [Linux Foundation - 2018].

The Linux Foundation Open Source Networking Stack, shown in Figure 3.5 [Linux Foundation - 2018], provides a comprehensive view of where such components operate in the networking stack, giving also guidelines on how these could be aggregated to provide a cloud CO with fully virtualized infrastructure, control, and management.

3.4.1   Infrastructure

We start by describing the technological innovation at the lower layer of the network, e.g., the disaggregated hardware part shown in Figure 3.5. The aim is to provide an overview of how virtualization is changing the hardware level through a brief description of the main projects in this area.

3.4.1.1   Disaggregated Hardware

Legacy optical transport systems (for example OTN) are complex vertically integrated solutions that compel operators to deploy single vendor networks to ensure interoperability and manageability. The dense wavelength division multiplexing (DWDM) transponder hardware includes DSP ASICs and complex optoelectronic components. These components are a maturing technology, with recent advancements mainly in integrating functions to drive down the power, size, and cost. The software running in the transponder is traditionally bundled with the hardware, and innovation happens only at the system level. In addition, the higher layers of the network (e.g., the packet or electronic switching level) use the transport system as a black box. The concept of packet optical convergence targets this issue, aiming to provide the operators with the ability to adjust and find capacity across multiple vendor equipment domains, reconfigure networks dynamically and respond to failure or traffic demands in short time (e.g., in the order of seconds). Packet optical convergence can largely benefit from disaggregation of components of the transmission system (optical transponders and optical line systems) and providing open software to control a multi-vendor assembly of components. Disaggregation of transponders from the optical line system enables more rapid innovation of the transponders, frees network operators to mix and match technology from multiple suppliers, exercises greater control over their infrastructures and lessens the chances of vendor lock-in.

The Telecom Infra Project (TIP) [Telecom Infrastructure Project - 2018] Open Packet DWDM uses combined packet and DWDM technology for metro and long-haul fiber optic transport networks. It enables a clean separation of software and hardware. TIP have defined a packet transponder and open line transport system with open optical specifications that enables any interested party to contribute systems, components, or software. The OOPT project has formed four working groups: Optical Line System (OLS); Disaggregated Transponder & Chip (DTC); Common API (C-API), Network Management System and System Integration and Solution Development; Physical Layer Simulation Environment (PSE).

In parallel, the ONF's [Open Networking Foundation - 2018] Open and Disaggregated Transport Network (ODTN) project uses intent-based networking and converged topology graphs to define capacity and path availability programmatically. ODTN employs the Transport-API (TAPI) as a northbound interface for cross-domain orchestration and OpenConfig - [July. 2018] as a southbound interface between the ONOS controller and the optical link. Plugins mediate between OpenConfig and Openflow as well as communicating through legacy switch protocols such as TL1 and the path computation element protocol (PCEP). The ODTN automatically discovers disaggregated components and controls the entire transport network. TAPI integrates control and monitoring of the optical transport network with higher level applications such as end-to-end dynamic bandwidth services across a multi-domain carrier network and support of network slicing to enable connectivity for high bandwidth or ultra-low latency 5G services. The ODTN project builds on the TIP Open Optical and Packet Transport group's work on network planning tools and APIs. The TIP open line system includes Yang software data models of each component in the system, and an open northbound software interface ([NETCONF Enns - 2006; Thrift Paolucci et al. - 2018], etc.) to the control plane software. This allows multiple applications to run on top of the open software layer, enabling software innovations in DWDM system control algorithms and network management systems. The ODTN project also makes use of [OpenROADM MSA - 2016] Multi-Source Agreement (MSA) specifications for optical system interoperability, in particular reconfigurable optical add/drop multiplexers (ROADMs), transponders and pluggable optics. The ODTN software stack works with TIP-compliant hardware, such as the Voyager optical switch and transport platform. In future, the ODTN project may adopt some of the OpenROADM Yang data models to ensure interoperability across data planes of separate control domains.

3.4.1.2   I/O Abstraction and Data Path

CO data planes may be grouped into those that are handled by dedicated hardware (for instance ASIC-based) and those handled by a general purpose processor (that is CPUs). Traditionally, there is a trade-off between the stability afforded by the former and the flexibility afforded by the latter. In terms of dedicated hardware based data planes, the OpenDataPlane (ODP) [Open Data Plane - 2018] project defines programmatic interfaces (APIs) to network SoCs (system on chips) for the purpose of building network data planes. ODP applications are typically hardware agnostic, portable and can make use of underlying hardware accelerator techniques. OpenFlow data plane abstraction (OF-DPA) [Belter et al. - 2014] is a hardware abstraction layer between OpenFlow compliant controllers and Broadcom's StrataXGS and StrataDNX high performance switch architectures. Similar in functionality to OpenNSL (Network Switch

Library), OF-DPA requires the Indigo Openflow Agent to mediate with Northbound Openflow controllers. OF-DPA is released as a Broadcom Silicon Software Development Kit (SDK) and includes libraries, APIs and an abstract switch functionality that supports bridging, routing, data center tunnel gateways, an MPLS provider edge, label switch routing and QoS use case packet flows.

P4 (Programming Protocol-Independent Packet Processors) [Bosshart et al. - 2014] is a behavioural description language for expressing how packets are processed by the data plane of a programmable forwarding element, such as a hardware or software switch, network interface card or router. P4 defines how the data plane recognizes, parses and matches protocol data unit headers and what subsequent action should be taken. As well as implemented current protocols, P4 can implement new protocols and features that facilitate greater network control and visibility. A number of P4 compiler targets have been created, including XDP (eXpress Data Path), Netcope VHDL FPGA, Xilinx PX FPGA, P4GPU (Cuda), P4FPGA (Verilog) Netronome SmartNIC as well as Open-DataPlane. T4P4S [Laki et al. - 2016] is a P4 compiler that generates applications that abstract the interface to the Data Plane Development Kit (DPDK), Freescale NPU and OpenWRT [O. Team - 2016] without compromising performance of packet processing. Some equipment manufacturers such as Juniper have adopted P4 as the language that describes the interaction between the control plane and the data plane of their routers and switches. Stratum is a silicon-independent switch operating system that allows a switch to be controlled by a local or remote network OS (NOS) via P4, P4Runtime and OpenConfig. The objective of the Stratum project is to avoid vendor lock-in to legacy data plane architectures that depend on proprietary silicon interfaces and closed software APIs. DPDK [Intel - 2012] is a set of user space libraries that improve packet processing performance bypassing the Linux kernel, minimizing the time needed to process packets by using direct memory access, Poll Mode drivers, Huge Pages and optimized cache. FD.io (Fast Data input/output) - [2018] uses DPDK for the network IO layer (getting packets to/from (v)NICs and thread/cores). FD.io, based on the Cisco Vector Packet Processing (VPP) project, focuses on ensuring open source networking deployments have the highest throughput, lowest latency and most efficient IO services. FD.io components are typically used in conjunction with other projects such as Open-Daylight, OpenNFV, and OpenStack. The IO Visor Project focuses on dynamic run-time extensibility of data plane capabilities in the kernel. IO Visor aims at creating a repository of IO Modules that are portable across multiple possible data planes like eBPF [Miano et al. - 2018] in the Linux kernel and frameworks like FD.io.

Lastly, there is a range of tools and techniques aimed at bridging the approaches whereby the data plane is implemented in hardware or in software running on CPU. For instance, OpenFastPath [Open Fast Path - 2018] is a variant of ODP that uses DPDK to accelerate routing and forwarding, tunnelling and termination of protocols through the Linux kernel. PISCES [Shahbaz et al. - 2016] is a software switch created from Open vSwitch (OVS) and a hard-wired hypervisor switch, whose behaviour is customized using P4. PISCES is not hard-wired to specific protocols, which makes it easy to add new features. Also, SDN-enabled Broadband Access (SEBA) is a lightweight variant of R-CORD that is optimized to allow traffic to "fastpath" through to the backbone without requiring VNF processing on a server. SEBA supports a range of virtualized access technologies at the edge of the carrier network, which include PON, G.Fast, and eventually DOCSIS. SEBA supports both residential access and wireless backhaul.

3.4.1.3 Data Centre Switching Fabric

The ONF Trellis is a L2/L3 switching non-blocking fabric architecture for data centers using white box switching hardware and open source software. Unlike traditional networking approaches, the fabric itself does not run a control protocol (such as BGP, OSPF or RSTP). Instead, all the intelligence is moved into applications running on the clustered ONOS controller. In this way the fabric switches can be simplified, the entire fabric can be optimized by leveraging a holistic view of all activity, and new features and functionality can be deployed without upgrading the switches.

3.4.1.4 Optimized Infrastructure Projects

Most data centers purchase and use a large number of inexpensive general-purpose servers from different vendors. The Open Compute Project (OCP) [Heiliger - 2011] is a project formed by Facebook to design and build custom software, servers and other data center components in order to address its diverse range of infrastructure. The community formed around OCP has collaborated with other open source software projects to successfully develop energy-efficient servers, an air-side economizer and an evaporative cooling system to support its servers, in addition to designs for server racks and battery cabinets, integrated DC/AC power distribution scheme and an operating system agnostic switch (the Open Network Install Environment – ONIE).

3.4.2 Management and Control

This section describes the main software elements being developed for the control and management of the network, represented in the higher half of Figure 3.5.

3.4.2.1 Network Control

The past decade has seen the emergence and growth of a large number of Openflow and, in general, SDN control planes, ranging from basic standalone controllers (such as Floodlight [GitHub - July 2018], POX and RYU) that manage only groups of switches, through to full architectures that administer entire data centers and telecommunications networks and WANs. We are primarily interested in this latter category, which includes OpenContrail [Singla and Rijsman - 2018], OpenDayLight [Medved et al. - 2014] and ONOS [Berde et al. - 2014].

OpenContrail is a tactical SDN framework, which has been adopted by Juniper as a control framework for its SDN compatible equipment. Architecturally it is composed of four subsystems. vRouters handle network slicing, traffic steering and MPLS or VXLAN based overlay networks. The Configuration subsystem manipulates the high-level service data model into a form for consumption by the devices. The Controller component manages and monitors network state. Lastly, the Analytic subsystem collects and collates data about system performance. OpenContrail uses IF-MAP (interface for metadata access points), the open-standard IT Orchestration protocol, for model definition. In time IF-MAP will be replaced by the YANG-based configuration format.

OpenDayLight (ODL) is an open source SDN architectural framework, based on the Cisco Extensible Network Controller (XNC). The Service Provider variant of ODL has renderers for IETF's NetConf configuration, BGP and PCEP. Topology queries for the purposes of discovery, host tracking and inventory management are performed through a REST API. SDN models are defined using YANG-based MD-SAL (Model-Driven

Service Abstraction Layer), where applications are defined as a data model and the APIs required to access them can be auto-generated as part of the integration process.

ONOS is a network operating system specifically developed for service providers, driven and supported by the ONF, which also maintains the Openflow standards. ONOS is a specific ONF project with development and maintenance resources allocated to it by service providers such as AT&T and NTT. The ONOS project objectives are to provide an SDN platform with carrier-grade performance and availability and a number of use cases to demonstrate the carrier capability of the system have been outlined. These are, for example, an SDN IP Peering use case, a Network Function as a Service (NFVaaS) use case and a use case demonstrating failover using IETF Segment Routing (Spring Project). The NFVaaS use case demonstrates a virtual OLT (vOLT) solution for GPON. ONOS does not rely solely on Openflow as its SDN control plane technology, as demonstrated in the Segment Routing use case. The PCE use case looks at the issue of over-dimensioning of current packet optical cores so as to handle both network outages and peak bursts. The ONOS PCE application is used to configure, orchestrate and monitor the packet optical core to achieve much higher levels of utilization without compromising on redundancy.

3.4.2.2 Cloud and Virtual Management

The core component of the Cloud and Virtual Management function is the CORD project. Its objective is to transform the edge of the operator network into an agile service delivery platform. The CORD platform uses the paradigms of SDN, NFV and cloud technologies to disaggregate legacy network platforms and converge services, through the provision of agile data centers at the network edge. This enables the operator to deliver an optimal end-user experience and stimulate the creation of next-generation innovative services. The technologies used in the CORD access network range from GPON, [G.Fast B. Telecom - 2013] through XGPON and DOCSIS. CORD comes in a number of different flavors suiting different market segments: M-CORD is for mobile central office, E-CORD is for enterprise central office and R-CORD is for delivering broadband residential services. R-CORD implements the complete residential GPON solution as a collective of virtual machines and containers running on general-purpose data center infrastructure. While a traditional OLT is custom functionality built on monolithic hardware available from a restricted set of vendors, in R-CORD the OLT is disaggregated, in that only the OLT physical and MAC layers are created in dedicated hardware (called the OLT-MAC), and all other functions run on software distributed throughout the CORD cloud. The vOLT (virtual OLT) runs on commodity servers with GPON media access control (MAC), GPON OMCI [Effenberger et al. - 2007], 802.1ad-compliant VLAN bridging and ethernet MAC functions. Each physical port on the white box OLT terminates either 32 or 64 ONUs.

The VOLTHA [Walter - 2018] (virtual OLT hardware abstraction) function wraps the OLT-MAC as an Openflow manageable resource. On its south-bound side, VOLTHA communicates with PON hardware devices using vendor-specific protocols and protocol extensions through adapters. There are two pieces of software that work together to implement the vOLT functionality. The first is a vOLT agent running in a container or VM and facilitates a connection between ONOS and the hardware. The agent exposes a northbound Openflow interface, which enables it to be controlled by ONOS and maps Openflow messages to the native APIs of the hardware device. The vOLT Agent is comprised of an Indigo Hardware abstraction layer, Netconfd daemon [Sun Mi Yoo et al. - 2005], the proprietary APIs for the PON physical layer and an ONU management and control interface (OMCI) stack. The agent abstracts the entire PON system as a single switch to the controller. The agent is able to understand a limited subset of Openflow messages from the controller and configure the hardware appropriately. The second piece of software is a set of ONOS functions that facilitates subscriber attachment and authentication, and establishes and manages VLANs for connecting consumer devices to the central office switching fabric on a per-subscriber basis and manages other control plane functions of the OLT. Here the vOLT mimics functions of a traditional OLT such as 802.1X [Congdon et al. - 2003], IGMP Snooping [Wang et al. - 2002] and VLAN bridging. Subscriber traffic is identified by two VLAN tags within the central office. The inner tag (C-tag) identifies the specific subscriber within the PON. The outer tag (S-tag) identifies the PON. Taken together, these two tags can identify a subscriber uniquely across all OLT devices in the system. ONOS instructs the OLT which VLANs to use through OpenFlow messages.

The R-CORD virtualized customer premises equipment known as the virtual subscriber gateway runs these subscriber features in a Linux container on commodity hardware located in the CO. These include functions such as DHCP and NAT as well as optional services such as firewall, and parental control. Indeed, R-CORD supports basic Internet connectivity, as well as a collection of optional features, such as service suspension and resumption, parental control, bandwidth metering, firewalling and access diagnostics.

3.4.2.3 Orchestration, Management and Policy

Orchestration is the coordination of resources across a number of domains, disciplines and timescales so as to provision and maintain an end-to-end service. To this end, orchestrators have knowledge of the overall topology and capacity and the mechanisms to achieve particular service objectives across diverse vertical and horizontal domains, for example optical, wireless and computational. MANO [Ersue - 2013] is one such European Telecommunications Standards Institute (ETSI) project, for the management and orchestration of software-defined networks and network function virtualization. MANO deals with supporting multi-site deployments, onboarding of NFV, virtual network functions packaging, upgrading and installations on an SDN controller, creating development environments and service modelling.

Another orchestrator, ECOMP [Ersue - 2016] extends ETSI MANO and introduces the concept of the resource controller and policy component as well as the concept of metadata, for lifecycle management of the virtual environment enabling network agility and elasticity. ECOMP defines a master service orchestrator that is responsible for automating end-to-end service instances. The orchestration automates configuration processes, programmability rules and policy-driven operational management. To this end, ECOMP supports open cloud standards, such as OpenStack and OPNFV and uses Netconf, Yang configuration and management models, and Restful APIs.

Lastly, The application-based network operation (ABNO) [Aguado et al. - 2015] is an IETF SDN framework that is not solely restricted to using Openflow as a communication protocol to the data plane components. Instead the ABNO can communicate with MPLS and GMPLS multi-domain networks using PCE as the controlling agent and PCEP as the control protocol. The ABNO also has a policy manager, a I2RS (interface to routing system) client, a virtual network topology manager (VNTM) for multi-layer co-ordination and an application-layer traffic optimization server. Southbound communication with components such as Openflow are achieved using a provisioning manager. Statefulness is provided by an LSP-DB and TED database. As an example, the ABNO has been used [Napoli et al. - 2015] to demonstrate the multi-domain and multilayer configuration of commercial equipment (such as ADVA, Juniper nodes and OTN 400 Gbps channels) and the validation of the PCEP extensions to support remote GMPLS LSP set-up.

3.4.3 Cross-Layer Components

Alongside components working on a specific layer of the Linux Foundation Network stack, we can see in Figure 3.5 that there are a number of components that span multiple functional levels. ECOMP and Open-O Orchestration have been subsumed into the open source orchestrator project called Open Network Automation Platform (ONAP) [ONAP - 2018]. ONAP provides a unified architecture and offers a policy-driven software automation of VNFs and network capabilities that allows software, network and cloud providers to rapidly create and efficiently orchestrate new services. ONAP is primarily targeted at providing an open source automation and orchestration platform for service providers, particularly telecommunication vendors, to run SDN and offer virtual network functions.

Secondly, OPNFV [Price and Rivera - 2012] establishes a reference NFV platform for facilitating the development and evolution of multi-vendor NFV components. OPNFV is concerned with performance and use case based testing on current standards specifications and on work from open source communities for specific NFV use cases. The objective of OPNFV is to accelerate the development of emerging NFV products and services thereby ensuring headline performance targets and interoperability are met. OPNFV's work concentrates on NFV interfaces (NFVI) and an NFV virtualized infrastructure manager and relies on components from other open source projects such as OpenDaylight, ONOS, OpenStack, Ceph Storage, KVM, OpenvSwitch, DPDK and Linux.

3.5 Conclusions

This chapter has exposed the ongoing work on the convergence of networks, focusing primarily on the access-metro network, but also emphasizing the importance of integration with the mobile and cloud edge ecosystems. After describing the 5G requirements behind the drive towards metro convergence and virtualization, we first provided a description of convergence at the physical network architecture (showing the evolution of node consolidation concepts) and then provided details of the main software frameworks currently implementing converged network solutions.

Bibliography

A. Aguado, V. Lopez, J. Marhuenda, O. Gonzalez de Dios, and J. P. Fernandez-palacios. Abno: a feasible sdn approach for multivendor ip and optical networks [invited]. *IEEE/OSA Journal of Optical Communications and Networking*, 7(2):A356–A362, 2015.

G. fast B. Telecom. Release of bt cable measurements for use in simulations. Technical report, ITU-T SG15, 2013.

M. P. Belanger, M. O'Sullivan, and P. Littlewood. Margin requirement of disaggregating the dwdm transport system and its consequence on application economics. In *2018 Optical Fiber Communications Conference and Exposition (OFC)*, pages 1–3, March 2018.

B. Belter, A. Binczewski, K. Dombek, A. Juszczyk, L. Ogrodowczyk, D. Parniewicz, M. Stroi nski, and I. Olszewski. Programmable abstraction of datapath. In *2014 Third European Workshop on Software Defined Networks*, pages 7–12, 2014.

Pankaj Berde, Matteo Gerola, Jonathan Hart, Yuta Higuchi, Masayoshi Kobayashi, Toshio Koide, Bob Lantz, Brian O'Connor, Pavlin Radoslavov, William Snow, and Guru Parulkar. Onos: Towards an open, distributed sdn os. In *Proceedings of the Third Workshop on Hot Topics in Software Defined Networking*, HotSDN '14, pages 1–6, 2014. ISBN 978-1-4503-2989-7.

Pat Bosshart, Dan Daly, Glen Gibb, Martin Izzard, Nick McKeown, Jennifer Rexford, Cole Schlesinger, Dan Talayco, Amin Vahdat, George Varghese, and David Walker. P4: Programming protocol-independent packet processors. *SIGCOMM Comput. Commun. Rev.*, 44(3):87–95, 2014. ISSN 0146-4833.

P. Congdon, B. Aboba, A. Smith, G. Zorn, and J. Roese. Ieee 802.1 x remote authentication dial in user service (radius) usage guidelines(no. rfc 3580). Technical report, No. RFC 3580, 2003.

Bruno Cornaglia, Gavin Young, and Antonio Marchetta. Fixed access network sharing. *Optical Fiber Technology*, 26:2–11, 2015.

U. Dötsch, M. Doll, H. Mayer, F. Schaich, J. Segel, and P. Sehier. Quantitative analysis of split base station processing and determination of advantageous architectures for lte. *Bell Labs Technical Journal*, 18(1):105–128, 2013.

F. Effenberger, D. Cleary, O. Haran, G. Kramer, R. D. Li, M. Oron, and T. Pfeiffer. An introduction to pon technologies [topics in optical communications]. *IEEE Communications Magazine*, 45(3):S17–S25, 2007.

M. S. Elbamby, C. Perfecto, M. Bennis, and K. Doppler. Toward low-latency and ultra-reliable virtual reality. *IEEE Network*, 32(2):78–84, 2018.

A. Elrasad, N. Afraz, and M. Ruffini. Virtual dynamic bandwidth allocation enabling true pon multi-tenancy. In *2017 Optical Fiber Communications Conference and Exhibition (OFC)*, pages 1–3, March 2017.

Rob Enns. Netconf configuration protocol. no. rfc 4741. Technical report, Network Working Group, 2006.

M. Ersue. Etsi nfv management and orchestration-an overview. Ietf meeting proceedings, institution, 2013.

M. Ersue. Ecomp: the engine behind our software-centric network. Technical report, AT&T's, 2016.

GitHub. Floodlight sdn openflow controller, July. 2018. URL https://github.com/floodlight/floodlight. Accessed on July. 2018.

J. Hagel, J. S. Brown, T. Samoylova, and M. Lui. *From exponential technologies to exponential innovation*. Deloitte Consulting white paper, Report 2 of the 2013 Shift Index series, 2013.

Rachid El Hattachi and Javan Erfanian. *NGMN 5G white paper*. NGMN Alliance, NGMN Ltd, 2015.

J. Heiliger. Building efficient data centers with the open compute project. Technical report, Facebook Engineering Notes, 2011.

Fast Data input output, 2018. URL https://fd.io/. Accessed on July. 2018.

Intel. Packet processing on intel architecture. Data plane development kit, Intel® Network Builders, 2012.

2083-0 ITU-R M. Imt vision: Framework and overall objectives of the future development of imt for 2020 and beyond. Technical report, ITU-R, 2015.

Sándor Laki, Dániel Horpácsi, Péter Vörös, Róbert Kitlei, Dániel Leskó, and Máté Tejfel. High speed packet forwarding compiled from protocol independent data plane specifications. In *Proceedings of the 2016 ACM SIGCOMM Conference*, SIGCOMM '16, pages 629–630, 2016. ISBN 978-1-4503-4193-6.

Linux Foundation, 2018. URL https://www.linuxfoundation.org/. Accessed on June 25, 2018.

A. Lord. The evolution of optical networks in a 5g world[keynote talk]. In *2018 International Conference on Optical Network Design and Modeling (ONDM)*, May 2018.

Richard MacKenzie. *NGMN Overview on 5G RAN Functional Decomposition*. NGMN Alliance, NGMN Ltd, 2018.

S. McGettrick, L. Guan, A. Hill, D. B. Payne, and M. Ruffini. Ultra-fast 1+1 protection in 10 gb/s symmetric long reach pon. In *39th European Conference and Exhibition on Optical Communication (ECOC 2013)*, pages 1–3, Sept 2013a.

S. McGettrick, D. B. Payne, and M. Ruffini. Improving hardware protection switching in 10gb/s symmetric long reach pons. In *2013 Optical Fiber Communication Conference and Exposition and the National Fiber Optic Engineers Conference (OFC/NFOEC)*, pages 1–3, March 2013b.

S. McGettrick, F. Slyne, N. Kitsuwan, D. B. Payne, and M. Ruffini. Experimental end-to-end demonstration of shared n:1 dual homed protection in long reach pon and sdn-controlled core. In *2015 Optical Fiber Communications Conference and Exhibition (OFC)*, pages 1–3, March 2015.

S. McGettrick, F. Slyne, N. Kitsuwan, D. B. Payne, and M. Ruffini. Experimental end-to-end demonstration of shared n:m dual-homed protection in sdn-controlled long-reach pon and pan-european core. *Journal of Lightwave Technology*, 34(18):4205–4213, 2016.

J. Medved, R. Varga, A. Tkacik, and K. Gray. Opendaylight: Towards a model-driven sdn controller architecture. In *Proceeding of IEEE International Symposium on a World of Wireless, Mobile and Multimedia Networks 2014*, pages 1–6, June 2014.

S. Miano, M. Bertrone, F. Risso, M. Tumolo, and M.V. Bernal. Creating complex network service with ebpf: Experience and lessons learned. In *2018 IEEE International Conference on High Performance Switching and Routing (HPSR)*, 2018.

Open ROADM MSA. ROADM Network Model and Device Model. Open ROADM Multi-Source Agreement, http://www.openroadm.org/home.html, 2016.

Francesco Musumeci, Cristina Rottondi, Avishek Nag, Irene Macaluso, Darko Zibar, Marco Ruffini, and Massimo Tornatore. A survey on application of machine learning techniques in optical networks. *CoRR*, abs/1803.07976, 2018. URL http://arxiv.org/abs/1803.07976.

A. Nag, D. B. Payne, and M. Ruffini. N:1 protection design for minimizing olts in resilient dual-homed long-reach passive optical network. *IEEE/OSA Journal of Optical Communications and Networking*, 8(2):93–99, 2016.

A. Napoli, M. Bohn, D. Rafique, A. Stavdas, N. Sambo, L. Poti, M. Nölle, J. K. Fischer, E. Riccardi, A. Pagano, A. Di Giglio, M. S. Moreolo, J. M. Fabrega, E. Hugues-Salas, G. Zervas, D. Simeonidou, P. Layec, A. D'Errico, T. Rahman, and J. P. F. P. Giménez. Next generation elastic optical networks: The vision of the european research project idealist. *IEEE Communications Magazine*, 53(2):152–162, 2015.

O. Team. Openwrt: A linux distribution for wrt54g. Technical report, O. Team, 2016.

ONAP. *Open Network Automation Platform (ONAP) Architecture white paper*. ONAP, ONAP a Series of LF Project, 2018.

Open Data Plane, 2018. URL http://www.opendataplane.org. Accessed on July. 2018.

Open Fast Path, 2018. URL http://www.openfastpath.org. Accessed on July. 2018.

Open Networking Foundation, 2018. URL https://www.opennetworking.org/. Accessed on July. 2018.

OpenConfig, July. 2018. URL http://www.openconfig.net/. Accessed on July. 2018.

F. Paolucci, A. Sgambelluri, F. Cugini, and P. Castoldi. Network telemetry streaming services in sdn-based disaggregated optical networks. *Journal of Lightwave Technology*, 36(15):3142–3149, 2018.

D.B. Payne, and R.P. Davey. The future of fibre access systems? *BT Technology Journal*, 20(4):104–114, 2002.

L. Peterson, A. Al-Shabibi, T. Anshutz, S. Baker, A. Bavier, S. Das, J. Hart, G. Palukar, and W. Snow. Central office re-architected as a data center. *IEEE Communications Magazine*, 54(10):96–101, 2016.

T. Pfeiffer. Converged heterogeneous optical metro-access networks. In *36th European Conference and Exhibition on Optical Communication*, pages 1–6, 2010.

C. Price and S. Rivera. *Opnfv: An open platform to accelerate NFV white paper*. A Linux Foundation Collaborative Project, A Linux Foundation, 2012.

C. Raack, R. Wessälly, D. Payne, and M. Ruffini. Hierarchical versus flat optical metro/core networks: A systematic cost and migration study. In *2016 International Conference on Optical Network Design and Modeling (ONDM)*, pages 1–6, May 2016.

M. Ruffini. Access-metro convergence in next generation broadband networks. In *2016 Optical Fiber Communications Conference and Exhibition (OFC)*, pages 1–61, March 2016.

M. Ruffini. Multidimensional convergence in future 5g networks. *Journal of Lightwave Technology*, 35:535–549, 2017.

M. Ruffini and D.C. Kilper. From central office cloudification to optical network disaggregation. In *2018 IEEE Photonics Society Summer Topicals*, July 2018.

M. Ruffini, D. B. Payne, and L. Doyle. Protection strategies for long-reach pon. In *36th European Conference and Exhibition on Optical Communication*, pages 1–3, Sept 2010.

M. Ruffini, M. Achouche, A. Arbelaez, R. Bonk, A. Di Giglio, N. J. Doran, M. Furdek, R. Jensen, J. Montalvo, N. Parsons, T. Pfeiffer, L. Quesada, C. Raack, H. Rohde, M. Schiano, G. Talli, P. Townsend, R. Wessaly, L. Wosinska, X. Yin, and D. B. Payne. Access and metro network convergence for flexible end-to-end network design [invited]. *IEEE/OSA Journal of Optical Communications and Networking*, 9(6):524–535, 2017.

Muhammad Shahbaz, Sean Choi, Ben Pfaff, Changhoon Kim, Nick Feamster, Nick McKeown, and Jennifer Rexford. Pisces: A programmable, protocol-independent software switch. In *Proceedings of the 2016 ACM SIGCOMM Conference*, SIGCOMM '16, pages 525–538, 2016. ISBN 978-1-4503-4193-6.

A. Singla and B. Rijsman, 2018. URL http://www.opencontrail.org/opencontrail-architecture-documentation. Accessed on July. 2018.

CPRI Specification. Common public radio interface (cpri) v6.0: Interface specification. Technical report, CPRI Specification, 2013.

Sun Mi Yoo, Hong Taek Ju, and James Won Ki Hong. Web services based configuration management for ip network devices. In *2005 Management of Multimedia Networks and Services, 8th International Conference on Management of Multimedia Networks and Services(MMNS)*, pages 1–12, 2005.

T. Taleb, K. Samdanis, B. Mada, H. Flinck, S. Dutta, and D. Sabella. On multi-access edge computing: A survey of the emerging 5g network edge cloud architecture and orchestration. *IEEE Communications Surveys Tutorials*, 19 (3):1657–1681, 2017.

Telecom Infrastructure Project, 2018. URL https://telecominfraproject.com/. Accessed on July. 2018.

E. Walter. Ofc 2018: At&t's pon& edge compute vision. In *2018 Optical Fiber Communications Conference and Exposition (OFC)*, pages 1–21, March 2018.

Jun Wang, Limin Sun, Xiu Jiang, and ZhiMei Wu. Igmp snooping: a vlan-based multicast protocol. In *5th IEEE International Conference on High Speed Networks and Multimedia Communication (Cat. No.02EX612)*, pages 335–340, 2002.

4

Multicore Fibres for 5G Fronthaul Evolution

Ivana Gasulla* and José Capmany

iTEAM Research Institute, Universitat Politècnica de València, Valencia, 46022, Spain

4.1 Why 5G Communications Demand Optical Space-Division Multiplexing

Emerging information and communication technology scenarios that will have a profound impact on our future society, such as 5G wireless communications and the Internet of Things, are expected to pose a formidable challenge for existing telecommunication networks. Exigent and demanding requirements are envisaged in terms of cell coverage (from a few meters to several kilometers), number of connected devices (over 1 billion worldwide), transmission format diversity (from single-input single-output to multiple-input multiple-output), multiple spectral regions (from a few hundred MHz to 100 GHz), supported applications, increased bandwidth per end user (up to 10 GHz), smooth and adaptive integration of the fiber and wireless segments, and efficient energy management [Samsung Electronics Co - 2015; China Mobile Research Institute - 2011; Pizzinat et al. - 2015]. In view of this set of disruptive 5G capabilities, it is clear than neither photonic nor radiofrequency (RF) technologies can solve this problem on their own. Rather, a multidisciplinary technology is called upon that, in addition, can lend itself to resource and functionality sharing by suitable software definition.

Fibre-wireless communications, or what the experts named 40 years ago as microwave photonics (MWP), is actually such a technology. MWP can be considered nowadays as a mature interdisciplinary field that merges the worlds of RF, photonics and optoelectronics engineering, [Capmany et al. - 2013; Seeds - 2002; Capmany and Novak - 2007; Yao - 2009]. This technological fusion allows for the generation, processing, and distribution of microwave and millimeter-wave signals by photonic means, benefiting from the well-known advantages inherent to photonics, such as high bandwidth, a low loss level that is independent of the radio frequency, and electromagnetic interference immunity. In addition, MWP brings the fundamental added value of enabling the implementation of key features such as fast tunability and reconfigurability, which are either complex or not even possible using electronic approaches, [Capmany et al. - 2013].

* Corresponding Author: Ivana Gasulla; ivgames@iteam.upv.es

Optical and Wireless Convergence for 5G Networks, First Edition.
Edited by Abdelgader M. Abdalla, Jonathan Rodriguez, Issa Elfergani, and Antonio Teixeira.
© 2020 John Wiley & Sons Ltd. Published 2020 by John Wiley & Sons Ltd.

These advantageous properties are behind the increasing interest from the research community experienced over the last four decades in two main areas of application.

- Telecommunication distribution networks, where broadband RF signals are delivered between a central office and a set of remote end users or base stations. MWP enables radio-over-fiber signal distribution, including multiple-input multiple-output (MIMO) antenna connectivity.
- RF signal processing systems. MWP enables different functionalities such as tuneable and reconfigurable microwave signal filtering, radio beam-steering in phased array antennas, arbitrary waveform generation, optoelectronic oscillation and multi-gigabit-per-second analogue-to-digital conversion, [Capmany et al. - 2013; Seeds - 2002; Capmany and Novak - 2007; Yao - 2009]. These functionalities, in turn, are required in a variety of information and communication technology applications.

Not only does MWP bring considerable added value to traditional RF systems in both civil and defence scenarios, but it also holds a promising future in a myriad of emerging fields where wireless and wired signals must coexist. These include, among others, the Internet of Things, smart cities, medical imaging, sensing, optical coherence tomography, as well as converged fiber-wireless radio access networks. The potential for the extension of microwave photonics to these emerging areas under the umbrella of 5G depends on the reduction of size, weight and power consumption while assuring seamless broadband versatility, reconfigurability, multifunctionality and performance stability.

This challenge requires the development of innovative photonic technologies that must go beyond mere fiber-based distribution and connectivity functionalities. Instead, they should embrace – at the same time – high-performance microwave and millimeter-wave signal processing functionalities that will be critical in 5G smart radiating systems. However, as Figure 4.1 shows, the present fiber-wireless paradigm embraces:

- Static, inefficient and replicating distribution architectures where bundles of fibers are deployed between central nodes (hosting shared-cost equipment) and several remote antennas or user terminals.
- Bulky, heavy and power-consuming signal processing systems that make use of discrete opto-electronic or fiber-based devices.

In particular, integrated microwave photonics, which aims to integrate the highest number of photonic components either in a monolithic or hybrid platform, has been proposed recently to address processing systems in isolation [Marpaung et al. - 2013]. Nevertheless, both processing and distribution functionalities must be addressed as a whole, since many applications require parallelization in a distributed nature of the same processing functionality that is usually carried by "brute-force" system replication. There is one revolutionary approach that has been left untapped in finding innovative ways to address this challenge: exploiting space – the last available degree of freedom for optical multiplexing. Space-division multiplexing (SDM) has been recently touted as a solution for the capacity bottleneck in digital communications by establishing independent light paths in a single fiber via multicore fibers (MCFs) or few-mode fibers [Richardson et al. - 2013]. Although SDM was initially envisioned in the context of core and metro

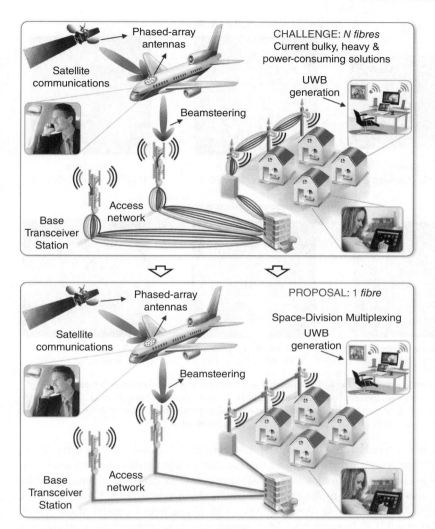

Figure 4.1 Application of space-division multiplexing fiber solutions to a representative 5G communications scenario.

optical networks, its full potential in next-generation fiber-wireless communications has not been fully unveiled yet.

4.2 Multicore Fibre Transmission Review

The addition of the space dimension to the portfolio of optical multiplexing technologies has been welcomed by the optical communications community as a promising solution to the saturation of conventional single-mode fiber (SMF) capacity, [Richardson et al. - 2013]. Enlargement of the transmission capacity per cross-sectional area of the

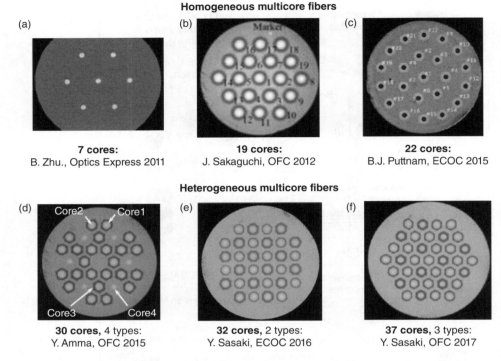

Figure 4.2 Representative uncoupled multicore fibers reported in the literature with both homogeneous cores: (a) as reported in Zhu et al. [2011], (b) as reported in Sakaguchi et al. [2012], and (c) as reported in Puttnam et al. [2015]; and heterogeneous cores: (d) as reported in Amma et al. [2015], (e) as reported in Sasaki et al. [2016], and (f) as reported in Sasaki et al. [2017].

optical fiber over a fixed bandwidth, which for single-core SMFs has reached a maximum of 100 Tb s^{-1}, is achieved by increasing the number of light paths that are transmitted within the same optical fiber. Different SDM approaches have been investigated for the last few years, including multicore fibers [Richardson et al. - 2013; Koshiba - 2014; Matsuo et al. - 2016], few-mode fibers [Ryf et al. - 2015] and even a combination of both [Sakaguchi et al. - 2015]. In particular, MCF solutions aim to increase the transmission capacity by spatially multiplexing N different signals into the N different cores (either single-mode or few-mode) that comprise the fiber.

As a first step, MCFs can be classified into uncoupled and coupled transmission. In the case of uncoupled transmission, each core must be suitably arranged to keep the inter-core crosstalk sufficiently small for long-distance applications. A variety of core arrangements have been reported for uncoupled MCFs, including homogeneous MCFs with multiple identical cores, quasi-homogeneous MCFs with slightly different cores, and heterogeneous MCFs with several types of different cores. Representative homogeneous and heterogeneous uncoupled structures reported in the literature are gathered in Figure 4.2.

4.2.1 Homogeneous MCFs

MCF links allow for the transmission of a multiplex of different data signals along the different cores that comprise the fiber while assuring a low level of crosstalk. Most of the

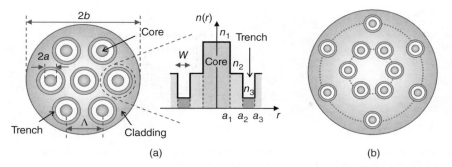

Figure 4.3 Cross-sectional diagrams and refractive index profile of representative trench-assisted homogeneous MCF structures: (a) 7-core hexagonal, (b) 12-core dual ring.

research activity on MCF transmission has focused on the so-called homogeneous multicore fiber, where all the N cores are, in theory, identical. The highest capacities demonstrated so far featured crosstalk values between -72 and -22 dB/100 km (uncoupled transmission) and cladding diameters ranging from 125 to 260 µm, [Koshiba - 2014]. Early MCF transmission was based on hexagonal structures of seven cores disposed with a core separation (or pitch) $\wedge$ around 46.8 µm, as illustrated in Figure 4.2a, large enough to obtain a crosstalk level below -40 dB, [Zhu et al. - 2011].

Subsequent optimization of the refractive index profile resulted in trench-assisted configurations, where each core is surrounded by a trench of refractive index n_3 lower than the cladding index n_2, as detailed in Figure 4.3a. Thanks to better confinement of the optical field, this technique makes the crosstalk insensitive to the fiber bending radius increase and enables reduced pitch values down to 35 µm, for instance, a close-packed 19-core fiber as the one depicted in Figure 4.2b, [Sakaguchi et al. - 2012]. A crosstalk reduction down to -50 dB led to very high capacities, as the recent record 2.15 Pbit s^{-1} (over 31 km) reached in a 22-core fiber with a cladding diameter of 260 µm [Puttnam et al. - 2015], as shown in Figure 4.2a. Another representative example is the long-haul transmission of more than 1 Eb s^{-1} km^{-1} (bidirectional 344 Tbit s^{-1} over 1500 km) reported using a 12-core two-ring structure, see Figure 4.3b [Kobayashi et al. - 2013].

4.2.2 Heterogeneous MCFs

Heterogeneous MCFs were proposed by [Koshiba et al. - 2009] to increase the density of cores as compared to homogeneous MCFs. They are composed of dissimilar cores that are characterized by different effective indices and arranged in the fiber cross-section area so that the crosstalk between neighbouring cores is reduced as the phase matching condition is prevented. As seen in Figure 4.4a, a typical heterogeneous MCF is composed of two or more interleaved triangular lattices of identical cores that are separated by a distance d. The overall arrangement results in a layout where neighboring cores are characterized by different propagation constants and are separated by a pitch value given by $\wedge = d/\sqrt{3}$. Initial designs of heterogeneous 19-core fibers based on this triangular arrangement were able to reduce the core separation or pitch down to 23 µm for a 125 µm cladding diameter by accommodating three different core compositions [Koshiba et al. - 2009]. The use of high values of the refractive index contrast

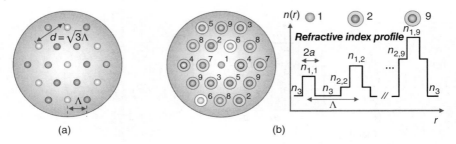

Figure 4.4 Cross-sectional diagrams and refractive index profile of representative heterogeneous MCF structures: (a) 19-core hexagonal step index, (b) 19-core hexagonal.

Δ(Δ = 1.15%, 1.20% and 1.25%), increased the packing density accommodating up to 19 cores with a core pitch $\wedge$ = 23 μm. [Kokubun and Watanabe - 2011] found in addition that a double-cladding structure could accommodate up to nine different equivalent refractive indexes (see Figure 4.4b).

The incorporation of trench-assisted profiles, [Tu et al. - 2013; 2016; Hayashi et al. - 2011], led to larger core multiplicities as, for instance, the 30-core fiber comprising four different types of trench-assisted cores arranged in a cladding diameter of 228 μm as illustrated in Figure 4.2d, [Amma et al. - 2015], the $32 - core$ fiber reported in [Sasaki et al. - 2017] with a cladding diameter of 243 μm and only two types of cores, and the 37-core fiber reported by [Sasaki et al. - 2017] comprising three types of trench-assisted cores in a 248 μm cladding diameter (see Figure 4.2*f*) [Sasaki et al. - 2017]. Further progress incorporating few-mode transmission in each core achieved a record spatial channel over 100, for instance by combining 36 heterogeneous cores and 3-mode transmission inside a cladding diameter of 306 μm, [Sakaguchi et al. - 2015].

4.3 Radio Access Networks Using Multicore Fibre Links

Next generation global IT paradigms, such as 5G and the Internet of Things, require evolving existing wireless radio access platforms in combination with entirely new technologies. This involves more efficient use of the existing bandwidth by exploiting spatial diversity through massive MIMO and beamforming, as well as the incorporation of carrier aggregation and new RF bands. As the density of urban cell sites increase, emission power requirements will be reduced, and much higher spectral reuse will be possible. Broad bandwidth, high reliability, and low latency will be essential and will require smooth matching between radio access and wired segments via flexible and adaptive RF–photonic interfaces. Centralized radio access networks (C-RANs), also known as cloud RANs, have been proposed to meet the above requirements for 5G converged fiber-wireless access networks [China Mobile Research Institute - 2011; Pizzinat et al. - 2015; Chanclou et al. - 2013; Saadani et al. - 2013]. They are based on the principle of hosting all the resources managed by the baseband processing units (BBUs) corresponding to different base stations in a shared central office. This implies the introduction of a fronthaul fiber segment between the remote radio head (RRH), which is placed at the base station, and the corresponding BBU. The C-RAN concept brings considerable benefits, noteworthy among which are:

- Both operational and capital expenditure savings.
- Low energy consumption.
- Improved radio performance derived from the potentially reduced latency between base stations, enabling as a consequence the implementation of coordinated multi-point (CoMP) protocols.

As illustrated in Figure 4.5, a fronthaul fiber segment is required for the remote feeding of different RRHs from the central office. Digital radio-over-fibre (DRoF) was proposed initially to transport the baseband data signal directly from a particular BBU located in the central office to the corresponding RRH, and vice versa, using one of the protocols defined in the common public radio interface (CPRI).However, C-RANs face several relevant challenges that can be summarized as follows:

- The first challenge relates to bit rate requirements. CPRI protocols involve very high symmetric bit rates as compared to the real data rate that is required at the user end. To give an example, the delivery of five contiguous LTE-A channels featuring a 20-MHz bandwidth will require a bit rate of 6.144 Gbit s^{-1} per antenna sector. This number could, in addition, grow up to almost 50 Gbit s^{-1} per antenna sector if, for instance, spatial diversity through 8x8 MIMO is exploited. To solve this limitation, the transmission of analogue radio-over-fibre (RoF) in the fronthaul segment has gained increasing interest [Mitchell - 2014; Liu et al. - 2014].
- The second one relates to fiber availability. If we consider, in principle, three-sector antennas, we will need at least six optical fibers to feed every single antenna. If we

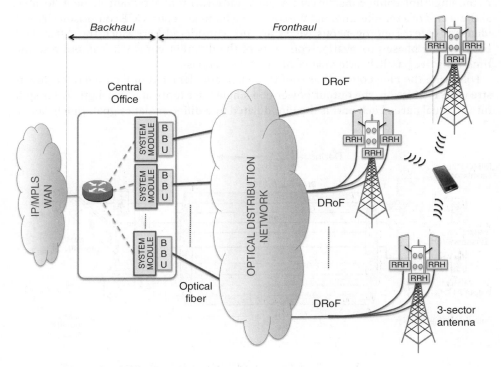

Figure 4.5 Centralised radio access network architecture.

include MIMO configurations with multiple radiant elements, then we should consider as well some multiplexing, such as wavelength-division multiplexing (WDM) and SDM, to cope with the constraint of not increasing the fiber count.

- C-RANs should support additional features that will be key in 5G communications, such as carrier aggregation, dynamic capacity allocation as well as centralized supervision and management.
- Finally, C-RANs must be flexible enough to integrate current and/or evolved versions of additional distribution networks such as passive optical networks (PONs).

In order to solve the second challenge, without losing track of the rest, researchers have proposed the use of MCFs to support C-RAN architectures, [Galvé et al. - 2016]. MCF-based C-RAN solutions are potentially flexible enough to support both digital and analog RoF architectures, and capacity upgrades by carrier aggregation and massive MIMO, as well as true cloud operation. More importantly, the spatial diversity inherent to MCF transmission is capable of enabling the above mentioned features employing electronic spatial switching placed at the shared central office, allowing software defined networking (SDN) as well as network function virtualization (NFV). In addition, MCF-based C-RAN configurations will be compatible with WDM and will support a PON overlay.

4.3.1 Basic MCF Link Between the Central Office and Base Station

Figure 4.6 depicts the basic building block of an MCF-based C-RAN architecture, that is, the single link connecting the central office and an array of N radiant elements located at the sector of a remote antenna. We consider a homogeneous MCF comprising $2N + 1$ identical cores. Heterogeneous MCFs are not considered here since the differences between the phase propagation constants of the different cores will compromise the time-sensitive parallel performance of MIMO antennas.

Following the blue coloured area of the schematic depicted in Figure 4.6 for downstream transmission, the output power of a single laser (optical wavelength λ_D) is split into N equal parts. Each part is then modulated by a different data signal, which can be

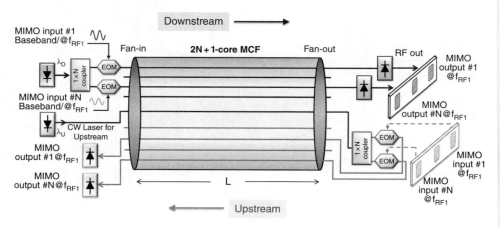

Figure 4.6 Link between the central office and base station sector based on homogeneous MCF comprising $2N + 1$ cores, Source: © [2018] IEEE. Reprinted, with permission, from Galvé et al. [2016].

either baseband I-Q, intermediate frequency or radiofrequency. The figure illustrates a representative MIMO transmission scenario, where each modulated signal is injected into a different fiber core. We use, in this case, subcarriers with the same RF, f_{RF1}. At the other fiber end (base station), each data signal is detected by a different receiver and processed by the RRH before being distributed to the antenna and radiated. We must note that the link is versatile enough to accommodate different RF signals centered at subcarriers with different RF frequencies. Of special relevance is the fact that a single core can be dedicated to distributing, for instance, the CW laser (optical wavelength λ_U) to the BS, (as required for upstream delivery) or even a control and supervision separate channel. In the upstream direction (green coloured), the distributed common optical carrier (CW signal) is split into N parts. Each of the N antenna radiating elements receives its particular signal, the RRH processes it and modulates the common carrier accordingly. The received data channels are then sent to the central office through N cores of the fiber.

4.3.2 MCF Based RoF C-RAN

The single central office to base station MCF link described in the previous section can serve as the essential connection between the BBU and a given antenna sector in 5G radio access networks. Figure 4.7 shows a representative C-RAN distribution architecture where the central office is connected to the sectors conforming the different base stations by means of $(2N + 1)$-core MCFs, (N cores for downstream and N cores for upstream). For simplicity, three-sector antennas have been considered in Figure 4.7. The ith input core to (from) antenna sector j in base station BS_k is labelled as $s_i^{k,j}$, as shown in the MCF link expanded in the lower part of Figure 4.7. A centralized switch placed in the central office provides the dynamic mapping of (to) downstream (upstream) subcarrier channels and bands to (from) the spatial ports. The frequencies of the RF subcarriers are

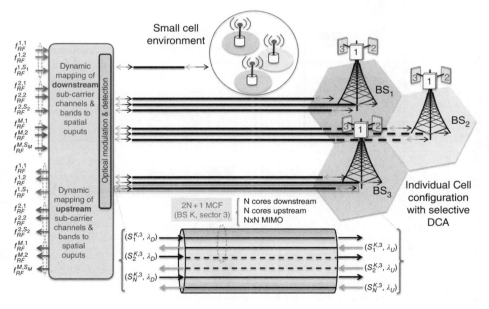

Figure 4.7 MCF-based C-RAN configuration for RoF operation, Source: © [2018] IEEE. Reprinted, with permission, from Galvé et al. [2016].

labelled as f_{RF}^{m,S_m}, for the mth band, $1 \leq m \leq M$, M being the total number of bands and S_m the number of subcarriers in the mth band. The fabric comprises as well an internal electronic core where SDN switching is carried before the modulation of the λ_D optical carrier in the downstream and after detection of the data signal in the upstream.

Several advantages arise from the incorporation of spatial diversity in the central office. First, each antenna sector within a given base station can be independently addressed and configured with single optical fiber, avoiding fiber bundling. Furthermore, the central office can carry out resource allocation electronically, allowing dynamic carrier aggregation and MIMO schemes. Finally, the number of MIMO radiant elements in a given antenna sector can be independently and dynamically set from 1 to N at the central office, enabling, if required, CoMP protocols between sectors in neighboring base stations.

To better understand the capabilities of this scheme, consider for instance the resource allocation table illustrated in Figure 4.8. Here, we show a representative switching configuration involving three base stations: base station BS_1 (orange coloured) provides

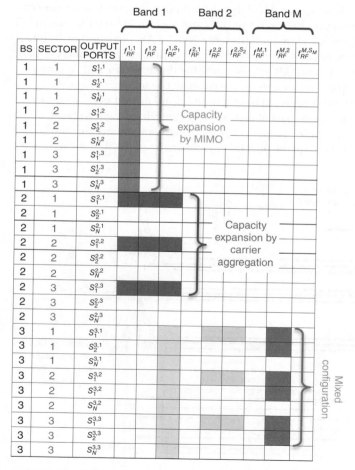

Figure 4.8 Example of a resource allocation table for the MCF-based RoF C-RAN configuration.

capacity expansion by realizing NxN MIMO operation in all the antenna radiator elements over one RF allocated in band 1 ($f_{RF}^{1,1}$); base station BS_2 (red coloured) implements capacity expansion by RF carrier aggregation using three RF carriers in band 1 ($f_{RF}^{1,1}, f_{RF}^{1,2}, f_{RF}^{1,S_1}$) in the three sectors activating only one radiating element per sector; and base station BS_3 (green coloured) features capacity expansion, a mixed configuration that implies the use of NxN MIMO in band 1 (f_{RF}^{1,S_1}), RF carrier aggregation in band 2 ($f_{RF}^{2,2}, f_{RF}^{2,S_2}$) as well as $2x2$ MIMO in band M ($f_{RF}^{M,2}$).

4.3.3 MCF Based DRoF C-RAN

Figure 4.9 illustrates the configuration of the proposed MCF-based C-RAN architecture when considering DRoF operation. In the downstream direction, the input to the electronic switch (or in the upstream direction, the output from the electronic switch) located at the central office corresponds to a BBU pool where a set of virtual BBUs are defined by software to service base stations BS_1 to BS_M. Each virtual BBU_m ($m = 1$ to M) is tied to a specific set of fiber cores and can be reconfigured dynamically to allocate different capacities. An overall manager can allocate resources to the virtual BBUs.

Two representative examples of resource reconfiguration that can be given in a typical DRoF scenario are gathered in the resource allocation table shown in Figure 4.10. In both examples, we assume equal capacity per sector at each base station. Situation 1

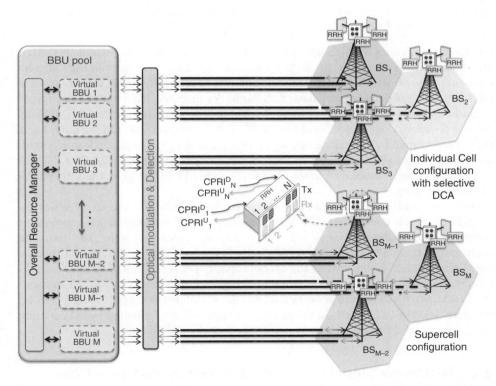

Figure 4.9 MCF based C-RAN configuration for DRoF operation, Source: © [2018] IEEE. Reprinted, with permission, from Galvé et al. [2016].

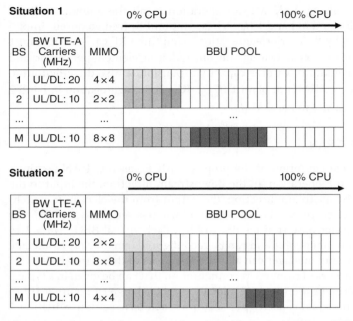

Figure 4.10 Example of a resource allocation table for the MCF-based DRoF C-RAN configuration, Source: © [2018] IEEE. Reprinted, with permission, from Galvé et al. [2016].

represents carrier aggregation operation over a multi-band non-contiguous (800-MHz and 900-MHz bands) $10 + 10$ MHz bandwidth for LTE-A implementing different configurations of MIMO. When only one antenna is active, the CPRI bit rate is 1.536 Gbit s^{-1}, which is depicted as a basic CPU unit. For instance, base stations BS$_1$ (green coloured) and BS$_2$ (yellow coloured) implement independent 4x4 and 2x2 MIMO with an overall capacity per sector of 6.14 and 3.07 Gbit s^{-1}, respectively. On the other hand, the set conformed by base stations BS$_{M-2}$, BS$_{M-1}$, and BS$_M$ (red coloured) implements a supercell with 8x8 MIMO and an overall capacity per sector of 12.28 Gbit s^{-1}. The cumulative capacity employed in terms of percentage of CPU usage of the BBU pool is illustrated in grey.

Let us see study a case (situation 2) where the capacity is re-assigned. We consider a single-band non-contiguous (900-MHz band) $(5 + 5)$-MHz bandwidth implementing several MIMO configurations as well. When only one antenna is active, the basic unit corresponds to a CPRI bit rate of 0.768 Gbit s^{-1}. For base station BS$_1$, the bit rate per core remains 1.536 Gbit s^{-1} while the overall capacity per sector is 3.07 Gbit s^{-1} and the number of radiators is reduced to 2; for base station BS$_2$ the number of radiant elements increases up to 8, the bit rate per core is 0.768 Gbit s^{-1} while the overall capacity per sector increases to 6.14 Gbit s^{-1}. In the case of the supercell implemented by base stations BS$_{M-2}$, BS$_{M-1}$ and BS$_M$, the capacity per sector reaches 3.07 Gbit s^{-1}.

4.4 Microwave Signal Processing Enabled by Multicore Fibers

At the heart of most applications demanded in microwave photonics signal processing, we can find the true time delay line (TTDL), an optical subsystem that provides a frequency independent and tuneable delay within a given frequency range, [Capmany and Novak - 2007; Yao - 2009]. This key building block enables essential functionalities such as controlled RoF distribution, tuneable and reconfigurable RF signal filtering, radio beam-steering in phased array antennas, optoelectronic oscillation, arbitrary waveform generation, and multi-gigabit-per-second analogue-to-digital conversion, [Capmany et al. - 2013; Seeds - 2002; Capmany and Novak - 2007; Yao - 2009]. These functionalities, in turn, are required in a variety of information technology applications, such as broadband wireless communications, satellite communications, distributed antenna systems, signal processing, sensing, medical imaging, and optical coherence tomography.

Different technology approaches have been reported for the implementation of this microwave photonics subsystem over the past 40 years, where the required diversity (different group delays area applied to a set of samples of the RF signal) is realized by exploiting either the time or the wavelength multiplexing domains. Configurations built upon standard single-mode fibers, which include configurations based on switched fiber links or dispersive fibers [Wilner and van den Heuvel - 1976], the inscription of fibre bragg gratings (FBGs) [Capmany et al. - 1999; Zeng and Yao - 2005; Wang and Yao - 2013], and the exploitation of nonlinear phenomena such as stimulated Brillouin scattering [Morton and Khurgin - 2009], demonstrated operation bandwidths between 0.1 and 40 GHz for delays between 0.4 and 8 ns. On the other hand, integrated photonics technologies, including the use of ring cavities in silicon on insulator [Marpaung et al. - 2013], racetrack resonators in Si_3N_4 [Marpaung et al. - 2013], photonic crystal structures [Sancho et al. - 2012], and semiconductor optical amplifiers Ohman et al. - 2007] based on indium phosphide [Ohman et al. - 2007], featured operation bandwidths between 2 and 50 GHz for delays in the range between 40 and 140 ps.

A radically different approach lies in the exploitation of the inherent parallelism of SDM fibers to implement TTDLs, leading to the so-called "fiber distributed signal processing", a concept that exhibits great potential in the context of converged fiber-wireless communications. Despite MCFs being originally envisioned in the context of core (and metro) networks, we must keep in mind that they can also be applied to a wide range of fields, including not only radio access networks and multiple antenna connectivity as described in the previous section, but also RF signal processing and multiparameter sensing, [Capmany et al. - 2013; Seeds - 2002; Capmany and Novak - 2007; Yao - 2009]. In particular, MWP signal processing can benefit from the use of MCFs regarding volume and weight, but also in terms of performance versatility and stability. Up to date, MWP approaches have relied on, on the one hand, discrete fiber-based subsystems or photonic integrated circuits to process the microwave signal, while, on the other hand, separate fiber links to implement the required distribution of the signal. The use of SDM technologies will enable the implementation of both parallel signal processing and distribution to the end user (wireless base station, indoor antenna, radar antenna, etc.) simultaneously.

Let us see how to implement a sampled discrete TTDL with a generic MCF. As Figure 4.11 depicts, the goal is to obtain at the output of the MCF-based link (or component) a given set of time-delayed replicas of the modulated signal. This series of samples must feature a constant differential delay between adjacent samples (named basic differential delay, $\Delta\tau$). If only one optical wavelength is implicated, as shown in Figure 4.11, the TTDL features 1D (one-dimensional) performance, since all the samples results from exploiting the fiber spatial diversity.

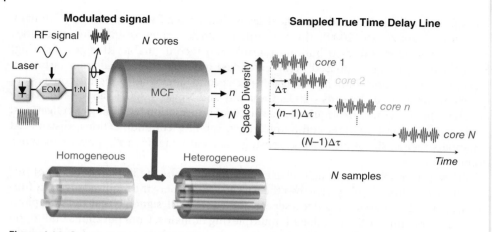

Figure 4.11 Generic 1D sampled TTDL based on an MCF (space diversity operation).

We can implement this scheme using a heterogeneous MCF if each core features different group delay and dispersion characteristics, [Gasulla and Capmany - 2012; Garcia and Gasulla - 2015; García and Gasulla - 2016]. The use of homogeneous MCFs, where all the cores are in principle identical, calls for the introduction of dispersive elements along the cores [Gasulla et al. - 2017].

This approach offers, in addition, unprecedented 2D (two-dimensional) operation if we combine the spatial diversity provided by the cores with the optical wavelength diversity provided by the use of multiple optical wavelengths. Figure 4.12 shows this concept for an array of M lasers. When using diversity in optical wavelength, the basic differential delay results from the propagation difference experienced by two adjacent wavelengths λ_{m+1} and λ_m in a given core n. On the other hand, the use of the spatial diversity creates the basic differential delay from the propagation difference experienced by two adjacent cores $n + 1$ and n for a particular optical wavelength λ_m.

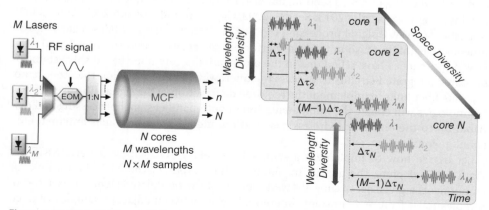

Figure 4.12 Generic 2D sampled TTDL based on an MCF (space and optical wavelength diversity operation).

4.4.1 Signal Processing Over a Heterogeneous MCF Link

As described in Section 2.2 of this chapter, most of the research on MCF links has focused on high-capacity digital communications. While digital signal distribution usually demands similar propagation characteristics in all the cores, the implementation of tuneable TTDLs requires different group delay values for a specific wavelength. This requirement calls for the development of customized heterogeneous MCFs where, while keeping low intercore crosstalk and considerable immunity to bends, we can tailor the chromatic dispersion of every single core. The group delay $\tau_n(\lambda)$ of core n can be expanded following a second-order Taylor series around a reference or anchor wavelength λ_0 as

$$\tau_n(\lambda) = \tau_n(\lambda_0) + D_n(\lambda - \lambda_0) + \frac{1}{2}S_n(\lambda - \lambda_0)^2 \tag{4.1}$$

where D_n is the chromatic dispersion and S_n the dispersion slope of core n at the reference wavelength. For proper tuneable time delay operability we need, first, a linear dependence of the spectral group delay with the optical wavelength and, second, a linear increment of D_n with the core number n. If all cores share the same group delay at the anchor wavelength, $\tau(\lambda_0)$, we can control the basic differential delay, $\Delta\tau = \tau_{n+1}(\lambda) - \tau_n(\lambda)$, to achieve continuous tunability from 0 up to tens (or even hundreds) of picoseconds per kilometer. This allows for the implementation of distributed signal processing on the fly for link lengths up to a few kilometers.

The design of the required heterogenous MCF involves the customization of the refractive index profile of every single core, where trench-assisted configurations are preferred since they provide more design versatility. Different designs of heterogeneous 7-core fibers behaving as sampled TTDLs have been reported [Gasulla and Capmany - 2012; Garcia and Gasulla - 2015; García and Gasulla - 2016]. In [García and Gasulla - 2016], an optimum design was presented in terms of both group delay tunability and minimum crosstalk for a fiber whose schematic cross section is depicted in Figure 4.13 (left). The computed parameters for each core are gathered in Figure 4.13 (right), where α_1 is the core radius, α_2 is the core-to-trench distance, w is the trench width, Δ_1 is the core-to-cladding relative index difference, and n_{eff} is the effective index. The core pitch is 35 μm while the cladding diameter is the standard 125 μm.

This 7-core fiber offers a linear spectral group delay with incremental values from core to core, as shown in Figure 4.14a. In other words, it fulfils the demand for tuneable delay line operation and can be used as the basic element to perform multiple MWP signal

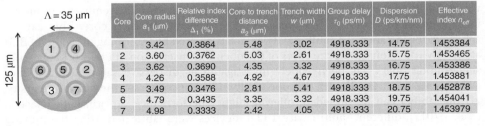

Core	Core radius a_1 (μm)	Relative index difference Δ_1 (%)	Core to trench distance a_2 (μm)	Trench width w (μm)	Group delay τ_0 (ps/m)	Dispersion D (ps/km/nm)	Effective index n_{eff}
1	3.42	0.3864	5.48	3.02	4918.333	14.75	1.453384
2	3.60	0.3762	5.03	2.61	4918.333	15.75	1.453465
3	3.62	0.3690	4.35	3.32	4918.333	16.75	1.453386
4	4.26	0.3588	4.92	4.67	4918.333	17.75	1.453881
5	3.49	0.3476	2.81	5.41	4918.333	18.75	1.452878
6	4.79	0.3435	3.35	3.32	4918.333	19.75	1.454041
7	4.98	0.3333	2.42	4.05	4918.333	20.75	1.453979

$\Lambda = 35$ μm

125 μm

Figure 4.13 Left: schematic cross-section of the designed heterogeneous MCF; right: design parameters and computed propagation characteristics for each core.

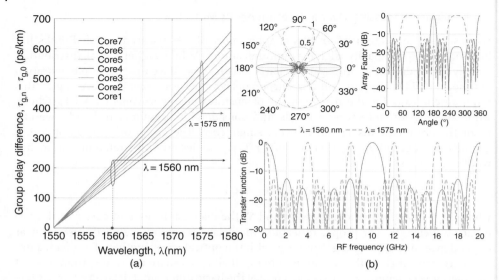

Figure 4.14 (a) Computed group delay (difference with respect to the group delay at the anchor wavelength of 1550 nm) as a function of the optical wavelength for the dispersion-engineered heterogeneous MCF. (b) Computed array factor of a phased-array antenna as a function of the beam pointing angle (upper part) and transfer function of a signal filter as a function of the radio frequency (lower part) for an optical wavelength of 1560 nm (blue solid lines) and 1575 nm (red dashed lines), Source: Reproduced with permission of MDPI.

functionalities, as signal filtering and radio beamsteering for phased array antennas. As a proof of concept, Figure 4.14b shows the computed frequency responses of the corresponding optical beamforming (i.e. array factor, upper part) and signal filtering (lower part) functionalities for a tuneable delay line implemented with 10 km of the designed MCF. By tuning the optical wavelength of the optical source, we can see how the filter free spectral range (FSR) and the direction of the phased-array antenna beam are modified. In particular, increasing the operation wavelength from 1560 up to 1575 nm reduces the FSR from 10 down to 4 GHz (lower figure) while the angle of the radiation varies from 180° down to 90° without nonlinear degradation due to higher-order dispersion.

4.4.2 RF Signal Processing Over a Homogeneous MCF Multi-Cavity Device

The implementation of a tunable true time delay lines built upon homogeneous MCFs, where all the cores share the same propagation characteristics, requires the in-line incorporation of additional dispersive optical elements in each core. This can be achieved, for instance, if a multi-cavity device is created by inscribing individual FBGs at selective positions along the cores of the fiber. The inscription of FBGs in single-core single-mode fibers has been widely investigated as dispersive sampled delay lines [Wang and Yao - 2013] with 1D operability. If we incorporate the spatial diversity, 2D operability can be provided, offering higher performance versatility. In [Gasulla et al. - 2017], different multi-cavity TTDL devices were fabricated using the moving phase mask technique, inscribing either the same grating in planes containing a set of cores or individual gratings in single cores. Figure 4.15a illustrates the schematic of a

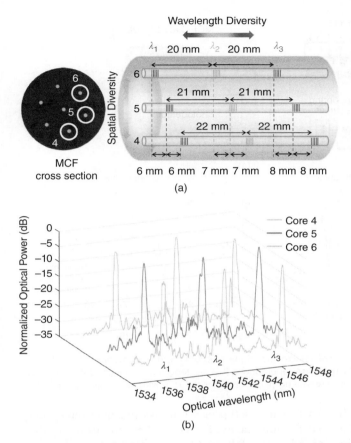

Figure 4.15 (a) Multi-cavity device implemented by selectively inscribing FBGs in three outer cores of a homogeneous 7-core fiber; (b) measured optical spectrum in reflection, Source: Reproduced with permission of MDPI.

multi-cavity fiber device that is built upon the inscription of independent FBGs in three of the outer cores, identified in the figure as cores 4, 5 and 6. The fiber employed is a commercial homogeneous 7-core fiber with a 125-μm cladding diameter and 35-μm core pitch. Each one of the selected three cores comprises an array of three uniform gratings centred at different optical wavelengths, (λ_1 = 1537.07 nm, λ_2 = 1541.51 nm, and λ_3 = 1546.26 nm) was inscribed at different longitudinal positions in cores 4, 5 and 6. To enable operation using wavelength diversity, the gratings in a given core were inscribed with incremental distances from one core to another, that is 20 mm for core 6, 21 mm for core 5 and 22 mm for core 4. On the other hand, the spatial diversity is produced by the incremental displacement between grates centred at the same wavelength but inscribed in adjacent cores, i.e. 6 mm for λ_1, 7 mm for λ_2 and 8 mm for λ_3. As we can see from the normalized optical spectra in reflection plotted in Figure 4.15a, all the gratings have similar strength levels with a maximum difference between gratings of 3 dB.

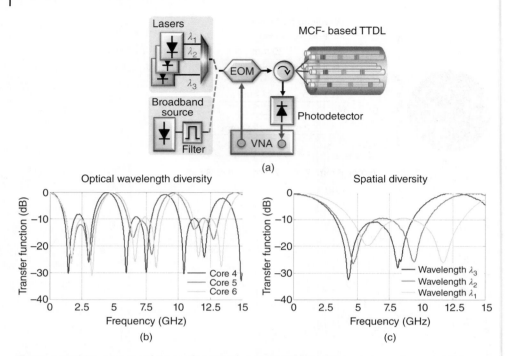

Figure 4.16 (a) Experimental setup of a microwave signal filter built upon a homogeneous MCF multi-cavity device; measured RF responses operating in (b) wavelength diversity (for different cores) and (c) spatial diversity (for different wavelengths), Source: Reproduced with permission of MDPI.

Figure 4.16a illustrates the experimental setup of a microwave signal filter implemented with the described MCF multi-cavity device. Either an array of three low-linewidth lasers or a broadband source followed by an optical filter with a 2 nm bandwidth were used. Note that we require the broadband optical source in order to avoid coherent interference when the samples coming from different cores are detected together (differential delay between samples is then lower than the coherence time of the low-linewidth laser). Figure 4.16b shows the measured frequency responses for the signal filtering provided when we exploit wavelength diversity, that is when one gathers the different samples coming from a given core. Filter reconfigurability is possible if we select another core. In this particular example, going from core 4 to core 6 increases the FSR from 4.45 to 4.97 GHz.

If we detect instead the signal samples that come from different cores and share the same optical wavelength, we can operate the delay line using spatial diversity provided by the fiber cores. This can be seen in Figure 4.16c, which gathers the experimental frequency responses produced by each of the three optical wavelengths. Tuning the wavelength from λ_3 to λ_1 increases the FSR from 12.50 up to 17.76 GHz. Although the experimental results are in good agreement with the theoretical results, we must highlight that slight discrepancies in the differential delay arise due to discrepancies between the theoretical and real values of the refractive index of the cores (which, in addition, are not exactly the same for all the cores) and small imbalances produced in the reflectivity strength of the gratings.

4.5 Final Remarks

Future fiber-wireless access networks will potentially benefit from SDM-based approaches in terms of compactness as compared to a set of parallel single-mode fibers, performance stability against mechanical or environmental conditions as well as operation versatility offered by the simultaneous use of the spatial- and wavelength-diversity domains. Moreover, the synergistic combination of photonic integrated circuits (understood as "vertical integration") and the SDM fiber technologies gathered in this chapter (understood as a "horizontal integration") will contribute to the reduction of size, weight and power consumption that will drive the future of microwave photonics.

The application of SDM fibers to microwave photonics links and systems lead to the term "fiber-distributed signal processing", where we can process the signal while it is being distributed throughout the radio access network. Regarding signal distribution, the use of MCFs to support C-RAN architectures will be capable of addressing the main future challenges envisioned for 5G communications, allowing SDN and NFV technologies while being compatible with both WDM and PON overlay expansion. SDN and NFV can be achieved by exploiting the fact that both in the DRoF and RoF approaches the traffic and capacity characteristics of each sector in each base station can be reconfigured by software in the electrical domain through an overall resources manager located in the electronic switch place at the central office. Furthermore, the proposed architecture is potentially integrable with MCF-based solutions for metro networks if architecture on demand gateway nodes [Amaya et al. - 2014] are introduced at the central office. In terms of RF signal processing, both dispersion-engineered heterogeneous MCF links and multi-cavity devices on commercial homogeneous MCFs can be exploited to implement several microwave photonics functionalities that will be specially demanded in the context of 5G communications. These include, among others, arbitrary waveform generation, reconfigurable signal filtering, optical beamforming networks for phased array antennas and analogue-to-digital conversion.

Bibliography

Y. Amma, Y. Sasaki, K. Takenaga, S. Matsuo, J. Tu, K. Saitoh, M. Koshiba, T. Morioka, and Y. Miyamoto. High-density multicore fiber with heterogeneous core arrangement. In *2015 Optical Fiber Communications Conference and Exhibition (OFC)*, pages 1–3, 2015.

N. Amaya, et al. (2014). Software defined networking over space division multiplexing optical networks: features, benefits and experimental demonstration. *Optics Express*, 22 (3): 3638–3647.

J. Capmany, et al. (1999). New and flexible fiber-optic delay-line filters using chirped Bragg gratings and laser arrays. *IEEE Transactions on Microwave Theory Technologies*, 47 (7): 1321–1326.

J. Capmany and D. Novak. Microwave photonics combines two worlds. *Nature Photon*, 1, 2007.

J. Capmany, J. Mora, I. Gasulla, J. Sancho, J. Lloret, and S. Sales. Microwave photonic signal processing. *IEEE/OSA Journal of Lightwave Technology*, 31, 2013.

P. Chanclou, A. Pizzinat, F. Le Clech, T. L. Reedeker, Y. Lagadec, F. Saliou, B. Le Guyader, L. Guillo, Q. Deniel, S. Gosselin, S. D. Le, T. Diallo, R. Brenot, F. Lelarge, L. Marazzi, P.

Parolari, M. Martinelli, S. O'Dull, S. A. Gebrewold, D. Hillerkuss, J. Leuthold, G. Gavioli, and P. Galli. Optical fiber solution for mobile fronthaul to achieve cloud radio access network. In *2013 Future Network Mobile Summit*, pages 1–11, 2013.

China Mobile Research Institute (2011). C-ran: The road towards green ran. White paper, China Mobile Research Institute, 2011.

J. M. Galvé, I. Gasulla, S. Sales, and J. Capmany. Reconfigurable radio access networks using multicore fibers. *IEEE Journal of Quantum Electronics*, 52 (1):1–7, 2016.

S. Garcia and I. Gasulla. Design of heterogeneous multicore fibers as sampled true-time delay lines. *Opt. Lett.*, 40(4):621–624, Feb 2015.

S. García and I. Gasulla. Dispersion-engineered multicore fibers for distributed radiofrequency signal processing. *Opt. Express*, 24(18): 20641–20654, 2016.

S. García, D. Barrera, J. Hervas, S. Sales, and I. Gasulla. Dispersion-engineered multicore fibers for distributed radiofrequency signal processing. *Photonics*, 4(49), 2017.

I. Gasulla and J. Capmany. Microwave photonics applications of multicore fibers. *IEEE Photonics Journal*, 4(3):877–888, 2012.

I. Gasulla, D. Barrera, J. Hervas, and S. Sales. Spatial division multiplexed microwave signal processing by selective grating inscription in homogeneous multicore fibers. nature, Scientific Reports, 2017.

T. Kobayashi, H. Takara, A. Sano, T. Mizuno, H. Kawakami, Y. Miyamoto, K. Hiraga, Y. Abe, H. Ono, M. Wada, Y. Sasaki, I. Ishida, K. Takenaga, S. Matsuo, K. Saitoh, M. Yamada, H. Masuda, and T. Morioka. 2 × 344 tb/s propagation-direction interleaved transmission over 1500-km mcf enhanced by multicarrier full electric-field digital back-propagation. In *39th European Conference and Exhibition on Optical Communication (ECOC 2013)*, pages 1–3, 2013.

Y. Kokubun and T. Watanabe. Dense heterogeneous uncoupled multi-core fiber using 9 types of cores with double cladding structure. In *17th Microoptics Conference (MOC)*, pages 1–2, 2011.

M. Koshiba. Design aspects of multicore optical fibers for high-capacity long-haul transmission. In *Microwave Photonics (MWP) and the 2014 9th Asia-Pacific Microwave Photonics Conference (APMP) 2014 International Topical Meeting on*, pages 318–323, 2014.

M. Koshiba, K. Saitoh, and Y. Kokubun. Heterogeneous multi-core fibers: proposal and design principle. *IEICE Electronics Express*, 6(2):98–103, 2009.

C. Liu, J. Wang, L. Cheng, M. Zhu, and G. K. Chang. Key microwave-photonics technologies for next-generation cloud-based radio access networks. *Journal of Lightwave Technology*, 32(20):3452–3460, 2014.

D. Marpaung, C. Roeloffzen, R. Heideman, A. Leinse, S. Sales, and J. Capmany. Integrated microwave photonics. *Lasers Photonics Review*, 7, 2013.

S. Matsuo, K. Takenaga, Y. Sasaki, Y. Amma, S. Saito, K. Saitoh, T. Matsui, K. Nakajima, T. Mizuno, H. Takara, Y. Miyamoto, and T. Morioka. High-spatial-multiplicity multicore fibers for future dense space-division-multiplexing systems. *IEEE/OSA Journal of Lightwave Technology*, 34(6):1464–1475, 2016.

J. E. Mitchell. Integrated wireless backhaul over optical access networks. *Journal of Lightwave Technology*, 32(20):3373–3382, 2014.

P. A. Morton and J. B. Khurgin. Microwave photonic delay line with separate tuning of the optical carrier. *IEEE Photonics Technology Letters*, 21(22): 1686–1688, 2009.

F. Ohman, K. Yvind, and J. Mork. Slow light in a semiconductor waveguide for true-time delay applications in microwave photonics. *IEEE Photonics Technology Letters*, 19(15):1145–1147, 2007.

A. Pizzinat, P. Chanclou, F. Saliou, and T. Diallo. Things you should know about fronthaul. *Journal of Lightwave Technology*, 33(5):1077–1083, 2015.

B. J. Puttnam, R. S. Luís, W. Klaus, J. Sakaguchi, J. M. Delgado Mendinueta, Y. Awaji, N. Wada, Y. Tamura, T. Hayashi, M. Hirano, and J. Marciante. 2.15 Pb/s transmission using a 22 core homogeneous single-mode multi-core fiber and wideband optical comb. In *2015 European Conference on Optical Communication (ECOC)*, pages 1–3, 2015.

D. J. Richardson, Fini J. M., and L. E. Nelson. Space-division multiplexing in optical fibres. *Nature Photonics*, 7, 2013.

R. Ryf, H. Chen, N. K. Fontaine, A. M. Velzquez-Bentez, J. Antonio-Lpez, C. Jin, B. Huang, M. Bigot-Astruc, D. Molin, F. Achten, P. Sillard, and R. Amezcua-Correa. 10-mode mode-multiplexed transmission over 125-km single-span multimode fiber. In *2015 European Conference on Optical Communication (ECOC)*, pages 1–3, 2015.

A. Saadani, M. El Tabach, A. Pizzinat, M. Nahas, P. Pagnoux, S. Purge, and Y. Bao. Digital radio over fiber for lte-advanced: Opportunities and challenges. In *2013 17th International Conference on Optical Networking Design and Modeling (ONDM)*, pages 194–199, 2013.

J. Sakaguchi, B. J. Puttnam, W. Klaus, Y. Awaji, N. Wada, A. Kanno, T. Kawanishi, K. Imamura, H. Inaba, K. Mukasa, R. Sugizaki, T. Kobayashi, and M. Watanabe. 19-core fiber transmission of $19 \times 100 \times 172 - gb/s$ sdm-wdm-pdm-qpsk signals at 305tb/s. In *OFC/NFOEC*, pages 1–3, 2012.

J. Sakaguchi, W. Klaus, J. M. D. Mendinueta, B. J. Puttnam, R. S. Luis, Y. Awaji, N. Wada, T. Hayashi, T. Nakanishi, T. Watanabe, Y. Kokubun, T. Takahata, and T. Kobayashi. Realizing a 36-core, 3-mode fiber with 108 spatial channels. In *2015 Optical Fiber Communications Conference and Exhibition (OFC)*, pages 1–3, 2015.

Samsung Electronics Co. 5G vision. White paper, Samsung Electronics Co, 2015.

J. Sancho, J. Bourderionnet, J. Lloret, S. Combrie, I. Gasulla, S. Xavier, S. Sales, P. Colman, G. Lehoucq, D. Dolfi, J. Capmany, and A. De Rossi. Integrable microwave filter based on a photonic crystal delay line. *Nature Communications*, 3, 2012.

Y. Sasaki, R. Fukumoto, K. Takenaga, K. Aikawa, K. Saitoh, T. Morioka, and Y. Miyamoto. Crosstalk-managed heterogeneous single-mode 32-core fibre. In *ECOC 2016; 42nd European Conference on Optical Communication*, pages 1–3, 2016.

Y. Sasaki, K. Takenaga, K. Aikawa, Y. Miyamoto, and T. Morioka. Single-mode 37-core fiber with a cladding diameter of 248μ*m*. In *2017 Optical Fiber Communications Conference and Exhibition (OFC)*, pages 1–3, 2017.

A. J. Seeds. Microwave photonics. *IEEE Transactions on Microwave Theory and Techniques*, 50, 2002.

J. Tu, K. Saitoh, M. Koshiba, K. Takenaga, and S. Matsuo. Optimized design method for bend-insensitive heterogeneous trench-assisted multi-core fiber with ultra-low crosstalk and high core density. *IEEE/OSA Journal of Lightwave Technology*, 31(15):2590–2598, 2013.

J. Tu, K. Long, and K. Saitoh. An efficient core selection method for heterogeneous trench-assisted multi-core fiber. *IEEE Photonics Technology Letters*, 28(7):810–813, 2016.

C. Wang and J. Yao. Fiber bragg gratings for microwave photonics subsystems. *Opt. Express*, 21:22868–22884, Sep 2013.

K. Wilner and A. P. van den Heuvel. Fiber-optic delay lines for microwave signal processing. *Proceedings of the IEEE*, 64(5):805–807, 1976.

J. Yao. Microwave photonics. *IEEE/OSA Journal of Lightwave Technology*, 27, 2009.

F. Zeng and J. Yao. All-optical microwave filters using uniform fiber bragg gratings with identical reflectivities. *Journal of Lightwave Technology*, 23(3):1410–1418, 2005.

B. Zhu, T.F. Taunay, M. Fishteyn, X. Liu, S. Chandrasekhar, M. F. Yan, J. M. Fini, E. M. Monberg, and F. V. Dimarcello. 112-tb/s space-division multiplexed dwdm transmission with 14-b/s/hz aggregate spectral efficiency over a 76.8-km seven-core fiber. *Opt. Express*, 19(17):16665–16671, Aug 2011.

5

Enabling VLC and WiFi Network Technologies and Architectures Toward 5G

Isiaka Ajewale Alimi[,1], Abdelgader M. Abdalla[1], Jonathan Rodriguez[1,3], Paulo Pereira Monteiro[1,2], Antonio Luís Teixeira[1,2], Stanislav Zvánovec[4], and Zabih Ghassemlooy[5]*

[1] *Instituto de Telecomunicações, 3810-193, Aveiro, Portugal*
[2] *Department of Electronics, Telecommunications and Informatics (DETI), Universidade de Aveiro, 3810-193, Aveiro, Portugal*
[3] *University of South Wales, Pontypridd, Wales, UK*
[4] *Department of Electromagnetic Field, Faculty of Electrical Engineering, Czech Technical University in Prague, Prague 16627, Czech Republic*
[5] *Optical Communications Research Group, Northumbria University, Newcastle upon Tyne NE1 8ST, United Kingdom*

5.1 Introduction

Machine-to-machine (M2M) communications have been observed to be a key part of the emerging Internet of Things (IoT) within the context of smart environments such as healthcare, security, industry, etc. With these technologies, information can be exchanged between autonomous devices such as actuators, mobile phones, sensors, and radio-frequency identification tags being pervasively deployed within the network without the need for human interaction. Furthermore, more intelligent and efficient wireless IoT devices capable of offering reliable interconnections to the internet are anticipated in the next generation IoT applications. IoT is envisaged to offer a platform in multitude of environments and applications that will support several heterogeneous devices with embedded elements such as sensors and actuators. The main aims of these technologies are based on sharing, gathering, and forwarding information adaptively in accordance with the obtainable data. For instance, it has been estimated that IoT devices will grow beyond 30 billion globally by 2020, resulting into a massive connection to the internet [Shahin et al. - 2018; Lv et al. - 2018; Parne et al. - 2018]. Consequently, the anticipated numerous devices will not only have notable impact on data traffic in networks but will also contribute significantly to the problem of transmission channel availability and latency. In addition, they will have a huge influence on the design and implementation of the fifth generation (5G) and beyond-5G (B5G) wireless networks [Lv et al. - 2018].

Moreover, for efficient operation of the devices, stringent requirements regarding massive connectivity and low latency will be demanded of the associated networks

* Corresponding Author:Isiaka Ajewale Alimi iaalimi@ua.pt

Optical and Wireless Convergence for 5G Networks, First Edition.
Edited by Abdelgader M. Abdalla, Jonathan Rodriguez, Issa Elfergani, and Antonio Teixeira.
© 2020 John Wiley & Sons Ltd. Published 2020 by John Wiley & Sons Ltd.

[Lv et al. - 2018]. Meanwhile, there has been a remarkable development in wireless communication technology regarding ubiquitous connectivity provision between people across the world. However, further efforts are required for heterogeneous objects to communicate with each other by means of wireless communication networks. So, devices and users with the support for information exchange, which can be from device-to-device (D2D)/terminal-to-terminal (T2T), person-to-person (human-to-human (H2H)) and/or person-to-device (H2D) are enabled to communicate in the IoT network [Parne et al. - 2018].

It is worth mentioning that, by nature, M2M requires further analysis in order to comprehend the fundamental human demand it serves. For instance, wireless local area network (WLAN) or ZigBee could be implemented for a small deployment of M2M devices at very low-data rates. However, for a large deployment of M2M communications, cellular networks should be employed for connectivity. Furthermore, IoT and M2M communications present a large-scale network that has the ability to support various applications and a huge number of heterogeneous devices such as vending machines, sensors, and vehicles that are interconnected for automatic data transmission [Lo et al. - 2013]. Moreover, M2M communications have many applications in different fields like computers, smartphones, healthcare monitoring system, manufacturing floors, energy grids, warehouses, healthcare, cloud based systems, transportations, traffic lights, intelligent tracing, and tracking systems. It is also applicable to other smart systems such as smart cities, smart transportations, smart electricity grids, and smart meters [Parne et al. - 2018]. Based on the requirements of each application, the M2M system demands a number of technologies to ensure the last mile connectivity.

Wireless technologies are promising solutions due to merits such as mobility support, ubiquitous coverage, and plug-and-play features [Lo et al. - 2013]. However, it has been observed that the existing RF wireless spectrum is not only highly regulated by the associated authorities but also extremely congested. The emergence of IoT and M2M communications is not helping the situation and makes it even more demanding. Consequently, spectrum congestion in high-density areas may lead to limited access to the network. Likewise, the link performance of the existing RF based systems is hindered by multipath effects, most especially in a dense urban scenario. The problem is more demanding for instance in the indoor scenario where the available bandwidth is not sufficient for efficient operation of heterogeneous users' equipment. This is due to the fact that, over 70% of wireless data traffic occurs in indoor settings such as offices and homes [Ghassemlooy et al. - 2016]. The pressure on the current base stations can be lessened by offloading mobile data traffic to effective microwave technologies such as wireless fidelity (WiFi) and Femtocells [Alimi et al. - 2018]. Nevertheless, densely deployed WiFi hotspots have been found to be one of the bottlenecks for system capacity improvement [Wu et al. - 2017]. Therefore, to ensure seamless wireless communication, it is imperative to employ highly reliable and low-cost technologies [Ghassemlooy et al. - 2016].

In addition, to attend to the network demand, effective concepts can be adopted to improve the RF based schemes performance. It should be noted that, irrespective of the employed technology, the following options can be implemented to enhance the capacity of wireless radio systems [Ghassemlooy et al. - 2016].

1. Additional spectrum (channel or band) allocation, allotment, and assignment, result in a more exploitable bandwidth.
2. Deployment of more radio access nodes (i.e. overlay of small cells such as femtocells, picocells, and microcells on the conventional macrocells), which helps in spectrum reuse and provides not only effective network capacity but also seamless wireless coverage.
3. Improve the spectral efficiency by employing innovative schemes such as effective frequency reuse, resource scheduling, spectrum allocation, data/channel aggregation, and compression techniques to increase the network capacity [Tayade - 2016].
4. Significantly higher efficiency in terms of power (energy), resources utilization, and cost.

However, the aforementioned schemes come with different stringent trade-offs. For instance, acquisition and deployment of a new spectrum is cost-ineffective [Ghassemlooy et al. - 2016]. Likewise, based on a small-cell concept, more densely deployed radio access nodes can be employed. Nevertheless, apart from the associated cost, a dense network presents additional challenges regarding inter-cell/inter-tier interference and management issues with spectral resources reuse as anticipated in the 5G and B5G cellular networks [Alimi et al. - 2018]. Moreover, there have been concerted efforts in improving the spectral efficiency of RF based wireless communication systems over the year; however, with different evolving bandwidth-intensive applications and services, the efforts seem to be insufficient [Ghassemlooy et al. - 2016]. In addition, the current RF based wireless systems fall short of the required bandwidth to meet up with the ever-growing demand. Consequently, with the enumerated associated challenges of the RF based schemes, innovative, cost-effective and viable technologies are required [Ghassemlooy et al. - 2016; Alimi et al. - 2018, 2017a]. This can be achieved by adopting optical wireless scheme as an alternative and/or complementary technology as expatiated in Section 5.2. This will not only help in addressing the problem but also aids in alleviating the relapsing pressure on the RF spectrum.

5.2 Optical Wireless Systems

An optical wireless communication (OWC) system is an innovative and promising broadband access technology that offers a scalable, ultra-high-speed, improved capacity, cost-effective, and easier to deploy solution while still maintaining the inherent advantages of optical fiber based solutions. Therefore, the OWC can address the bandwidth demands of different services and applications of the next-generation networks at comparatively low-cost in certain applications. Furthermore, the fundamentally unlimited bandwidth presented by the OWC can be accredited to diverse bands such as ultraviolet, visible (VL) and infrared (IR) sub-bands being used for communication purposes.

The sub-bands have huge bandwidths that have mostly remained unexploited. Consequently, to attend to the bandwidth demand, research efforts have been focused on OWC. This is mainly owing to its inherent advantages compared to the traditional RF based communication systems. For instance, OWC offers exceptional features such as ease of deployment, ultra-high capacity/bandwidth (THz), lower power consumption (i.e. resulting in high energy efficiency), more compact/low-mass equipment, reduced time-to-market, immunity to RF electromagnetic interference (EMI), better protection against interference, cost-effectiveness due to no licensing fee, inherent/high-degree of security against eavesdropping, and a high-degree of spatial confinement bringing practically unconstrained frequency reuse [Alimi et al. - 2017d,e,c]. Moreover, OWC systems are highly attractive for various applications such as remote-sensing/monitoring/surveillance, illumination, data communications, disaster recovery, radio astronomy, high-definition TV transmission, metropolitan area network extension, sharing of medical imaging in real-time, and fronthaul/backhaul of wireless cellular networks because of their salient features [Ghassemlooy et al. - 2016; Alimi et al. - 2018, 2017a; Ghassemlooy et al. - 2012].

In addition, OWC systems are able to comply with the following 5G Infrastructure Public Private Partnership (5G PPP) identified key performance indicators PPP in [Horizon 2020 - 2014].

1. Provision of 1000 times higher wireless area capacity and further diverse service capabilities in relation to 2010.
2. Conservation of about 90% of energy per service provided with major attention in mobile communication systems where the main energy consumption is due to radio access networks.
3. Minimization of average service provision time cycle from 90 hours to 90 minutes.
4. Creation of a secure and dependable internet with a "zero perceived" downtime for services delivery.
5. Facilitation of ultra-dense deployments of wireless communication links towards connecting over seven trillion wireless devices that will be serving over seven billion people.
6. Enabling advanced user controlled privacy.

OWC can be generically grouped into two technologies, which are the indoor and outdoor optical wireless schemes. For the indoor OWC schemes, IR (780–950 nm) or VL (380–780 nm) system is employed for an in-building (indoor) wireless solution [Ghassemlooy et al. - 2015]. An indoor OWC scheme is of high significance mainly in scenarios where it is challenging to provide network connectivity via physical wired connections. In addition, the indoor-based OWC schemes can be grouped into diffused, tracked, directed line of sight (LOS) and non-directed LOS configurations. In outdoor based OWC systems, an optical carrier is employed for transmitting information from one node/point to another via an unguided channel. The transmission channel can be an atmosphere or free space. Therefore, the outdoor-based (terrestrial point-to-point) OWC is also known as a free-space optical (FSO) communication system. It should be noted that FSO systems normally operate at the near IR frequencies. Also, they are categorized into terrestrial and space optical links, which comprise building-to-building, ground-to-satellite, satellite-to-ground, satellite-to-satellite, and satellite-to-airborne platforms [Ghassemlooy et al. - 2016; Alimi et al. - 2018, 2017a; Ghassemlooy et al. - 2012]. The OWC system classification is depicted in the tree diagram in Figure 5.1. This chapter focuses on visible light communication (VLC) technologies within the context of 5G.

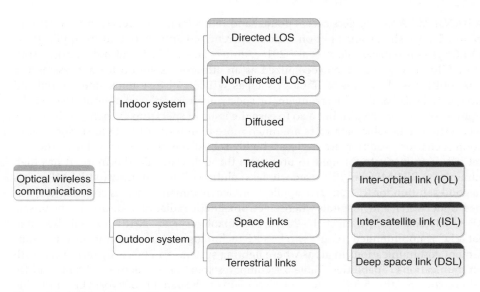

Figure 5.1 OWC system classification.

5.3 Visible Light Communication (VLC) System Fundamentals

As stated in Section 5.1, the current RF spectrum is heavily regulated and congested. Also, spectrum congestion in high-density areas can cause poor/limited access to the network users. This is even more demanding for cell-edge users due to the experienced very-low SNR/data rate, which consequently leads to poor quality of service (QoS). This effect can be attributed to factors such as high-path losses, inter-cell interference, and fading effects that are due to wireless channel conditions. Consequently, innovative and feasible technologies are required [Ghassemlooy et al. - 2016; Alimi et al. - 2018, 2017a]. As expatiated in Section 5.2, an alternative and/or a complementary technology based on an optical wireless scheme such as a VLC system can be adopted not only to address the problem but also to help in alleviating the so-called RF "spectrum crunch". This is due to the fact that OWC uses different bands of the electromagnetic spectrum and is capable of supporting different 5G deployment/usage scenarios such as enhanced mobile broadband (eMBB) that is for mobile and fixed wireless; massive machine type communications (mMTC), which is for the IoT; as well as ultra-reliable and low latency communications (URLLC), which is for mission-critical services as discussed in Section 5.4.

The existing improvement in the light emitting diode (LED) chip design that has the ability of swift nanosecond switching times as well as wide deployment of LEDs for energy efficiency are parts of factors that pave the way for the adoption of a VLC system [Sevincer et al. - 2013; Rajagopal et al. - 2012]. Consequently, the VLC system seems to be an appealing technology for addressing challenges such as bandwidth limitation, energy efficiency, electromagnetic radiation, and safety in the wireless communication systems [Ying et al. - 2015; Yang and Gao - 2017]. Based on this, it is an applicable solution for a number of short- and medium-range communications such as M2M communications, wireless personal area networks (WPANs), wireless body area networks

(WBANs), WLANs, wireless access points, and vehicular and underwater networks. In Section 5.4, we shed some light on various current and anticipated future applications of VLC systems [Ghassemlooy et al. - 2016; Alimi et al. - 2017a; Uysal and Nouri - 2014].

In addition, VLC has been recognized as an attractive solution for addressing the issues of RF-based wireless technology such as WiFi. In VLC systems, unexploited and unregulated sub-bands of electromagnetic spectrum are employed not only to offer the required system bandwidth but also to ease the issue of spectrum crunch. For instance, VLC system achievable data rates are much more than those of WiFi for indoor application scenarios. Additionally, it gives a highly robust physical security due to the fact that it is very intricate for light to propagate through walls. Furthermore, it has high immunity to RF EMI from RF-based wireless solutions. The immunity to RF EMI makes it a good solution for RF-sensitive applications/environments such as scan centers, airlines, underwater, and hospitals where electromagnetic radiation is deemed to be unsafe (RF-restricted areas) [Alimi et al. - 2017a; Papanikolaou et al. - 2018]. It should be noted that VLC is not intended to oust wireless RF technologies; however, it is meant to supplement communications in areas where lighting systems can be easily retrofitted with communications capabilities. Table 5.1 compares some of the features of VLC and RF technologies. Section 5.5 gives further discussion on the benefits of hybrid LiFi and WiFi systems.

Furthermore, the global governmental policies to replace incandescent light bulbs with more energy-efficient lighting approaches laid the foundations for extensive future

Table 5.1 Comparison of VLC and RF technologies [Ghassemlooy et al. - 2015; IEEE 802.15.7 VLC Task Group - 2008].

	Parameter	VLC	RF
	Available spectrum	400 THz to 800 THz	3 KHz to 300 GHz
	Devices size	Small	Large
	Power consumption	Low	Medium
	Security	High	Low
	Distance	Short	Short–long
	Mobility	Limited	Good
	Bandwidth	Unlimited	Limited
	Line of sight	Yes	No
	Standard	IEEE 802.15.7	Matured
	Hazard	No	Yes
	RF EMI	No	High
Infra	Visibility (security)	Yes	No
to	Infra	LED illumination	Access point
mobile	Mobility	Limited	Yes
	Coverage	Narrow	Wide
Mobile	Visibility (security)	Yes	No
to	Power consumption	Relatively low	Medium
mobile	Distance	Short	Medium

application of LEDs. Consequently, LEDs have been widely employed in different places such as homes, offices, industries, and municipalities for several applications like illuminations, advertising displays, streetlights, outdoor lamps, car tail lights/headlights, traffic signs, and intelligent transport systems. Consequently, these are paving the way for connecting mobile devices to the internet [Ghassemlooy et al. - 2016; Uysal and Nouri - 2014; Haas et al. - 2016]. Therefore, with wide deployment of LEDs in different environments, the VLC system can leverage current lighting infrastructures for high-speed wireless communication and effective service provisioning.

In addition, the global research community via bodies like the IEEE standardization body, the Wireless World Research Forum, the VLC Consortium, the European OMEGA project, and the UK research council, have embraced VLC as an attractive solution due to its inherent exceptional features [Ghassemlooy et al. - 2015].

Basically, a VLC system employs LED lamps as access points (APs) and a photodetector or an image sensor based receiver based on an intensity modulation and direct detection (IM/DD) scheme. It is remarkable that, the VLC IM/DD scheme consist of comparatively inexpensive optoelectronic devices at both transceivers (LEDs and photodiodes). The emanating ultra-high speed light from the LED serves as the medium for conveying information streams and the streamed data is encoded by means of IM. For photon detection, the DD method is implemented. This entails the use of a photodiode at the receiving aperture (dongle) for converting the slightest changes in the optical signal amplitude into an electric current. The signal is then processed and the subsequent data stream is employed by wireless devices. Figure 5.2 depicts the basic operation of a VLC system.

It is noteworthy that different modulation techniques can be employed by VLC systems. Since VLC depends on electromagnetic radiation for information transmission, with basic modifications, modulation techniques that are commonly employed for RF communications are also applicable to VLC. Likewise, it offers several distinctive and exclusive modulation formats owing to VL being employed for wireless communication. Hence, it supports single-carrier modulations such as pulse amplitude modulation, pulse-position modulation (PPM) and on-off keying, that have been extensively considered in wireless infrared communication systems. These modulation formats are easy to implement and have typical achievable rates of several kbps to Mbps.

In addition, multi-carrier modulations that are based on orthogonal frequency division multiplexing (OFDM) and multi-band carrier-less amplitude and phase modulation (mCAP) are also applicable. For instance, formats such direct-current-biased optical OFDM and asymmetrically clipped optical OFDM can also be employed for high data rates [Haas et al. - 2016; Yamazato - 2017]. VLC specific modulation such as color shift keying proposed in IEEE 802.15.7 for improving the data rate can also be employed [Haas et al. - 2016; Khan - 2017], using an alternative modulation format mCAP VLC the largest improvements in the data rate and bandwidth efficiency of 10% and 40%, respectively have been achieved [Chvojka et al. - 2017].

5.4 VLC Current and Anticipated Future Applications

There are a number of feasible applications of optical wireless schemes in communication systems. The adopted technique depends mainly on the specified application. The

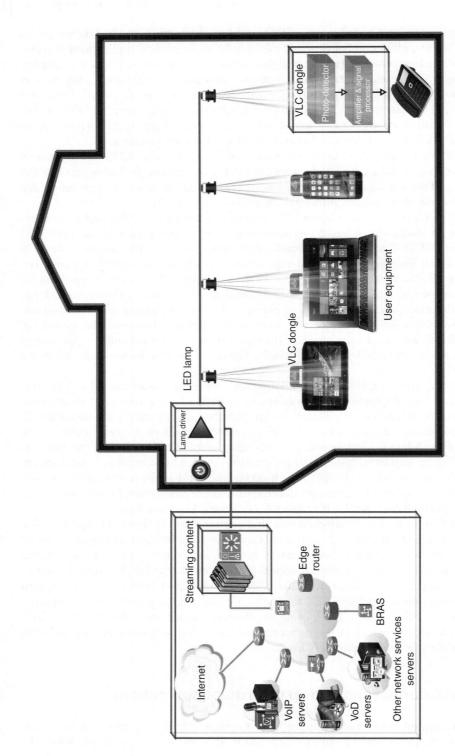

Figure 5.2 VLC system.

5G envisaged high data rate metrics required for different deployment/usage scenarios such as eMBB, mMTC, URLLC, and others can be achieved with the aid of OWC systems. eMBB indicates an improved support of the conventional MBB via enhanced coverage, capacity, and average/peak/cell-edge data rates. Moreover, URLLC signifies the demand for supporting evolving critical applications like smart grids, industrial internet, remote surgery, infrastructure protection, and intelligent transportation systems. In addition, mMTC implies the necessity to support the anticipated 5G IoT scenario with tens of billions of envisaged connected sensors and devices [Teyeb et al. - 2017].

The 5G technology use cases, along with typical required user data rates, are depicted in Figure 5.3 [Teyeb et al. - 2017; 5G Americas - 2017]. It is worth mentioning that OWC systems are capable of conveniently supporting the use cases. Take, for instance, that the contemporary OWC systems have the ability to support 10 Gbps ethernet. Comparatively, the supported bandwidth surpasses that of the 60 GHz RF wireless ethernet systems that offer 1.25 Gbps. Besides, it also exceeds the supported bandwidth of personal communication system Gigabit infrared that operates at data rates of 512 Mbps and 1.024 Gbps [Ghassemlooy et al. - 2016].

Similarly, the Wireless Gigabit Alliance (WiGig) offers about 7 Gbps data rate while it has been established that a single micro-LED can attained high-speed data transmission in the Gbps range (3 Gbps), but at the moment over a very short transmission range, which is expected to increase with maturity of devices (i.e. mostly LEDs). Moreover, broad illumination can be accomplished by combining laser LEDs with an optical diffuser. The resulting configuration is capable of delivering about 100 Gbps [Haas et al. - 2016; Wang and Haas - 2015]. In addition, due to the offered concurrent support for communication and illumination by the LED-based VLC, LED applications are progressively dominating a number of use cases globally. This encourages effective offloading into the light spectrum and hence reduces the pressure on the RF spectrum. Consequently, these are opening the way for connecting massive mobile devices to the internet, as anticipated by the 5G and B5G wireless networks deployment scenarios such as eMBB and mMTC among other things. Figure 5.4 illustrates some of the potential applications of VLC systems. In the following, we present different use cases of VLC technology for addressing the last-mile transmission bottleneck of wireless networks. These substantiate its current and future applications for wireless systems.

5.4.1 Underwater Wireless Communications

Underwater wireless communications entail data transmission in unguided water environments by means of wireless carriers for military or diver-to-diver communications [Alimi et al. - 2017a]. In underwater communication, the usually employed technique is based on acoustic transmission, mostly because it can cover a several kilometer range. Nevertheless, the acoustic technology is affected by factors such as extremely low bandwidth and high latency, which are as a result of low-propagation speed. Hence, underwater acoustic communication data rates are restricted to only a few hundreds or thousands of kilobits per second [Ghassemlooy et al. - 2016; Uysal and Nouri - 2014].

Likewise, it should be noted that radio waves cannot be effectively employed for underwater applications. This is due to significant radio signal absorption by water within a few feet of transmission [Johri - 2016; Kuppusamy et al. - 2016; Soni et al. - 2016]. In recent years we have seen growing research interest in underwater

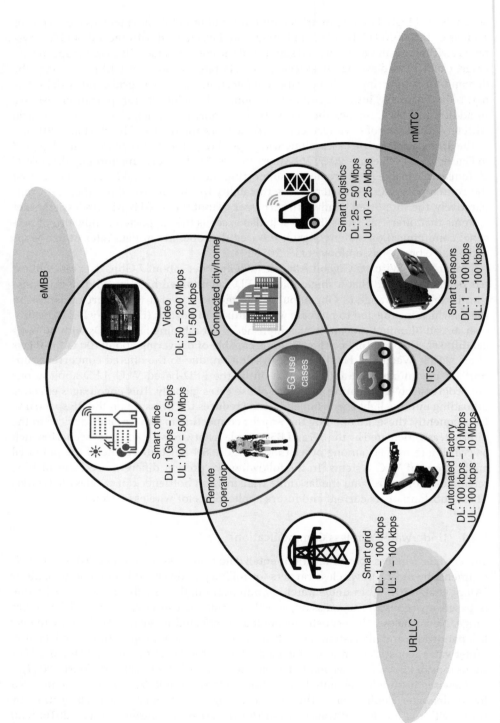

Figure 5.3 5G technology use cases.

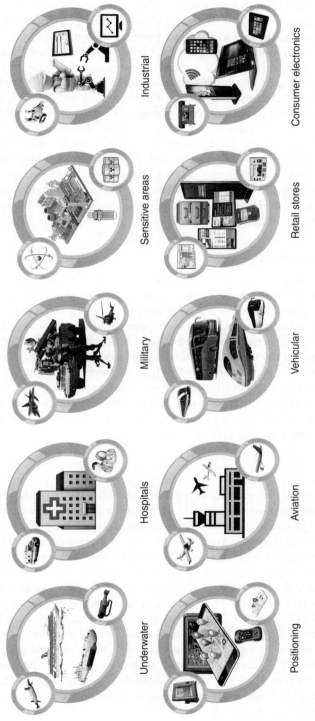

Figure 5.4 Potential applications of the VLC system [5G Americas - 2017].

optical wireless communications, which offers higher transmission bandwidth compared to the acoustic and RF based wireless technologies [Zeng et al. - 2017]. Therefore, VLC systems (mainly in the blue-green wavelength band) are a promising solution for underwater communication, which offer higher data rates over a short to moderate transmission range. This is due to the relative water transparency to the visible spectrum band resulting in minimum light absorption. Consequently, an underwater VLC based system can offer hundreds of megabit per second high data speeds for short ranges. This can be employed to complement acoustic communication that has a longer transmission range [Ghassemlooy et al. - 2016; Uysal and Nouri - 2014].

In addition, VLC can help in the deployment of an underwater wireless sensor network (UWSN). It can help UWSN entities such as autonomous underwater vehicles, remotely operated underwater vehicles (ROVs), relay buoys, and seabed sensors, in having effective connection among themselves. Beside the communication aspect, this can help not only in preventing the ROV vehicles from being struck but also enables better exploration [Alimi et al. - 2017a].

5.4.2 Airline and Aviation

Another potential application is in the aviation industries. Airlines are currently offering internet access and multimedia content such as movies and games to passengers on board. This deployment usually requires enormous amounts of wiring that normally results in high cost and an increase in the aircraft weight. VLC schemes can be employed for delivering broadband access on aircrafts. Besides the capability for high data rates, the implementation can offer advantages such as reduced weight, reduced system complexity, elimination of wiring, improved reliability, and safety by preventing RF interference on critical avionics. Also, in airports, there are strict requirements concerning the use of RF equipment that can interfere with aircraft communications systems or the control tower. Therefore, a VLC based system offers safer and alternative communication systems for this scenario [Firefly LiFi - 2018].

5.4.3 Hospitals

There have been notable concerns about the usage of WiFi in a lot of healthcare environments such as operation rooms. This can be attributed to RF interference with sensitive electronics devices or equipment like magnetic resonance imaging scanners, as well as other medical and monitoring instruments. So, the use of devices such as cell phones are normally prohibited in such sectors. VLC technology offers an effective means of addressing the interference issue and, consequently, presents a safer communication system for healthcare devices in places such as hospitals, workplaces, and homes. Also, it is a cost-effective technology that can provide more reliable communications between equipment (M2M) or between health workers/patients and monitoring stations (H2M). Moreover, it can be employed to monitor/track/sense/detect equipment or the patient wearing a device. When the equipment/patient (wearable device) position is compromised or out of the field of view, a notification signal can be sent to the nursing station with the aid of alarms [Firefly LiFi - 2018]. Some medical testing equipment like a cardio stress test could be re-designed in order to facilitate LED integration on the sensor units. This will enable VLC systems to offer better performance [Ghassemlooy et al. - 2016;

Uysal and Nouri - 2014]. Presently, the RF based wireless is the only technology used in electroencephalography (EEG) or clinical applications, which suffers from lack of frequency spectrum and electromagnetic interference [Dhatchayeny et al. - 2015]. Note, the latter also has adverse effects on patient's health and the medical equipment, especially in RF restricted zones like intensive care unit (ICU)s [Dhatchayeny et al. - 2015]. The VLC together with optical camera communications can be used for transmission, detection, analysis, and classification of EEG signals. It is remarkable that the envisaged massive applications, such as 5G wireless systems, wireless IoT, wearable devices, and other wireless devices with data-gathering medical actuators/sensors will really benefit from VLC systems. Moreover, applications are expected to be designed to comply with current exposure limits in order to prevent the likely health hazards. Hence, the VLC based solution offers an efficient approach and a more secure and safer communication system than the RF based technologies [GSMA - 2017].

5.4.4 Vehicular Communication Systems

Vehicular-to-everything communication is an important system that has the ability to enable connected, safe, reliable, and efficient automated transportation services with high potential of averting automotive traffic accidents. There are a number of potential 5G vehicular network applications such as vehicle-to-vehicle, vehicle-to-infrastructure, vehicle-to-pedestrian, and vehicle-to-network communications for VLC systems in automobiles [Luo et al. - 2017]. For instance, a VLC based system can be deployed in the vehicle headlights and tail lights to enable them to communicate with one another. Also, it can facilitate vehicle communications with the road side infrastructure (RSI) such as traffic lights and street lamps. Furthermore, LED-based RSI technology can be employed for signaling in addition to broadcasting safety-related information in vehicular networks. It is remarkable that, LED-based communication systems are capable of attending to the low latency demanded for safety functionalities such as emergency electronic brake lights, vehicle platooning, in-vehicle signage, pre-crash sensing, traffic signal violation warning, stop sign movement/left-turn assistant, curve speed/lane-change warning, and cooperative forward/intersection collision warning. Consequently, a VLC system can be employed not only to prevent road accidents but also to reduce traffic congestion through effective information exchange. In addition, the technology is capable of supporting the high speeds required in infotainment applications such as high-speed internet access, media downloading, point of interest notification, multi-player gaming, cooperative downloading, map downloads and updates [Ghassemlooy et al. - 2016; Uysal and Nouri - 2014].

In addition, charging and communication systems can also be provided by VLC based technology. This implementation enables VLC to connect vehicles to charging stations or base stations. Furthermore, another significant application is that the VLC system can be employed within vehicles to minimize or eliminate wiring required for offering internet access and multimedia services as anticipated by 5G and B5G. This implementation presents quite a lot of advantages such as reduction in manufacturing costs, weight, and fuel consumption. It is also applicable in train stations and other mass transit locations for communication with trolleys, trains, and other equipment. This helps in preventing RF interference as well as congestion issues [Firefly LiFi - 2018].

5.4.5 Sensitive Areas

EMI makes WiFi unsuitable for sensitive or hazardous locations such as industrial gas/nuclear/power production plants, military, chemical, or petroleum storage plants. The control rooms and the associated equipment in these facilities offer significant functions and any kind of failure might endanger millions of lives. In these scenarios where other transmission mediums can be hazardous, VLC can offer safer wireless connectivity since it does not interfere with the associated electronic devices [Kuppusamy et al. - 2016; Soni et al. - 2016; Firefly LiFi - 2018; Alimi et al. - 2017a].

5.4.6 Manufacturing and Industrial Applications

VLC can provide M2M wireless connectivity within industrial and automated work cell areas. There are a huge number of devices in industrial environments that need adequate monitoring and control; for these applications, VLC implementation offers a better solution compared with RF and wired methods of communication. The wired solutions are not only expensive but might also be unviable because of certain equipment that is usually in motion. Similarly, RF implementations are susceptible to interference as well as congestion issues in addition to lower levels of security. Consequently, VLC wireless systems can enhance system reliability, safety, and production efficiency in a large number of industrial scenarios [Firefly LiFi - 2018]. For instance, in the fourth industrial (Industry 4.0) revolution, robust wireless connectivity is required for interconnecting digitized physical assets like machines for a virtual seamless experience in the digital ecosystems. This enables the production entities such as machines, conveyors, and logistic supply robots to adapt seamlessly and automatically to the specified production steps [Verzijl et al. - 2014; PricewaterhouseCoopers - 2016].

5.4.7 Retail Stores

A large number of LED lights have been deployed in different retail shops. The lights are intended mostly for beautification and lighting purposes. Besides these applications, they can be employed for internet access and other connectivity purposes. Based on this, there can be effective and real-time communications between the devices such as touchscreen kiosks, interactive touch screens, counters, video screens, digital signage/television monitors, computers, scanners, and printers in the store. Moreover, VLC motion detection features can be exploited for monitoring the movement of people in the store [Firefly LiFi - 2018; Soni et al. - 2016].

5.4.8 Consumer Electronics

The unprecedented growth in the number of consumer electronics such as tablets, digital camcorders, fitness monitors, smartphones, headphones, laptops, and smart TVs demands seamless, cost-effective, and high rate data transfer between devices (D2D) and cloud storage without RF interference or congestion concerns. Apart from being capable of addressing these requirements, VLC also offers high security for sensitive data. In addition, it is remarkable that LED technology has already been integrated into the user interface of various consumer devices. Therefore, this makes the deployment of the VLC connectivity solution easy and inexpensive. Moreover, the embedded LED/LCD screens and sensors/cameras of various consumer electronics can be employed as part of VLC connectivity without the usual intricate network configurations [Firefly LiFi - 2018]. For instance, a smart phone camera can be used for communication purposes. In this scenario, the embedded image sensor (phone camera) can be employed for a number of M2M communications such as phone-to-TV, phone-to-phone, and phone-to-vending machine [Ghassemlooy et al. - 2016; Uysal and Nouri - 2014].

5.4.9 Internet of Things

The IoT is a system with a plethora of interconnected and uniquely identified entities such tablets, smartphones, refrigerators, thermostats, etc. that are capable of spontaneous data transfer between themselves by means of the internet. In such a system, low-cost and interference free communications are required between the devices. RF and wired connectivity are unviable and cost inefficient, and VLC is a promising solution that can be employed for massive data transmission between such *things* and the cloud [Firefly LiFi - 2018]. Consequently, LED lights that dominate everywhere, such as classrooms, stores, streets, outdoor lamps, markets, windows, walls, signage, traffic signs, and vehicle tail lights/headlights, can be exploited for accessing the internet, thus making VLC based IoT a reality [Kuppusamy et al. - 2016].

5.4.10 Other Application Areas

VLC can be employed to offer high bandwidth and low latency for intra-chip and inter-chip communications. It can also be used for musical signal transmissions. Similarly, VLC transceivers can be integrated into wearable devices and clothing as a part of WBAN. Consequently, the VLC can ensure wireless connection of sensors to the central unit and for subsequent connection to the internet. The VLC system is also applicable in WPAN connectivity. Moreover, RF based systems like WiFi have been the major indoor WLAN solution; however, with the advent of VLC based system like LiFi, there is high tendency for the indoor infrared WLAN solution to be the dominant scheme only in certain applications in the future. This will be facilitated by the need for energy efficient LED lighting technologies with high tolerance to humidity and long life expectancy [Ghassemlooy et al. - 2016; Uysal and Nouri - 2014]. In addition, VLC can serve as a good indoor localization/positioning system in order to address the RF-based global positioning system challenges like high-attenuation and multi-path, which cause relatively low accuracy in enclosed environments such as indoor and under ground (e.g., tunnels) [Li et al. - 2018]. It is remarkable that OWC links can be easily installed and deployed. This salient feature makes them highly relevant during natural disasters such as cyclones, earthquakes, hurricanes or tsunami, and in disastrous environments where local infrastructures are undependable. Likewise, the OWC links are also applicable in research and the tactical field by defense companies and military organizations [Ghassemlooy et al. - 2016; Uysal and Nouri - 2014; Kuppusamy et al. - 2016]. Owing to the huge number of different luminous devices such as indoor/outdoor lamps, traffic signs, car headlights/tail lights, commercial displays, TVs, and the like being employed everywhere, VLC can be used for ubiquitous computing and a broad range of network connectivity in smart devices, smart homes, smart cities, smart factories, and smart grids [Ghassemlooy et al. - 2015; Alimi et al. - 2017a].

5.5   Hybrid VLC and RF Networks

As stated earlier, the VLC system employs energy efficient LED based transmitters that offer concurrent support for data communications, indoor localization and illumination. Thus, data transmission speed depends on the illumination intensity [Alimi et al. - 2017a; Papanikolaou et al. - 2018]. Nevertheless, the major shortcoming of the VLC networking solution is the limited coverage feature. This can be noticed in the indoor scenario that is prone to optical radiation blockage and link/channel interruption (co-channel interference) that might be as a result of movement, rotation or blockage of the receiver. On the other hand, WiFi, which is based on RF, is capable of offering ubiquitous and hence more reliable coverage. However, WiFi has a number of challenges regarding efficiency, capacity, and security. Therefore, to support the ever increasing demand and enhance the system performance with high QoS among users, there is a need for VLC and RF convergence. Consequently, a hybrid VLC and RF system exploits the advantages of both schemes and simultaneously addresses their weaknesses [Alimi et al. - 2017a; Papanikolaou et al. - 2018].

A hybrid VLC/RF network has been considered to be an emerging technology for the next generation indoor wireless communications. The network exploits the inherent high-speed data transmission of the VLC and the ubiquitous coverage offered by the RF systems [Wu et al. - 2017]. Moreover, a hybrid VLC and RF system can be achieved by WiFi overlay on the VLC network. It should be noted that there is no interference between the VLC and RF systems because they operate on different spectra. This aids a hybrid VLC/RF network in accomplishing the aggregate throughput of both VLC and RF standalone networks. Thus, a hybrid VLC/RF system not only enhances the QoS but also substantially improves user throughput and quality of experience. In addition, the configuration aids in reducing the contention and subsequent spectrum efficiency losses by the WiFi system. It also helps in reducing the pressure on the RF spectrum through effective offloading into the free light spectrum. On the other hand, the VLC system is able to exploit the coverage at dead spots in the network. This helps in a guaranteed high throughput at all locations [Ghassemlooy et al. - 2016; Haas et al. - 2016].

In an indoor scenario, a huge quantity of VLC based APs can be deployed so as to realize a high area spectral efficiency within the network. This enhances the sum throughput of the hybrid system significantly. It should be noted that the VLC based LiFi attocell covers a considerably smaller area compared to the WiFi AP owing to the small size of the LiFi attocells. This implies that, depending on the location, any slightest movement by the user can result in several handovers among the LiFi attocells. So, there is high tendency that the excessive/relapsing handover can significantly impair the system throughput. This problem can be alleviated with the implementation of dynamic load balancing. This approach ensures that the quasi-static users in the network are served by the LiFi attocells, whereas mobile users are served through the WiFi AP. Nevertheless, as a result of the user movement, local overload conditions could happen. This situation can hinder handover and consequently initiates a lower throughput [Wang and Haas - 2015]. In Section 5.6, we discuss the different challenges of VLC based systems and proffer potential solutions by means of making the VLC system a better scheme for the next-generation network.

5.6 Challenges and Open-Ended Issues

As stated in Section 5.4, there are a number of practical applications of VLC systems; however, there are some associated challenges that demand significant attention to make them compatible and effective for the 5G and B5G wireless network deployment scenarios such as eMBB, mMTC, and others. In this section, we present the major challenges of the VLC system and proffer viable solutions to address the system implementation concerns.

5.6.1 Flicker and Dimming

The concurrent implementation of light sources for illumination and data communication create a number of challenges that demand attention. The main challenges of a VLC system are flicker and dimming. The existing lighting systems are equipped with dimming control to allow the users to regulate the average brightness of the light sources to the desired level. The variation in the intensity of light perceived by the human naked eye is known as flicker. It is noticeable when the light source is switching on and off incessantly during data transmission. It is highly imperative to mitigate any potential flicker due to the harmful physiological changes that it can cause in humans. Thus, the VLC system has to support efficient flicker mitigation and dimming control [Rajagopal et al. - 2012; Jan et al. - 2015]. Flicker can be mitigated by ensuring that the variations in the intensity are within the maximum flickering time period (MFTP). The MFTP represents the maximum time period in which the light brightness can be changed without human eye perception [Rajagopal et al. - 2012; Alimi et al. - 2017a].

Furthermore, flicker and dimming require significant consideration in the VLC modulation scheme design. It is worth mentioning that VLC modulation has to be well effected to prevent flickering. Moreover, it is important to ensure that the chosen dimming level for a specific modulation is supported by the illuminating LEDs [Khan - 2017]. Consequently, a number of modulation formats have been proposed for VLC considering dimming control and flicker mitigation. For example, owing to its remarkable capability to control dimming, variable pulse position modulation (VPPM) was proposed in the IEEE 802.15.7 standard. The notable aspect of the VPPM is that it integrates the pulse width modulation as well as PPM so as to support communication with dimming control [Rajagopal et al. - 2012; Jan et al. - 2015; Alimi et al. - 2017a].

5.6.2 Data Rate Improvement

The VLC is an emerging technology in which LEDs are widely employed. However, LEDs are initially intended and designed for lighting. Consequently, their implementation for communication purposes is limited. One of the major concerns is that, as they are not designed for high bandwidths, their modulation bandwidth is limited to a few megahertz (MHz). It is noteworthy that this signifies a hardware constraint that hinders high data rate VLC link deployment. Nevertheless, the limiting factor can be practically addressed by employing techniques such as transmitter and/or receiver equalizations, bandwidth-efficient/high-order modulation schemes, multiple access schemes, and frequency reuse [Bawazir et al. - 2018; Jha et al. - 2017].

Moreover, the information rate can be enhanced by redesigning LEDs. For instance, micro LEDs (μLEDs) can be designed with significantly smaller size and faster rise time at an extremely low-power level. Then, coding, modulation, and pre-equalization schemes can be employed to increase the data rates. Subsequently, there can be spatial reuse of the resource and various levels of spatial multiplexing exploitation [Jha et al. - 2017]. Furthermore, the capacity of wireless communication systems can be improved by implementing multiple-input-multiple-output (MIMO) schemes. This could also be applied to the VLC system by using multiple transceiver apertures in order to exploit the considerable multiplexing gain offered by the MIMO schemes [Alimi et al. - 2017b; Jha et al. - 2017; O'Brien et al. - 2008]. In addition, another potential method of attending to the transmitter low-bandwidth is by blocking the phosphor component at the receiver. This can be achieved by employing a blue filter. However, the price for the approach is a slight reduction in the received power that can be attributed to filter losses [O'Brien et al. - 2008].

Moreover, other challenges are blocking, multipath induced intersymbol interference, high path losses, handover based mobility issues, and artificial light induced interference. Additionally, LED electro-optic response nonlinearity requires significant attention [Ghassemlooy et al. - 2015; Alimi et al. - 2017a].

5.7 Conclusions

This chapter has presented an OWC system as a viable and emerging technology that is capable of attending to the RF based wireless communication interference and spectrum congestion issues in efficient ways. Unregulated and unlicensed visible light communications with several salient features have also been presented not only for concurrent support for high-speed data communication and broad illumination but also for a range of feasible and exciting applications. Owing to the potential broad applications, the VLC system is progressively becoming an attractive area of research and, consequently, it is envisaged to dominate various aspects of the 5G and B5G wireless communication systems. This can help in reducing the pressure on the heavily regulated and congested RF spectrum through effective offloading into the free light spectrum. Furthermore, a hybrid VLC and RF system has been considered so as to exploit advantages of both schemes and simultaneously address their weaknesses. A number of technical challenges and viable solutions have also been discussed.

Acknowledgments

This work is supported by Ocean12-H2020-ECSEL-2017-1-783127 and FCT and the ENIAC JU (THINGS2DO - GA n. 621221) projects.

Bibliography

5G Americas. 5G Network Transformation. Technical report, 5G Americas, December 2017. URL http://www.5gamericas.org/files/3815/1310/3919/5G_Network_Transformation_Final.pdf.

I. Alimi, A. Shahpari, A. Sousa, R. Ferreira, P. Monteiro, and A. Teixeira. Challenges and opportunities of optical wireless communication technologies. In Pedro Pinho, editor, *Optical Communication Technology*, chapter 2. InTech, Rijeka, 2017a. doi: 10.5772/intechopen.69113. URL https://doi.org/10.5772/intechopen.69113.

I. A. Alimi, A. M. Abdalla, J. Rodriguez, P. P. Monteiro, and A. L. Teixeira. Spatial interpolated lookup tables (luts) models for ergodic capacity of mimo fso systems. *IEEE Photonics Technology Letters*, 29(7):583–586, April 2017b. ISSN 1041-1135. doi: 10.1109/LPT.2017.2669337.

I. A. Alimi, A. L. Teixeira, and P. P. Monteiro. Toward an efficient c-ran optical fronthaul for the future networks: A tutorial on technologies, requirements, challenges, and solutions. *IEEE Communications Surveys Tutorials*, 20(1): 708–769, Firstquarter 2018. doi: 10.1109/COMST.2017.2773462.

Isiaka Alimi, Ali Shahpari, Vítor Ribeiro, Artur Sousa, Paulo Monteiro, and António Teixeira. Channel characterization and empirical model for ergodic capacity of free-space optical communication link. *Optics Communications*, 390:123 –129, 2017c. ISSN 0030-4018. doi: https://doi.org/10.1016/j.optcom.2017.01.001. URL http://www .sciencedirect.com/science/article/pii/S0030401817300019.

Isiaka A. Alimi, Paulo P. Monteiro, and António L. Teixeira. Analysis of multiuser mixed rf/fso relay networks for performance improvements in cloud computing-based radio access networks (cc-rans). *Optics Communications*, 402:653 –661, 2017d. ISSN 0030-4018. doi: https://doi.org/10.1016/j.optcom.2017.06.097. URL http://www .sciencedirect.com/science/article/pii/S0030401817305734.

Isiaka A. Alimi, Paulo P. Monteiro, and António L. Teixeira. Outage probability of multiuser mixed rf/fso relay schemes for heterogeneous cloud radio access networks (h-crans). *Wireless Personal Communications*, 95(1):27–41, Jul 2017e. ISSN 1572-834X. doi: 10.1007/s11277-017-4413-y. URL https://doi.org/10.1007/s11277-017-4413-y.

S. S. Bawazir, P. C. Sofotasios, S. Muhaidat, Y. Al-Hammadi, and G. K. Karagiannidis. Multiple access for visible light communications: Research challenges and future trends. *IEEE Access*, 6:26167–26174, 2018. doi: 10.1109/ACCESS.2018.2832088.

P. Chvojka, K. Werfli, S. Zvanovec, P. A. Haigh, V. H. Vacek, P. Dvorak, P. Pesek, and Z. Ghassemlooy. On the m-cap performance with different pulse shaping filters parameters for visible light communications. *IEEE Photonics Journal*, 9(5):1–12, 2017.

D. R. Dhatchayeny, A. Sewaiwar, S. V. Tiwari, and Y. H. Chung. Experimental biomedical eeg signal transmission using vlc. *IEEE Sensors Journal*, 15(10): 5386–5387, 2015.

Firefly LiFi. Li-Fi. Technical report, Firefly Wireless Networks, December 2018. URL https://www.fireflylifi.com/applications.html.

Z. Ghassemlooy, W. Popoola, and S. Rajbhandari. *Optical Wireless Communications: System and Channel Modelling with MATLAB ®*. Taylor & Francis, 2012. ISBN 9781439851883.

Z. Ghassemlooy, S. Arnon, M. Uysal, Z. Xu, and J. Cheng. Emerging optical wireless communications-advances and challenges. *IEEE Journal on Selected Areas in Communications*, 33(9):1738–1749, Sept 2015. ISSN 0733-8716. doi: 10.1109/JSAC.2015.2458511.

Z. Ghassemlooy, M. Uysal, M. A. Khalighi, V Ribeiro, F. Moll, S. Zvanovec, and A. Belmonte. An overview of optical wireless communications. In M. Uysal, C. Capsoni, Z. Ghassemlooy, A. Boucouvalas, and E. Udvary, editors, *Optical Wireless Communications: An Emerging Technology*, Signals and Communication Technology, chapter 1, pages 1–23. Springer International Publishing, Springer International Publishing, 2016. ISBN 9783319302010.

GSMA. 5G, the Internet of Things (IoT) and Wearable Devices What do the new uses of wireless technologies mean for radio frequency exposure? Technical report, GSMA, September 2017. URL https://www.gsma.com/publicpolicy/wp-content/uploads/2017/10/5g_iot_web_FINAL.pdf.

H. Haas, L. Yin, Y. Wang, and C. Chen. What is lifi? *Journal of Lightwave Technology*, 34(6):1533–1544, March 2016. ISSN 0733-8724. doi: 10.1109/JLT.2015.2510021.

IEEE 802.15.7 VLC Task Group. Visible Light Communication- Tutorial. Technical report, IEEE P802.15 Working Group for Wireless Personal Area Networks (WPANs), March 2008. URL http://ieee802.org/802_tutorials/2008-03/15-08-0114-02-0000-VLC_Tutorial_MCO_Samsung-VLCC-Oxford_2008-03-17.pdf.

S. U. Jan, Y. D. Lee, and I. Koo. Comparative analysis of DIPPM scheme for visible light communications. In *2015 International Conference on Emerging Technologies (ICET)*, pages 1–5, Dec 2015. doi: 10.1109/ICET.2015.7389192.

Pranav Kumar Jha, Neha Mishra, and D. Sriram Kumar. Challenges and potentials for visible light communications: State of the art. *CoRR*, abs/1709.05489, 2017. URL http://arxiv.org/abs/1709.05489.

R. Johri. Li-fi, complementary to wi-fi. In *2016 International Conference on Computation of Power, Energy Information and Commuincation (ICCPEIC)*, pages 015–019, April 2016. doi: 10.1109/ICCPEIC.2016.7557216.

Latif Ullah Khan. Visible light communication: Applications, architecture, standardization and research challenges. *Digital Communications and Networks*, 3(2):78 –88, 2017. ISSN 2352-8648. doi: https://doi.org/10.1016/j.dcan.2016.07.004. URL http://www.sciencedirect.com/science/article/pii/S2352864816300335.

P. Kuppusamy, S. Muthuraj, and S. Gopinath. Survey and challenges of li-fi with comparison of wi-fi. In *2016 International Conference on Wireless Communications, Signal Processing and Networking (WiSPNET)*, pages 896–899, March 2016. doi: 10.1109/WiSPNET.2016.7566262.

Y. Li, Z. Ghassemlooy, X. Tang, B. Lin, and Y. Zhang. A vlc smartphone camera based indoor positioning system. *IEEE Photonics Technology Letters*, 30(13): 1171–1174, 2018.

A. Lo, Y. W. Law, and M. Jacobsson. A cellular-centric service architecture for machine-to-machine (m2m) communications. *IEEE Wireless Communications*, 20(5):143–151, October 2013. ISSN 1536-1284. doi: 10.1109/MWC.2013.6664485.

P. Luo, H. M. Tsai, Z. Ghassemlooy, W. Viriyasitavat, H. Le Minh, , and X. Tang. Car-to-car visible light communications. In Zabih Ghassemlooy, Luis Nero Alves, Stanislav Zvanovec, and Mohammad-Ali Khalighi, editors, *Visible Light Communications: Theory and Applications*, chapter 3, pages 253–282. CRC Press, CRC Press, 2017.

T. Lv, Y. Ma, J. Zeng, and P. T. Mathiopoulos. Millimeter-wave noma transmission in cellular m2m communications for internet of things. *IEEE Internet of Things Journal*, 5(3):1989–2000, June 2018. doi: 10.1109/JIOT.2018.2819645.

D. C. O'Brien, L. Zeng, H. Le-Minh, G. Faulkner, J. W. Walewski, and S. Randel. Visible light communications: Challenges and possibilities. In *2008 IEEE 19th International Symposium on Personal, Indoor and Mobile Radio Communications*, pages 1–5, Sept 2008. doi: 10.1109/PIMRC.2008.4699964.

V. K. Papanikolaou, P. P. Bamidis, P. D. Diamantoulakis, and G. K. Karagiannidis. Li-fi and wi-fi with common backhaul: Coordination and resource allocation. In *2018 IEEE Wireless Communications and Networking Conference (WCNC)*, pages 1–6, April 2018. doi: 10.1109/WCNC.2018.8377176.

B. L. Parne, S. Gupta, and N. S. Chaudhari. Segb: Security enhanced group based aka protocol for m2m communication in an iot enabled lte/lte-a network. *IEEE Access*, 6:3668–3684, 2018. doi: 10.1109/ACCESS.2017.2788919.

PPP in Horizon 2020. Advanced 5G Network Infrastructure for the Future Internet: Creating a Smart Ubiquitous Network for the Future Internet. Technical report, PPP/European Horizon 2020 framework, February 2014. URL https://cordis.europa.eu/docs/projects/cnect/5/317105/080/deliverables/001-D14Annex1Advanced5GNetworkInfrastructurePPPinH2020FinalNovember2013pdf.pdf.

PricewaterhouseCoopers. Industry 4.0: Building the digital enterprise: 2016 Global Industry 4.0 Survey. Technical report, PwC, September 2016. URL https://www.pwc.com/gx/en/industries/industries-4.0/landing-page/industry-4.0-building-your-digital-enterprise-april-2016.pdf.

S. Rajagopal, R. D. Roberts, and S. K. Lim. IEEE 802.15.7 visible light communication: modulation schemes and dimming support. *IEEE Communications Magazine*, 50(3):72–82, March 2012. ISSN 0163-6804. doi: 10.1109/MCOM.2012.6163585.

A. Sevincer, A. Bhattarai, M. Bilgi, M. Yuksel, and N. Pala. LIGHTNETs: Smart LIGHTing and mobile optical wireless NETworks- A survey. *IEEE Communications Surveys Tutorials*, 15(4):1620–1641, Fourth 2013. ISSN 1553-877X. doi: 10.1109/SURV.2013.032713.00150.

N. Shahin, R. Ali, and Y. T. Kim. Hybrid slotted-csma/ca-tdma for efficient massive registration of iot devices. *IEEE Access*, 6:18366–18382, 2018. doi: 10.1109/ACCESS.2018.2815990.

N. Soni, M. Mohta, and T. Choudhury. The looming visible light communication li-fi: An edge over wi-fi. In *2016 International Conference System Modeling Advancement in Research Trends (SMART)*, pages 201–205, Nov 2016. doi: 10.1109/SYSMART.2016.7894519.

Sudhanshu N. Tayade. Spectral efficiency improving techniques in mobile femtocell network: Survey paper. *Procedia Computer Science*, 78:734 –739, 2016. ISSN 1877-0509. doi: https://doi.org/10.1016/j.procs.2016.02.046. URL http://www.sciencedirect.com/science/article/pii/S187705091600048X. 1st International Conference on Information Security & Privacy 2015.

Oumer Teyeb, Gustav Wikström, Magnus Stattin, Thomas Cheng, Sebastian Faxér, and Hieu Do. Evolving LTE to fit the 5G future. Technical report, Ericsson, January 2017. URL https://www.ericsson.com/assets/local/publications/ericsson-technology-review/docs/2017/etr_evolving_lte_to_fit_the_5g_future.pdf.

M. Uysal and H. Nouri. Optical wireless communications- an emerging technology. In *2014 16th International Conference on Transparent Optical Networks (ICTON)*, pages 1–7, July 2014. doi: 10.1109/ICTON.2014.6876267.

Diederik Verzijl, Kristina Dervojeda, Jorn Sjauw-Koen-Fa, Fabian Nagtegaal, Laurent Probst, and Laurent Frideres. Smart Factories: Capacity optimisation. Technical report,

European Union, September 2014. URL https://www.pwc.com/gx/en/industries/industries-4.0/landing-page/industry-4.0-building-your-digital-enterprise-april-2016.pdf.

Y. Wang and H. Haas. Dynamic load balancing with handover in hybrid li-fi and wi-fi networks. *Journal of Lightwave Technology*, 33(22):4671–4682, Nov 2015. ISSN 0733-8724. doi: 10.1109/JLT.2015.2480969.

X. Wu, M. Safari, and H. Haas. Access point selection for hybrid li-fi and wi-fi networks. *IEEE Transactions on Communications*, 65(12):5375–5385, Dec 2017. ISSN 0090-6778. doi: 10.1109/TCOMM.2017.2740211.

T. Yamazato. Overview of visible light communications with emphasis on image sensor communications. In *2017 23rd Asia-Pacific Conference on Communications (APCC)*, pages 1–6, Dec 2017. doi: 10.23919/APCC.2017.8304093.

F. Yang and J. Gao. Dimming control scheme with high power and spectrum efficiency for visible light communications. *IEEE Photonics Journal*, 9(1): 1–12, Feb 2017. ISSN 1943-0655. doi: 10.1109/JPHOT.2017.2658025.

K. Ying, Z. Yu, R. J. Baxley, H. Qian, G. K. Chang, and G. T. Zhou. Nonlinear distortion mitigation in visible light communications. *IEEE Wireless Communications*, 22(2):36–45, April 2015. ISSN 1536-1284. doi: 10.1109/MWC.2015.7096283.

Z. Zeng, S. Fu, H. Zhang, Y. Dong, and J. Cheng. A survey of underwater optical wireless communications. *IEEE Communications Surveys Tutorials*, 19(1): 204–238, 2017.

6

5G RAN: Key Radio Technologies and Hardware Implementation Challenges

Hassan Hamdoun[*1], *Mohamed Hamid*[2], *Shoaib Amin*[3], *and Hind Dafallah*[4]

[1] *British Telecommunications (BT), Applied Research, BT Technology, Adastral Park, Ipswich, Suffolk, IP5 3RE, United Kingdom*
[2] *Ericsson, System & Technology, Lund, Business area Networks, SE - 223 62, Lund, Sweden*
[3] *Qamcom Research and Development AB, RF and Signal Processing, Linköping, SE - 583 30, Linköping, Sweden*
[4] *Ericsson, RFIC Product Development Unit, Lund, Business area Networks, SE - 223 62, Lund, Sweden*

6.1   Introduction

The rapid growth in consumer applications and services requiring high data rates, low latency and always-on connectivity is driving the innovation in technology across the whole mobile communications ecosystem, from user devices to the network infrastructure, protocols, air interface, and core network. In particular the innovations in the radio access network (RAN) are key as they enable the capability to reliably shift the volume of data required and meet the ever evolving consumer applications and services requirements.

As it has been the main theme of the evolution of different cellular communications generations, cell densification is the most effective and straightforward solution for meeting the ever increasing capacity demands [Dohler et al. - 2011]. This concept of densification is motivated by spectrum reusability and is clearly demonstrated when comparing a cell size of an order of hundreds of square kilometers in the first generation during the early 1980s and cell spacing of about two hundred square meters in today's deployments in dense urban areas [Andrews et al. - 2014]. A way forward for such densification in 5G systems is to approach further toward nested small cells such as femtocells and picocells. A similar concept from a coverage and capacity standpoint is to increase the radio resource reusability in space, exploiting large scale antenna arrays providing narrow beams transmitted from a cell having a single processing unit [Andrews et al. - 2014].

This chapter aims to shed light on the key RAN radio-enabling technologies that the authors believe will be instrumental for what the 5G new radio (NR) is likely to look like. Given the continued 3GPP standardization efforts and intensive efforts from industry vendors, mobile network operators (MNOs), original equipment manufacturers and the current intense race towards being first in the 5G market, it is timely to discuss the key RAN radio-enabling technologies and their implementations.

* Corresponding Author: Hassan Hamdoun Email: hassan.hamdoun@bt.com.

Optical and Wireless Convergence for 5G Networks, First Edition.
Edited by Abdelgader M. Abdalla, Jonathan Rodriguez, Issa Elfergani, and Antonio Teixeira.
© 2020 John Wiley & Sons Ltd. Published 2020 by John Wiley & Sons Ltd.

The authors, coming from wireless networking and RAN, and signal processing and radio frequency integrated circuit (RFIC) product development and design expertise, worked together to present in this chapter a 5G-focused RAN and hardware implementation discussion highlighting the main challenges and corresponding ongoing research topics to be addressed in order to satisfy the use cases' high data rate and low latency requirements.

The authors believe that the main distinction of this chapter is combining the theoretical and hardware aspects of the 5G NR key technologies, which, to the best of the authors' knowledge, have mostly been considered separately. Henceforth, the chapter concisely explains how hardware implementation and associated impairments can limit the performance of 5G NR technology. Furthermore, this chapter discusses mitigation techniques for some of these impairments and hardware design challenges.

The rest of this chapter is organized as follows: Section 6.2 presents the main uses cases that are likely to be implemented first and launched on commercial 5G networks. Section 6.3 discusses massive MIMO (M-MIMO), distributed MIMO (D-MIMO), dual connectivity, device-to-device (D2D) communications, carrier aggregation (CA), and licensed assisted access (LAA) spectrum technologies. Section 6.4 presents the transmitter, receiver, and transceiver hardware impairments, and discuss their impact on the effective implementation of the above-mentioned radio-enabling technologies with specific focus on the M-MIMO and on the millimeter wave (mmWave) frequency bands.

6.2   5G NR Enabled Use Cases

The early drop of 5G NR new RAT specs (phase 1) for enhanced mobile broadband (eMBB) non-standalone mode (NSA) was completed in Rel-15 in December 2017 [3GPP - 2017], with standalone (SA) standardization completed recently in June 2018. Such standardization development is paving the way for vendors and MNOs to start planning and preparing for 5G roll-out and the implementation of various aspects of the 5G ecosystem. Phase 2 in Rel-16 is expected to be completed in 2019. The 3GPP 5G NR road-map in [3GPP - 2017] defines the main use cases for 5G NR, focusing on eMBB and and ultra reliable and low latency communication (uRLLC), for which we focus on key RAN radio-enabling technologies and spectrum utilization enhancements in this chapter.

6.2.1   eMBB and uRLLC

The massive growth in mobile data, the number of connected devices and the dramatic change in user expectations of what the network should offer, are key market drivers shaping the IMT-2020 5G vision [ITUR-REC-M.2083 - 2015; Rumney - 2018]. Being naturally evolved from preceding cellular generations, eMBB is seen as the most mature and important case for 5G that enables MNOs to meet the expectations that the customers already have. For example, the provision of high data rates 10–20 Gbps peak, 100 Mbps whenever needed and support for high mobility. Key radio-enabling capabilities and functionalities of eMBB is the ability to provide MNOs with 100× cell densification, 100× reduction in network energy compared to Rel-14 long term evolution (LTE).

The eMBB is already, in various forms, implemented for early commercial 5G networks. These can be considered as a sub use case of eMBB. For example, fixed wireless access (FWA) is currently deployed by Verizon who even have a different air interface standard [Verizon 5G 3GP - 2016]. Notably, the main differences between eMBB and FWA lie in that there are no strict mobility and beam-tracking requirements for FWA, and that FWA successful trials and deployment are mainly in ≥3.5 GHz and in the mmWave band (24.25–52.6 GHz, with 24–28 GHz proven more beneficial) as demonstrated by Verizon, AT&T, Huawei and KT trials. Overall, FWA trial results have had varying degrees of success and several technical recommendations have emerged; however, few research questions are yet to be addressed. These questions are the same for the eMBB use case and are discussed in Section 6.3.

uRLLC is another complementing use case for the eMBB where the air interface latency target is 1 ms with 5 ms end-to-end latency (compared to 4G LTE latency of 40–70 ms, of which 20 ms is in air interface) and highly reliably communication reliability.

From the users' perspective eMBB is what allows for improving the quality of experience thanks to its promises of seamless network experience, incredibly fast real-time communication and ubiquitous all-things communication with highly reliable connectivity. For the MNO, the challenge lies in the technologies pieces of the puzzle that need to fit together in order to meet the users expectations while at the same time evolve the RAN and core networks for cost and design efficiency. Also, and most importantly, the challenge of designing and provisioning these technologies in a way such that the network is able to offer and drive new products and service, and increase revenue.

6.2.1.1   mMTC

The trend of low cost of computing, storage, and advances in intelligent machines, coupled with the massive progress in machine learning and artificial intelligence, has accelerated the adoption of both passive and smart sensors and devices used in various consumer and industrial applications. The cellular mobile industry is also taking a share of such evolution. Mobile technology is well suited to supporting cheap sensors and massive machine type communication (mMTC) involving high density of devices $(2 \times 10^5 \text{ km}^{-2})$ covering a long range with a low cost of connectivity, long battery life, and supporting asynchronous access [Bockelmann et al. - 2016].

3GPP has pushed the completion of standards for mMTC to Rel-16 as the current narrow band Internet-of-Things technology is sufficient for current requirements. Hence, this chapter does not focus on mMTC; more focus is devoted to the RAN enabling aspects for the low latency requirements of uRLLC.

6.2.2   Migration to 5G

Since mobile broadband provided by LTE is the base for the eMBB use case in 5G, then the 3GPP plan is to expand the LTE platform to converge to eMBB by improving its efficiency [Dahlman et al. - 2016]. Therefore, there are two stages towards the migration journey to 5G NR as follows. The first stage is NSA NR where the aim is to boost cellular network mobile broadband capacity. As the term implies, NSA NR uses an LTE core network and radio anchor with different configurations that allow NR technologies adoption. The second stage is NR SA where a 5G core and radio anchor will overlay LTE and operate independently. In SA NR, the aim is to expand the whole wireless ecosystem where the new use cases of 5G will be integrated. Moreover, NSA and SA are evolving towards edge cloud enhancements, for which the implementations are yet to be comprehensively studied, and this is discussed in Chapters 1, 3, 5, 10, and 12. Edge cloud brings together the latency benefits of a more distributed core processing at edge, i.e. mobile edge computing with more centralized radio processing facilitated by centralized-RAN (C-RAN).

6.3   5G RAN Radio Enabling Technologies

Over the last five decades, the capacity of mobile cellular networks has grown by a factor of one million. This was distributed as follows: a factor of 25 was due to spectrum increase, factor of 20 due to improvements in spectral efficiency, and a factor of 2000 due to the use of small cells, as discussed in [Rumney - 2018]. The analysis in [Rumney - 2018] showed that in order to achieve a data density growth of 125 billion to one, for various parameters of coverage, mobility, CAPEX/cell, and OPEX/cell/year, the cell densification dominates how mobile cellular systems will be best exploited/designed. We argue in this chapter that, while cell size is a major contributor to data rate growth, the combination and optimization of the three parameters is more suitable when discussing 5G NR RAN radio-enabling radio technologies as depicted below.

$$\text{Throughput} = \text{Available spectrum} \cdot \text{Cell density} \cdot \text{Spectral efficiency.}$$

$$\text{b s}^{-1}\text{ km}^{-2} \qquad\qquad \text{Hz} \qquad\qquad \text{Cell km}^{-2} \qquad\qquad rmb/s/Hz/Cell$$

Note that the 5G NR specifications for user equipment (UE) RRM, UE RRM core and UE radio core and UE demodulation are yet to be completed by the 3GPP, and hence more technical analysis, research, and contributions are to be made for which we aim to shed some light in this section. Hereafter, we discuss the primary radio technologies that will be adopted to support the needs for 5G networks, primarily the eMBB and FWA use cases, namely M-MIMO with specific focus on its role in mmWave frequency bands, D-MIMO, CA, dual connectivity and D2D communications.

6.3.1   Massive MIMO (M-MIMO)

With even smaller cells, there is a dire need for new techniques to effectively exploit the inter-site distance and manage interference effectively. Historically the evolution followed the trajectory from SISO, to receive diversity (SIMO) and transmit diversity (MISO) or a combination of the two. The evolution to MIMO systems is where n independent streams between transmitter and receiver are sent at the same time and frequency. Such spatial multiplexing with pre-coding or beam-forming has the potential to increase data rates by a factor of $x = \min(n, m)$ where n and m are the number of transmitter (Tx) and receive (Rx) antennas, respectively, at either end of the link.

In an $n \times m$ MIMO system with n Tx and m Rx antennas, several possibilities can be implemented. SU-MIMO, when n Tx antennas – on one device – and m Rx antennas – on another device – are helping to deliver m times the peak data rates when $n \geq m$. MU-MIMO extends this to uncoordinated m Rxs with the same data rate benefits. Coordination of Rx antennas across *n* cells, though requiring centralized base-band and remote radio head and symbol level coordination, helps improve the reliability of delivery of the above-mentioned data rate benefits.

Cellular mobile networks radio resource are RF interference limited and hence the techniques that allows for reuse of time-frequency radio resource is a key enabling technology of the 5G NR. Combating interference is feasible by preprocessing the transmitted signal in a way that a signals becomes disseminated within narrow beams toward its intended receiver and nowhere else. Subsequently, space is added as a dimension to the radio resource landscape and the time-frequency radio resource grid can be usable over different space points. This technique of transmitting narrow beam signals in order to reuse radio resources over space is known as beamforming and is a key ingredient exploited in the M-MIMO technology, as depicted in Figure 6.1.

M-MIMO was introduced in Rel-13 under the name full dimension MIMO with up to 64 antenna elements. However, different names have been used for M-MIMO such as hyper MIMO, very large MIMO, beamforming, and advanced antenna systems (AAS) (Section 6.4 below). M-MIMO and beamforming in the context of eMBB and FWA 5G use cases will be used interchangeably throughout this chapter.

Beamforming can be carried out by means of phase shifting and *possibly* re-scaling signals fed to different antenna elements of the array. Beamforming can be done either digitally or in the analog domain. Digital beamforming is performed in the baseband by means of ordinary mathematical operations. On the other hand, analog beamforming is accomplished using phase shifters in the analog domain. While digital beamforming has the advantage of being able to beam form different allocations in the same OFDM symbol in different directions, analog beamforming requires less fronthaul interfaces as compared to digital beamforming. The most promising approach for reducing the training overhead and hardware cost in M-MIMO is the use of a hybrid multiple-antenna architecture [Molisch et al. - 2017]. This architecture utilizes a combination of analog pre- and post-processing with lower dimensional digital processing [Ratnam et al. - 2018]. The realization of such architecture in hardware is the main research challenge that we elaborate on and discuss further in Section 6.4.

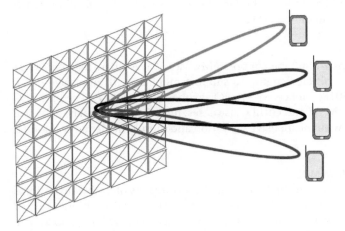

Figure 6.1 The concept of beamforming using a large scale antenna array (M-MIMO). Different colors represent different beams.

Efficient beamforming requires knowledge of the channel state information (CSI), which is used to assign complex weights to different signals sent to different antenna ports of the antenna array. This process of weight assignment is referred to as precoding. A precoder and a channel can be seen as two systems that produce narrow beams toward the intended users when cascaded. CSI can be reported aperiodically when needed or periodically, and it contains one or several pieces of information such as [Dahlman et al. - 2016]:

- Rank indicator. An indicator of the number of uncorrelated channels that can be used for downlink and implies the number of layers in the downlink transmission.
- Precoder matrix indicator (PMI). A PMI is an index to the precoding matrix chosen out of a set of matrices referred to as a codebook.
- CSI resource indicator (CRI). If the device is configured to monitor multiple beams, the CRI indicates the preferred beam at a specific time.

For more comprehensive coverage of CSI signaling information, the corresponding computational resource requirements, and overhead trade-offs arising from CSI and its role in beamforming, the reader is referred to [Dahlman et al. - 2016].

6.3.1.1 M-MIMO in mmWave

Traditionally, the mmWave spectrum has been used for backhauling for several reasons. First, its severe radio propagation losses makes mmWaves unsuitable for conventional cellular macro base stations where coverage concerns dominate. Moreover, there was no need for the high bandwidth available in the mmWave spectrum for fronthaul links, instead high capacity backhaul microwave links were – and of course still are – the links that require such a bandwidth, keeping in mind that operation in mmWave is more expensive compared to lower frequencies. In contrast, the available spectrum at mmWave is indeed an attraction for 5G ultra dense networks. Furthermore, the need for massive cell densification in 5G networks turns the propagation characteristics of mmWave into a beneficial advantage as it increases the radio spectrum re-usability. As for M-MIMO in 5G, mmWave frequencies have small bandwidth and, subsequently, a large number of antenna elements can be fabricated in small arrays facilitating the realization and implementation of M-MIMO

6.3.1.2 M-MIMO in sub 6 GHz

mmWave M-MIMO systems are suitable for both eMBB and FWA where high data rates are needed. However, it is still challenging to support mobility with mmWave based systems [Björnson et al. - 2018]. Hence, M-MIMO at sub 6 GHz is a key piece in the 5G NR puzzle. Moreover, for mMTC and URLLC use cases, high data rates can be achieved using sub 6 GHz spectrum with inexpensive operation and more reliable radio channels compared to mmWave.

6.3.1.3 Distributed MIMO (D-MIMO)

D-MIMO systems are radio systems that combine the advantages of both conventional distributed antenna systems and MIMO systems, i.e. both spatial multiplexing and macro-diversity are obtained simultaneously [Li et al. - 2017]. D-MIMO is also known as a cell-free MIMO where distributed antennas connected to the same base-band and processing unit are deployed in spots with poor coverage/capacity within a specific

Figure 6.2 The concept of D-MIMO where distributed antennas are connected to a single baseband and processing unit to enhance coverage/capacity within a specific area.

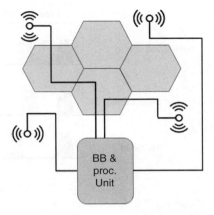

area (i.e. shopping mall, sport arena, etc). Figure 6.2 illustrates the concept of D-MIMO. Usually D-MIMO radio units are simpler and smaller scale compared to the ones used for ordinary cells, which imposes several challenges in hardware impairment mitigation and correction.

D-MIMO can be seen as a reduced scale C-RAN where the same concept of a centralized processing unit with distributed radio units is applied. Yet, the scales differences imposes different challenges for C-RAN and D-MIMO.

6.3.2 Carrier Aggregation and Licensed Assisted Access to an Unlicensed Spectrum

In Rel-10, exploiting a higher bandwidth by means of CA was introduced. CA implies using multiple carriers (each known as a component carrier (CC)) to transmit both uplink and downlink data. Carrier components can be of different bandwidths and not necessarily contiguous. In this context, there exist three CA cases as follows:

- Intraband CA with contiguous CCs.
- Intraband CA with non-contiguous CCs.
- Interband CA with non-contiguous CCs.

Figure 6.3 illustrates these three different CA cases.

CA in Rel-10 requires all CCs to have the same duplex scheme, FDD or TDD, while different uplink/downlink configurations are possible in Rel-11. Furthermore, in Rel-12, CA utilizing both FDD and TDD CCs simultaneously is supported with the maximum number of aggregated CCs, i.e. 32, in Rel-13 onwards [Dahlman et al. - 2016]. CA forms a base for LAA to the unlicensed spectrum standardized in Rel-13. With LAA, some CCs are within the operators licensed spectrum and some fall in an unlicensed spectrum. Therefore, the benefits of using an exclusive spectrum and accessing more spectra are combined without a compromise.

The main aim of LAA is to enable deploying reduced scale plug-and-play indoor access points known as femtocells with unlicensed spectrum access capabilities such as dynamic frequency selection and listen-before-talk mechanisms [Dahlman et al. - 2016]. Femtocells, together with a macro base station serving the same geographical area, form a network topology in a two tier heterogeneous network. The potentials of

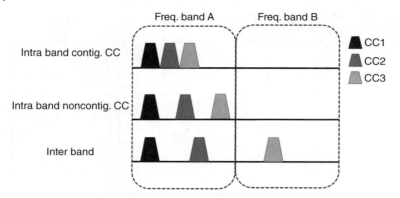

Figure 6.3 The three cases of CA.

cognitive femtocells are extensively investigated in the literature [Adhikary et al. - 2011; Buddhikot - 2010].

6.3.3 Dual Connectivity

Traditionally, in cellular networks a device is connected to a single cell handling its uplink and downlink for both control and data planes. This simple approach is suitable for low data rate applications while allowing a device to be connected to multiple cells would increase the data rate experienced by the device; this was realized in Rel-12. According to Rel-13, when a device is in dual connectivity mode, one eNodeB is a primary while the other is a secondary. Below are some scenarios where dual connectivity can be used [Dahlman et al. - 2016].

- User plane aggregation. Here, a device transmits/receives data to/from different cells.
- Control/user plane separation. A device control plane is handled by a cell while its data transmission is handled by another cell.
- Uplink/downlink separation. In this setup of dual connectivity, uplink transmission is handled by one cell while downlink transmission is handled by a different cell.

Figure 6.4 depicts the three above-mentioned dual connectivity scenarios

6.3.4 Device-to-Device (D2D) Communication

One of the potential technologies to meet the huge data rate demands in next generations wireless networks is D2D communications [Doppler et al. - 2009]. With D2D communications, devices in proximity to each other can exchange data directly among each other through short-range low-power radio links instead of double hop communications through base stations. On the other hand, control signals are handled through the serving base station as in conventional cellular systems. Accordingly, mode selection between cellular and D2D mode functionality is incorporated into the system when D2D communication is enabled [Doppler et al. - 2009]. Figure 6.5 explains graphically the D2D communication concept.

With the ultimate aim of efficiently utilizing the operator's available radio resource, radio resource allocation for D2D communications attracts much of research work and is well documented in the literature. In this regard, the major proposal is to adopt

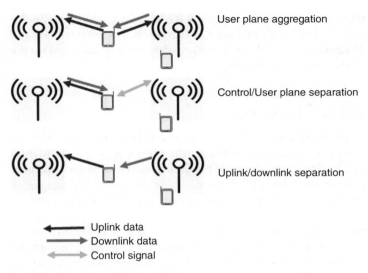

Figure 6.4 Dual connectivity scenarios.

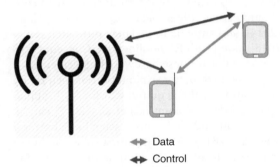

Figure 6.5 D2D communication.

inband-underlay spectrum sharing by D2D communications [Asadi et al. - 2014]. In D2D communications underlay cellular networks, both cellular and D2D users share the same spectrum, with D2D users being responsible for keeping the interference to cellular users under a certain limit.

Under this scheme, most of the research contributions aim at mitigating the interference between D2D and cellular users [Asadi et al. - 2014]. For example the major consensus follows the inband-overlay spectrum sharing approach proposed in [Pei and Liang - 2013] where part of the resources in cellular bands are reserved for D2D communications. Other works have proposed D2D communications to operate outband to eliminate the effects of D2D communications on cellular users [Asadi and Mancuso - 2013].

6.4 Hardware Impairments

In this section, we discuss the various hardware impairments that might slow and make the realization of the high data rate and low latency requirements of eMBB and

FWA use cases – in relation to the above discussed RAN radio enabling technologies – difficult to achieve. To align with various RAN architectures, interference and radio performance challenges discussed above, these impairments are discussed for the transmitter, receiver and transceiver separately.

6.4.1 Hardware Impairments – Transmitters

A massive number of arrays along with AAS are envisioned to be used in both the sub 6 GHz range and mmWave communication systems. The potential gain of M-MIMO comes from intelligently focusing the transmitted power in space, enabling spatial user separation, making it possible to serve multiple users within the same time-frequency resource, and help contribute to enhancing the system capabilities and capacity [Gustavsson et al. - 2014].

M-MIMO is comprised of up to hundreds of Tx paths, each with its own set of active components. Hence, to reduce the cost and complexity, a highly integrated design solution is required such as removing isolators between power amplifiers (PAs) and antenna, using shared local oscillators (LOs) and reducing the number of digital-to-analogue conversion channels via employing hybrid beamforming with sub-arrays. However, such highly integrated hardware is vulnerable to IQ imbalance [Khan et al. - 2017], cross-talk [Amin et al. - 2014], and antenna impedance mismatch [Gustavsson et al. - 2014; Hausmair et al. - 2018], along with PA nonlinearity. Consequently, M-MIMO performance will largely be reduced due to these hardware impairments.

When TDD M-MIMO systems are implemented, better exploitation of reciprocity to estimate the channel response and utilization of the acquired CSI for both uplink Rx and the downlink Tx precoding of payload data [Kolomvakis et al. - 2017] can be achieved. However, the IQ imbalance in the front-end analog processing has a harmful effect on the reciprocity [Kolomvakis et al. - 2017]. Also, IQ imbalance combined with the nonlinear behavior of the PAs negatively affects the modulation quality due to the generation of inband and out-of-band distortion [Chung et al. - 2018; Khan et al. - 2017]. Moreover, mutual coupling or cross-talk between antenna ports [Amin et al. - 2014; Gustavsson et al. - 2014; Hausmair et al. - 2018] combined with the nonlinear behavior of the PAs generates additional distortion. Still, TDD M-MIMO offers an alternative implementation to the FDD deployment, which is quite promising for eMBB at the TDD mid-band (3–5 GHz).

Compensation techniques such as digital pre-distortion (DPD) have been extensively used to mitigate nonlinear distortions in SISO systems [Isaksson et al. - 2006]. However, the behavior of M-MIMO systems is different due to the combined effect of IQ imbalance, mutual coupling and PA nonlinear behavior (cf [Alizadeh et al. - 2017; Amin et al. - 2014; Gustavsson et al. - 2014; Khan et al. - 2017] and references within). Nevertheless, most of the compensation techniques are targeted towards 4×4 MIMO, where each transmitter chain has its own feedback path for the monitoring of PA output. Hence, scalability of these compensation techniques towards M-MIMO remains a major research question. Extensive over-the-air (OTA) measurement is a key technology for understanding potential solutions to M-MIMO scalability. For example [Hausmair - 2018] reports the simulation of an OTA based DPD for 64×64 MIMO for the mitigation of nonlinear distortion in the presence of antenna cross-talk. However, it would be of interest to see a measurement based OTA-DPD of M-MIMO in the presence of the above-mentioned hardware impairments.

The above discussion focused on the hardware impairments from M-MIMO implementations, so the remainder of this section will be devoted to discussing the hardware challenges for CA implementations. The first case of inter-band CA – as discussed in 6.3.2– is the most promising approach to increase the 5G NR cell capacity. Also, inter-band CA can be used to reduce the radio size, i.e. by integrating two or more bands within a single radio sharing common active RF components, e.g., PAs. A promising technique is to integrate multiple bands within a single radio and utilization of common active RF components reduces the cost (hardware and installation) as well as radio size. However, inter-band CA signal amplification using wide-band PAs (also known as concurrent multi-band PAs) introduces self- and cross-modulation distortion products and introduces complex trade-offs in the RFIC design and implementation [Amin et al. - 2015].

To compensate for nonlinear distortions produced by concurrent multi-band PAs, SISO linearization techniques are not adequate as they do not take into consideration the cross-modulation terms. Several models have been proposed that compensate for nonlinearities produced in concurrent dual-band PAs [Amin et al. - 2017a; Liu et al. - 2013]. A physically inspired concurrent dual-band PA model is reported in [Amin et al. - 2017b], introducing new cross-modulation terms that are not studied in any other previously published models. Numerous characterization techniques [Wood - 2014; Ghannouchi et al. - 2015] and physically inspired behavioral models [Morgan et al. - 2006] have been developed for in-depth analysis of nonlinearities in SISO PAs, leading to mature compensation techniques. Therefore, it is of great interest to develop characterization techniques for the detailed understanding of the cause of the above-mentioned hardware impairments for 5G radios.

6.4.2 Hardware Impairments – Receivers

One of the main hardware challenges when moving to mmWave frequency is the Rx performance [Andersson et al. - 2017]; the expected noise floor (NF) increase from 5.1 dB@2 GHz to 9.1 dB@30 GHz. Moreover, a study carried out by Ericsson AB also concluded that a more than tenfold increase in the carrier frequency corresponds to designing a 2 GHz Rx in 15 year old low-voltage technology. The need for low cost, energy efficiency, and high integration make the CMOS process a potential candidate for 5G transceivers. With the CMOS technology, scaling the NF@30 GHz will reduce from 9.1 dB to 7.7 dB in ten years [Andersson et al. - 2017]. Some techniques are being introduced to enhance the performance of low noise amplifiers and receivers for mmWave applications such as capacitive neutralization and transformer-based matching networks [Lianming et al. - 2016].

6.4.3 Hardware Impairments – Transceivers

FDD transceivers suffer from the leakage between the transmitter and receivers. Usually, a duplexer filter is utilized to provide sufficient isolation from the strong transmit signal. However, the requirements of low cost and the need for non-contiguous carrier aggregation makes those bulky and expensive filters. Moreover, these filters operate in fixed frequency band pairs, which are less attractive to be used in 5G multiband flexible transceivers, especially with the introduction of intra-band and inter-band carrier aggregation in 3GPP [Kiayani et al. - 2018]. It is even more problematic in the inband full duplex (IBFD) communications, where the Tx and Rx work simultaneously at the same carrier frequency, which gives rise to a very strong coupling from Tx to Rx. Using active RF cancellation to improve the isolation is one of the useful approaches to overcome the signal interference and the Tx leakage problems in both IBFD and FDD systems [Kiayani et al. - 2018].

It is worth reiterating that carrier aggregation is a key enabler for 5G to achieve higher peak data rates and efficient use of a fragmented spectrum. Nevertheless this will add to the complexity of transceiver design, more challenging radio environment in terms of spurious and self-blocking is needed, requiring advanced techniques for filtering. The antenna front-end needs to be designed to support multiple frequency band combinations and hence antenna design will become more challenging with [CA Technologies - 2014].

As for the transceiver implementation and promising architectures, there are two common typologies: the homodyne and heterodyne. Homodyne transceivers are a low-cost solution due to the utilization of fewer components. Additionally, homodyne transceivers lend themselves well to integrated circuit design in comparison to heterodyne transceivers. However, they are not applicable for very high frequency conversion due to challenges associated with DC offset, LO leakage, IQ imbalance and flicker noise [Gu - 2005]. The heterodyne topology lends itself more to the mmWave frequency band. Figures 6.6 and 6.7 show both architectures [Lianming et al. - 2016; Dafallah - 2016].

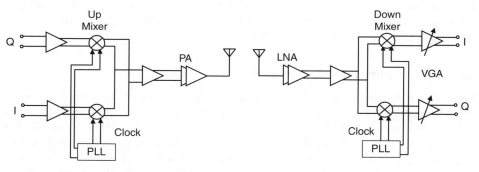

Figure 6.6 Homodyne transceiver.

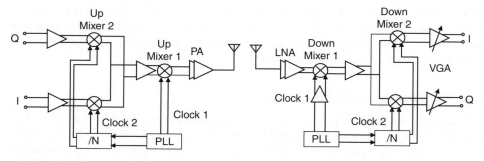

Figure 6.7 Heterodyne transceiver.

6.5 Technology and Fabrication Challenges

Despite the advantage of increased CMOS transistor speed due to size scaling, both the supply voltage and the ratio of the supply voltage to the transistor threshold voltage are also reduced. From the perspective of the mmWave circuit, this results in a low signal swing and low signal dynamic range. In addition, the insertion loss of passive devices increases at mmWave frequencies, which further degrades the power gain of the active devices, which leads to a low output power and high phase noise in the PA and oscillator, respectively. Current research efforts are focusing on developing techniques to overcome these limitations [Lianming et al. - 2016]. The semiconductor foundries need to scale the testing cost significantly to enable the low cost 5G networks as their experience with the high frequency fabrication has been for space applications where the high cost is acceptable [LAPEDUS - 2016]. Similar methods and techniques for testing cost scalability are required for 5G networks. A decision has to be made about which tests need to be done in an OTA chamber to reduce the cost. Chipset, devices and carriers all must have the same margin of error for certain performance parameters to reduce the number of OTA tests. Protocol tests which can be done without chambers will be beneficial in 5G [Khan - 2017]. Also, reducing variability in the manufacturing process by developing advanced fabrication parameters can reduce the cost through yield improvement [Resonant Inc. - 2018].

6.6 Conclusion

This chapter started by discussing the most promising and contributing 5G NR RAN radio-enabling technologies to meet the requirements for eMBB and FWA use cases. The optimization of various radio parameters for throughput maximization, spectral efficiency, spectrum and cell densification were discussed. To achieve efficient cell densification and reduce interference, M-MIMO with beamforming with hybrid multiple antenna architecture is a key promising RAN radio-enabling technology. The various challenges for the realization of M-MIMO, e.g. obtaining CSI and challenges of fabricating large antenna arrays for M-MIMO in mmWave bands, was discussed. D-MIMO is discussed as alternative implementation for M-MIMO that brings interesting advantages and hardware processing challenges.

Other RAN radio-enabling technology, licensed assisted access and its application for dual connectivity was presented, in particular for deployment scenarios using cognitive small cells (cognitive femtocells). The various scenarios for dual connectivity opens great possibilities for increasing downlink throughput with D2D adding more options to the table, provided intelligent interference management is put in place. The carrier aggregation, especially inband, was discussed as a key spectrum technology, with greater benefits when combined with licensed assisted access. Its associated hardware impairment challenges were also presented.

IQ imbalance and mutual coupling combined with the nonlinear distortion produced by PAs in M-MIMO were found to be the key challenges that degrade the transmitted signal quality, thus hindering efficient implementation and performance benefits of 5G systems and satisfying the eMBB and FWA use case requirements. The effect of these hardware impairments were discussed in detail. Moreover, the transceiver's architecture as well as leakage from transmitters to receivers was also investigated. The authors provided a discussion on current compensation techniques and predicted that their evolution will solve the nonlinearities in the PA. Also, scaling the transistors size was found to be promising in reducing the noise figure. Active RF cancellation was proven to also be a key technique that should be adopted. However, more research is still required to cover the interrelated hardware aspects in order to reach the required maturity level. The technology and fabrication challenges of the overall hardware architecture were also discussed and the authors suggested advanced fabrication models to reduce the cost of 5G network testing.

Bibliography

Verizon 5G TF V5G.201 v1.0 (2016-06. Verizon 5th Generation Radio Access; Physical layer; General description (Release 1). 3GPP, 2016. URL http://5gtf.org.

TS 38.202, NR; Services provided by the physical layer, Technical Specifications. 3GPP, 2017. URL https://portal.3gpp.org/.

3GPP. *5G Road Map*, 2017. URL http://www.3gpp.org/images/articleimages/.

A. Adhikary, V. Ntranos, and G. Caire. Cognitive femtocells: Breaking the spatial reuse barrier of cellular systems. In *2011 Information Theory and Applications Workshop*, pages 1–10, 2011.

M. Alizadeh, S. Amin, and D. Rönnow. Measurement and Analysis of Frequency-Domain Volterra Kernels of Nonlinear Dynamic 3×3 MIMO Systems. *IEEE Transactions on Instrumentation and Measurement*, 66(7):1893–1905, July 2017.

S. Amin, P. N. Landin, P. Händel, and D. Rönnow. Behavioral Modeling and Linearization of Crosstalk and Memory Effects in RF MIMO Transmitters. *IEEE Transactions on Microwave Theory and Techniques*, 62(4):810–823, April 2014.

S. Amin, W. Van Moer, P. Händel, and D. Rönnow. Characterization of concurrent dual-band power amplifiers using a dual two-tone excitation signal. *IEEE Transactions on Instrumentation and Measurement*, 64(10):2781–2791, Oct 2015. ISSN 0018-9456.

S. Amin, P. Händel, and D. Rönnow. Digital Predistortion of Single and Concurrent Dual-Band Radio Frequency GaN Amplifiers With Strong Nonlinear Memory Effects. *IEEE Transactions on Microwave Theory and Techniques*, 65(7):2453–2464, July 2017a.

Shoaib Amin, Per N. Landin, Peter Händel, and Daniel Rönnow. 2D Extended envelope memory polynomial model for concurrent dual-band RF transmitters. *International Journal of Microwave and Wireless Technologies*, 9(8):1619—1627, 2017b.

S. Andersson, L. Sundström, and S. Mattisson. Design considerations for 5g mm-wave receivers. In *2017 Fifth International Workshop on Cloud Technologies and Energy Efficiency in Mobile Communication Networks (CLEEN)*, pages 1–5, 2017.

J. G. Andrews, S. Buzzi, W. Choi, S. V. Hanly, A. Lozano, A. C. K. Soong, and J. C. Zhang. What will 5g be? *IEEE Journal on Selected Areas in Communications*, 32(6):1065–1082, 2014.

A. Asadi, Q. Wang, and V. Mancuso. A Survey on Device-to-Device Communication in Cellular Networks. *IEEE Commun. Surveys Tutorials*, 16 (4):1801–1819, 2014.

Arash Asadi and Vincenzo Mancuso. On the compound impact of opportunistic scheduling and d2d communications in cellular networks. In *Proc.16th ACM Int. Conf. Modeling, Analysis and Simulation of Wireless and Mobile Systems*, pages 279–288, New York, NY, USA, 2013. ACM.

Emil Björnson, Liesbet Van der Perre, Stefano Buzzi, and Erik G. Larsson. Massive MIMO in sub-6 ghz and mmwave: Physical, practical, and use-case differences. *CoRR*, abs/1803.11023, 2018. URL http://arxiv.org/abs/1803.11023.

C. Bockelmann, N. Pratas, H. Nikopour, K. Au, T. Svensson, C. Stefanovic, P. Popovski, and A. Dekorsy. Massive machine-type communications in 5g: physical and mac-layer solutions. *IEEE Communications Magazine*, 54(9): 59–65, 2016.

M. M. Buddhikot. Cognitive radio, dsa and self-×: Towards next transformation in cellular networks (extended abstract). In *2010 IEEE Symposium on New Frontiers in Dynamic Spectrum (DySPAN)*, pages 1–5, 2010.

A. Chung, M. B. Rejeb, Y. Beltagy, A. M. Darwish, H. A. Hung, and S. Boumaiza. Iq imbalance compensation and digital predistortion for millimeter-wave transmitters using reduced sampling rate observations. *IEEE Transactions on Microwave Theory and Techniques*, pages 1–10, Apr 2018.

Hind Dafallah. Highly Linear Attenuator and Mixer for Wide-Band TOR in CMOS. Master's thesis, Lund University, Sweden, 2016.

E. Dahlman, S. Parkvall, and J. Sköld. *4G, LTE-Advanced Pro and The Road to 5G*. Elsevier Science, 2016. ISBN 9780128046111. URL https://books.google.se/books?id=6J1FCgAAQBAJ.

M. Dohler, R. W. Heath, A. Lozano, C. B. Papadias, and R. A. Valenzuela. Is the phy layer dead? *IEEE Communications Magazine*, 49(4):159–165, 2011.

K. Doppler, M. Rinne, C. Wijting, C. B. Ribeiro, and K. Hugl. Device-to-device communication as an underlay to LTE-advanced networks. *IEEE Commun. Mag.*, 47(12):42–49, 2009.

S. Hong et al. Applications of self-interference cancellation in 5G and beyond. *IEEE Commun. Mag*, 52(2), Feb 2014.

F. M. Ghannouchi, O. Hammi, and M. Helaoui. *Behavioral Modeling and Predistortion of Wideband Wireless Transmitter*. John Wiley & Sons, West Sussex, UK, 2015.

Qizheng Gu. *RF System Design of Transceivers for Wireless Communications*. Springer, 2005.

U. Gustavsson, C. Sanchéz-Perez, T. Eriksson, F. Athley, G. Durisi, P. Landin, K. Hausmair, C. Fager, and L. Svensson. On the impact of hardware impairments on massive MIMO. In *2014 IEEE Globecom Workshops (GC Wkshps)*, pages 294–300, Dec 2014.

K. Hausmair. Modeling and Compensation of Nonlinear Distortion in Multi-Antenna RF Transmitters. *Chalmers University of Technology, PhD Thesis*, March 2018.

K. Hausmair, P. N. Landin, U. Gustavsson, C. Fager, and T. Eriksson. Digital Predistortion for Multi-Antenna Transmitters Affected by Antenna Crosstalk. *IEEE Transactions on Microwave Theory and Techniques*, 66(3):1524–1535, March 2018.

M. Isaksson, D. Wisell, and D. Ronnow. A comparative analysis of behavioral models for RF power amplifiers. *IEEE Transactions on Microwave Theory and Techniques*, 54(1):348–359, Jan 2006.

ITU R-REC-M.2083. Imt vision - framework and overall objectives of the future development of imt for 2020 and beyond. Technical Report M.2083-0 (09/2015), ITU, 2015. URL https://www.itu.int/rec/R-REC-M.2083.

Adnan Khan. Three 5G test challenges to overcome. Technical report, Anritsu, Dec 2017.

Z. A. Khan, E. Zenteno, P. Händel, and M. Isaksson. Digital Predistortion for Joint Mitigation of I/Q Imbalance and MIMO Power Amplifier Distortion. *IEEE Transactions on Microwave Theory and Techniques*, 65(1):322–333, Jan 2017.

A. Kiayani, M. Z. Waheed, L. Anttila, M. Abdelaziz, D. Korpi, V. Syrjälä, M. Kosunen, K. Stadius, J. Ryynänen,, and M. Valkama. Adaptive nonlinear rf cancellation for improved isolation in simultaneous transmit-receive systems. *IEEE Transactions on Microwave Theory and Techniques*, 66(5):2299–2312, 2018.

N. Kolomvakis, M. Coldrey, T. Eriksson, and M. Viberg. Massive mimo systems with iq imbalance: Channel estimation and sum rate limits. *IEEE Transactions on Communications*, 65(6):2382–2396, 2017.

MARK LAPEDUS. Waiting For 5G Technology. Technical report, Semiconductor Engineering, 2016. URL https://semiengineering.com/waiting-for-5g-technology/.

X. Li, X. Yang, L. Li, J. Jin, N. Zhao, and C. Zhang. Performance analysis of distributed mimo with zf receivers over semi-correlated *mathcalK* fading channels. *IEEE Access*, 5:9291–9303, 2017.

LI Lianming, NIU Xiaokang, CHAI Yuan, CHEN Linhui, ZHANG Tao, CHENG Depeng, XIA Haiyang, WANG Jiangzhou, CUI Tiejun, and YOU Xiaohu. The path to 5G: mmWave aspects. *Journal of Communications and Information Networks*, 1(2):4–15, Aug 2016.

Y. J. Liu, W. Chen, J. Zhou, B. H. Zhou, and F. M. Ghannouchi. Digital Predistortion for Concurrent Dual-Band Transmitters Using 2-D Modified Memory Polynomials. *IEEE Transactions on Microwave Theory and Techniques*, 61(1):281–290, Jan 2013.

A. F. Molisch, V. V. Ratnam, S. Han, Z. Li, S. L. H. Nguyen, L. Li, and K. Haneda. Hybrid Beamforming for Massive MIMO: A Survey. *IEEE Communications Magazine*, 55(9):134–141, Sep 2017.

D. R. Morgan, Z. Ma, J. Kim, M. G. Zierdt, and J. Pastalan. A generalized memory polynomial model for digital predistortion of rf power amplifiers. *IEEE Transactions on Signal Processing*, 54(10):3852–3860, 2006.

Y. Pei and Y. C. Liang. Resource Allocation for Device-to-Device Communications Overlaying Two-Way Cellular Networks. *IEEE Trans. Wireless Commun.*, 12 (7):3611–3621, 2013.

V. Ratnam, A. Molisch, O. Y. Bursalioglu, and H. C. Papadopoulos. Hybrid Beamforming with Selection for Multi-user Massive MIMO Systems. *IEEE Transactions on Signal Processing*, pages 1–1, 2018.

Resonant Inc. Rf innovation and the transition to 5g wireless technology. White paper, Resonant Inc., 2018.

Moray Rumney. Making 5g work: Capacity growth and rf aspects, 2018. URL https://event .on24.com/wcc/r/1655226/1965C3FC225758E6FD54E97E6B32480E.

Manuel Blanco Agilent Technologies. Carrier aggregation: Fundamentals and deployments. Technical report, Agilent Technologies, 2014.

J. Wood. *Behavioral Modeling and Linearization of RF Power Amplifier*. Artech House, Norwood, MA, 2014.

7

Millimeter Wave Antenna Design for 5G Applications

Issa Elfergani*[1], Abubakar Sadiq Hussaini[1,2], Abdelgader M. Abdalla[1], Jonathan Rodriguez[1,3], and Raed Abd-Alhameed[4]

[1] *Instituto de Telecomunicações, 3810-193, Aveiro, Portugal*
[2] *School of Information Technology & Computing, American University of Nigeria, 98 Lamido Zubairu Way, Wuro Hausa 640101, Yola, Nigeria*
[3] *University of South Wales, Pontypridd, CF37 1DL, United Kingdom*
[4] *Faculty of Engineering & Informatics, University of Bradford, Bradford, BD7 1DP, United Kingdom*

7.1   Introduction

The mobile industry is continuously innovating new handset devices to be aligned with future emerging market trends. Today's handsets support video streaming, multimedia and fast surfing internet that is enabled by the Long Term Evolution (LTE) standard, which is commonly referred to as 4G . The recent explosion in 4G communication technology has led to portable devices, such as notebooks and laptop computers, operating on the LTE bands, namely LTE 700 MHz, LTE 2400 MHz, and LTE 2600 MHz. Several LTE antenna designs have been reported recently [Hong et al. - 2014; Elfergani et al. - 2015a; Sethi et al. - 2017; Elfergani et al. - 2014, 2016; Belrhiti et al. - 2015], but are rigid in their design, to cover new design frequencies to cover these lower frequency operational designs.

The current communication regime is heading towards higher data rate connectivity as the market for rich multimedia content proliferates; this is supported by advances in nano-electronic devices and components that enable high processing power. In this context, the lower frequency bands that are being exploited by various communication networks are becoming congested, forcing the industry stakeholders to investigate alternative pools of the spectrum. One such option is the UWB unlicensed bands that range from 3.1 to10.6 GHz, which are being considered for indoor scenarios due to the shorter distances [REPORT and ORDER - 2002].

UWB antennas have gained huge consideration in academia and industry for applications in wireless transmission systems. Thus, several antennas designs with numerous approaches have been exploited to enhance the bandwidth coverage and enable the antenna to operate in the UWB frequency range [Elfergani et al. - 2017; Ray et al. - 2013; Gong et al. - 2016; Khan et al. - 2016; Akram et al. - 2015; Cruz et al. - 2017; Li et al. - 2006;

Corresponding Author: Issa Elfergani i.t.e.elfergani@av.it.pt

Optical and Wireless Convergence for 5G Networks, First Edition.
Edited by Abdelgader M. Abdalla, Jonathan Rodriguez, Issa Elfergani, and Antonio Teixeira.
© 2020 John Wiley & Sons Ltd. Published 2020 by John Wiley & Sons Ltd.

Abid et al. - 2015]. Moreover, as there exists some narrow band standards within the UWB frequency region, such as the wireless local area network (WLAN) for IEEE 802.11a operating at 5.15–5.825 GHz, the IEEE 802.16 WiMAX system operating at 3.3–3.7GHz, the C-band (4.4–5.0 GHz), and the X-band operating at 7.725–8.275 GHz for ITU applications, this has led to performance degradation for UWB systems due to co-channel interference. Thus, to protect UWB devices from such interference, a UWB antenna including filtering notch features is desired. The filtering notch performance may be accomplished by introducing a number of effective techniques as stated in [Abid et al. - 2015; Elfergani et al. - 2015b, 2012; Naser-Moghadasi et al. - 2013; Sung - 2013; Zhang et al. - 2013; Karmakar et al. - 2013; Abdollahvand et al. - 2010; Emadian and Ahmadi-Shokouh - 2015; Lin et al. - 2012; Elhabchi et al. - 2017; Reddy et al. - 2014]. However, although there are several advantages to the UWB standard, and in particular with UWB antennas, such as wide bandwidth, low power consumption, and high data-rate wireless connectivity among devices within a personal operating space, the above-mentioned antennas operating in such systems are limited to a carrier frequency ranging between 700 and 2600 MHz as in [Orellana and Solbach - 2008; Pan and Wong - 1997; Nakano et al. - 1984; Huang et al. - 1999; Boyle and Steeneken - 2007; Palukuru et al. - 2007; Hong et al. - 2014; Elfergani et al. - 2015a; Sethi et al. - 2017; Elfergani et al. - 2014, 2016; Belrhiti et al. - 2015][1-12], and from 3 to 10 GHz as in [Elfergani et al. - 2017; Ray et al. - 2013; Gong et al. - 2016; Khan et al. - 2016; Akram et al. - 2015; Cruz et al. - 2017; Li et al. - 2006; Abid et al. - 2015; Elfergani et al. - 2015b, 2012; Naser-Moghadasi et al. - 2013; Sung - 2013; Zhang et al. - 2013; Karmakar et al. - 2013; Abdollahvand et al. - 2010; Emadian and Ahmadi-Shokouh - 2015; Lin et al. - 2012; Elhabchi et al. - 2017; Gorai et al. - 2013; Reddy et al. - 2014]. Still the antennas on offer are not flexible enough to provide tuning/coverage for frequencies outside this range.

The next mobile era is heading towards 5G communications. The race towards 5G is driven by the market place that is forecasting mobile traffic to increase exponentially over the next decade. This is not only constitutes traditional mobile services such as high-speed data, but also future emerging use-cases such massive machine-type communication and the tactile internet. In synergy, these will place new design requirements on the mobile system that includes offering high capacity services, super low latency connectivity and high reliability. Clearly, current 4G systems that were designed preliminary for supporting medium-to-high data rate services are not engineered to support the expected diverse traffic types, nor able to have inherent capacity to support the growth in data; the existing cellular networks may need to deliver as much as a thousand times the capacity relative to current levels, and it is clear that legacy mobile systems will reach a saturation point.

To provide the fifth generation mobile network, the necessary tool to deliver next generation services, will require at first more spectrum since current capacity enhancement techniques can only deliver stepwise gains. In order to solve this problem, the 5G standardization community is looking towards the use of the higher frequency spectrum, the so-called millimeter wave bands (20–300 GHz), where a huge number of untapped spectrum pools are available. In fact, the millimeter wave bands have been recommended to be a significant and essential part of the 5G mobile network in order to provide multi-gigabit communication services such as high definition television (HDTV) and ultrahigh definition video (UHDV) [Elkashlan et al. - 2014]. It is believed that from both the regulatory and technological perspectives , the initial spectrum allocation for 5G will be between the bands of 24 and 57 GHz. It has been suggested/recommended that the mobile industry prioritize bands within the 25.25–29.5 GHz and the 36–40.5 GHz frequency blocks as primary targets for 5G (World Radio communication Conference 2019 – WRC-19) [Straight Path Communications Inc. - 2015]. In this context, the efficient deployment of 5G systems will require the design of even more compact yet efficient antennas. This provides the impetus for a new breed of handset design that in principle should be multi-mode in nature, energy efficient, and above all able to operate at the millimeter wave band, placing new design drivers on the antenna design.

There has been unprecedented activity within the antenna research community to come up with efficient antenna designs for use in the 5G paradigm, in particular designs that operate in dual 5G frequency bands, i.e. 28 and 38 GHz bands. The millimeter wave 5G antenna designs must take into consideration the high propagation loss due to atmospheric absorption at millimeter waves [Shubair et al. - 2015]. Furthermore, the 5G antennas are expected to have high gain, efficiencies, and bandwidth greater than 1 GHz. Several investigations have been carried out on wireless communications that employ a printed antenna technology for millimeter wave bands [Wong et al. - 2013; Chin et al. - 2011; Tong et al. - 2005; Wang et al. - 2012]. Microstrip antennas are exploited for various applications owing to their simple and low profile features. However, microstrip antennas suffer from narrow bandwidth, typically 2–5%, very low radiation efficiency (<10 %) and low gain (<0 dBi) [Jamaluddin et al. - 2009]. All these drawbacks may lead to a considerable reduction in the antenna gain and efficiency at millimeter wave frequencies. The printed monopole antenna has also been broadly used for mobile applications due to inherent attributes such as low profile, planar structure, multi-band properties, low cost, moderate to high gain, and ease of fabrication. Various antenna designs [Ullah et al. - 2017; Khalily et al. - 2016; Bisharat et al. - 2016; Choubey et al. - 2016; Park et al. - 2016; Dadgarpour et al. - 2016] have been proposed for 5G applications that operate in this new millimeter wave spectrum. Although these antenna designs cover the desired 5G allocated frequency bands, there are very few of them, which provides a wide frequency bandwidth. Moreover, the antennas that operate over a broad frequency range, but also have limited agile flexibility in their design to overcome interference [Ullah et al. - 2017; Khalily et al. - 2016; Bisharat et al. - 2016].

To mitigate interference and to have a more flexible inherent design, we re-engineer the printed monopole design, and propose a wide band monopole printed antenna encompassing simple and effective approaches towards enhancing the bandwidth, power gain, and efficiency, and reducing the interference at millimeter wave frequencies. This chapter first considers the design procedure based on a single band monopole antenna, which evolves through stepwise enhancements towards our final goal: an antenna that can operate with wide band characteristics. The enhancements consider notch features that were introduced by embedding an L-shaped slot on the right bottom corner of the radiating patch. To control the created rejected band, a lumped capacitor was positioned on the best location within the L-shaped slot and by changing its capacitance values; the notched-band can easily be shifted downwards over a wide and continues frequency range. The proposed antenna is demonstrated to be a suitable candidate antenna to operate over the 36–40.5 GHz frequency block, aligned with the frequencies range under consideration by WRC-19 [Straight Path Communications Inc. - 2015].

7.2 Antenna Design and Procedure

The millimeter wave printed monopole antenna structure and synthesis is shown in Figure 7.1. The proposed antenna is constructed on a $S_L \times S_W$, FR4 substrate with 0.8 mm thickness, dielectric constant $\varepsilon_r = 4.4$ and loss tangent tan $\delta = 0.017$. The dimensions of the radiating patch ($P_L \times P_W$) are 2.5×2.5 mm^2. The antenna is fed via a 50 Ohm microstrip transmission line having a width and length of $F_L \times F_W$. The other side of the substrate has a finite rectangular copper ground plane with dimensions $G_L \times G_W$, as depicted in Figure 7.1b.

This optimized ground plane shape and dimensions has helped in improving the impedance matching of the proposed antenna. The optimized parameters are summarized in Table 7.1. An embedded L-shaped slot dimension of $S_L \times S_W \times S_T$ was introduced on the right bottom corner over the patch and close to the feeding strip as shown in Figure 7.1b; this slot was added in an appropriate location in order to produce a rejected band at around 40 GHz.

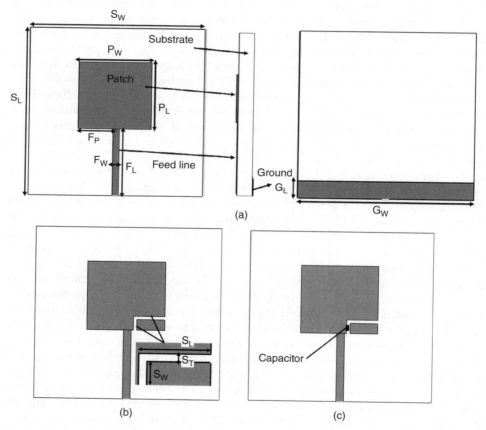

(a)

(b)   (c)

Figure 7.1 The main antenna structure: (a) antenna with defected ground plane, (b) antenna with L-slot, (c) antenna with attached capacitor.

7.3 Antenna Optimization and Analysis

To further investigate the influences of the key antenna parameters on performance, extensive parametric studies were carried out. The optimized values for the proposed design are tabulated in Table 7.1.Three parameters were chosen for this analysis, which include the ground plane length, feed position and type of substrate used; these three parameters were deemed as the most sensitive ones in defining the desired frequency bands along with best impedance matching. Each simulation was run with only one parameter varied, while the other parameters remain unchanged. The optimization analysis of the proposed antenna parameters were done with the aid of using a CST EM simulator [CST-Computer Simulation Technology AG - 2014].

7.3.1 The Influence of Ground Plane Length (G_L)

The influence of the length of the ground plane was analyzed by examining the variations in S_{11} against the ground plane length. The length of the ground plane was varied from 3.5 to 0.5 mm; when the ground plane length was set at 3.5 mm, the proposed antenna exhibited a narrow bandwidth at around 38 GHz.

On the other hand, this antenna demonstrates a wide frequency range from 30 GHz up to around 40 GHz when G_L was varied from 2.5 mm up to 1.5 mm. When the ground plane length decreased to 1 mm, the higher resonant frequency was increased to 42 GHz; this makes the antenna operate over a broad bandwidth from 30 to 42 GHz. However, when G_L was set at 0.5 mm, the bandwidth of the proposed antenna was further improved to cover the operational range from 30 to 45 GHz, as indicated in Figure 7.2, in which case this G_L value was selected to be the optimum value for the length of the ground plane for the proposed antenna design.

Table 7.1 The full dimensions of the proposed antenna.

Parameters	Volume (mm)
P_W	2.5
P_L	2.5
F_P	1.25
F_L	2
F_W	0.15
S_W	5
S_L	5
G_L	0.5
G_W	5
S_L	1
S_T	0.15
S_W	0.75

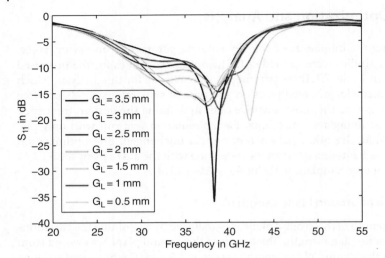

Figure 7.2 Simulated return loss (S_{11}) for different ground plane lengths of the first version of the proposed antenna.

7.3.2 The Effect of Feeding Strip Position (F_P)

The effects of the feed line position are illustrated in Figure 7.3. In this analysis, the position of the feed line is primarily set to the edge of the design structure, which corresponds to 1 mm, which is then shifted in steps of 0.25 mm closer to the end of the other edge, which is at 2 mm. One can clearly see that when the feeding strip is set at 0.5 and 1.5 mm, the proposed structure shows a good impedance matching of 23 dB return loss at 25 GHz. When the position of the feeding strip is moved over the values

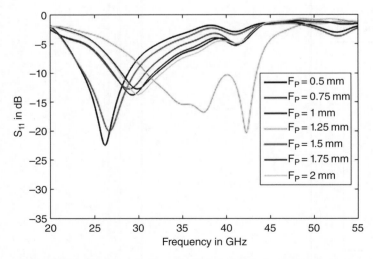

Figure 7.3 Simulated return loss (S_{11}) for different feeding line positions on the first version of the proposed antenna.

of 0.75, 1, 1.75, and 2 mm, the antenna resonant frequency is shifted upwards to around 30 GHz. Although the proposed antenna with the above-mentioned feeding position values show a good impedance matching at around 25 and 30 GHz, this is still considered as a narrow bandwidth resonator, which does not meet the targeted design value. However, when the position of the feeding line is set at 1.25 mm, the bandwidth of the proposed antenna was significantly improved to cover a wide frequency range from 30 to 45 GHz, as depicted in Figure 7.3. Therefore, this value of 1.25 mm was picked to be the best value for the feeding location.

7.3.3 The Influences of the Substrate Type

To estimate the effectiveness of the permittivity of the applied substrate, the variation in the permittivity of the substrates against the S_{11} response was investigated, as depicted in Figure 7.4. This examination was implemented in order to accomplish the best behavior for the proposed antenna at the targeted wide bandwidth range. Four substrates with different levels of permittivity were used in this study, where the wide band frequency response was analyzed, as indicated in Figure 7.4. These include (FR4 $\varepsilon_r = 4.4$), (RT/duriod 5880 $\varepsilon_r = 2.33$), (RT/duriod 5870 $\varepsilon_r = 2.33$) and (Roger RT 6006 $\varepsilon_r = 6.15$). It appeared that when the proposed radiator was printed over the RT/duriod 5870 $\varepsilon_r = 2.33$ and Roger RT 6006 $\varepsilon_r = 6.15$, it exhibited a resonant frequency only at 50 GHz, which does not meet the broad bandwidth target defined within this work. On other hand, when both the FR4 $\varepsilon_r = 4.4$ and Roger RT 6006 $\varepsilon_r = 6.15$ were utilized, the antennas achieved a very wide frequency range from 30 GHz up to around 45 GHz , as depicted in Figure 7.4, which is more suitable for 5G applications. However, as the permittivity of RT 6006 $\varepsilon_r = 6.15$ is quite high, this will affect and impair the performance of the antenna, particularly in terms of the antenna gain and efficiency. Thus, FR4 $\varepsilon_r = 4.4$ was selected to be used as the substrate material within this investigation.

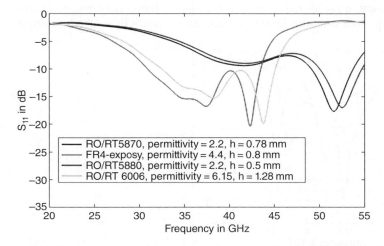

Figure 7.4 Simulated return loss (S_{11}) for different substrate materials on the first version of the proposed antenna.

7.4 Millimeter Wave Antenna Design with Notched Frequency Band

As was previously explained, to come up with the optimum wide band frequency millimeter wave antenna, several procedures were implemented. The first antenna design was implemented with a full ground plane, which resulted in good impedance matching at 37.5 GHz, as depicted in Figure 7.5. However, as the main target of this work was to design and develop a compact antenna that is able to operate over a wider frequency range within the millimeter wave spectrum, there was a need to look at alternative techniques such as the defected ground plane (DGP) approach.

In general, the conventional microstrip antenna structure also suffers from constraints and limitations such as single operating frequency, low impedance bandwidth, low power gain, and size increasing. Thus, several approaches have been taken into consideration in order to meet the above-mentioned limitations. These approaches include using stacking, different feeding methods, high permittivity dielectric substrates, frequency selective surfaces, electromagnetic band gaps (EBGs), photonic band gaps (PBGs), metamaterials, and so forth. However, none of these techniques can meet the aforementioned constraints. For example using a different feeding approach can improve the antenna bandwidth, but it might be considered complex and even costly; the use of a substrate with high permittivity will help in enhancing the antenna bandwidth, but will severely affect the antenna gain and efficiency; EBGs and PBGs are deemed as attractive avenues to improve the antenna performance, but both are not considered to be simple and low cost approaches.

The DGP has been given most attention among all the above-mentioned approaches, mainly because of its simple structural design. In fact, embedded single or multiple slots or defects over the ground plane of the microstrip circuits are referred to as defected ground plane/structure. The DGP is deemed as a promising procedure to effectively improve the performance of the patch antenna. This DGP approach was integrated within the antenna design, which led to several advantages including, improved radiation properties, antenna size reduction, suppression of harmonics, reduced cross polarization and improved isolation in MIMO antennas. The DGP is also an avenue

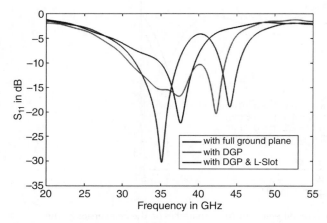

Figure 7.5 The S_{11} variations of the antenna with full ground plane, with DGP, and with DGP and L-slot.

that contributes in mitigating surface waves. This can be done by partially etching the ground structure of an antenna. Furthermore, DGP may enable antenna engineers to adjust the ground field shape, and since microstrip antennas suffer from severe surface wave losses, so in turn the DGP would improve the antenna performance.

In other words, the ground plane of the antenna is defected in order to upgrade the antenna performance. These defects can be a shape of periodic or non-periodic. Fundamentally, DGP is a resonant gap that is able to disturb the current distribution in the ground plane of the antenna. This alteration of the current distribution creates an impact on the characteristics of the antenna by disturbing the shunt capacitance and series inductance. As has been understood from [Breed - 2008], a DGP creates parallel-tuned circuits, where capacitance, inductance, and resistance change according to the structure of the DGP. It is a key factor that the DGP is tuned to different frequencies and provides multiband operation and enhanced bandwidth. All these advantages have made the DGP a promising, easy, and simple avenue to mainly improve the antenna bandwidth. Thus, within this chapter the DGP was implemented to enhance the antenna bandwidth. As can be seen from Figure 7.5, when the ground plane of the antenna was cut to form the DGP, the bandwidth was enhanced to cover a wide range from 30 to 45 GHz compared to the antenna with full ground plane that has only a narrow band around 37.5 GHz.

However, the antenna with DGP operates over a broad frequency range of the millimeter wave spectrum, i.e. 30 to 45 GHz, which is subject to expected severe interference among the co-channels. To mitigate or/and avoid such interference, a filter is needed to offer rejection to unwanted frequencies in the range of the targeted operating band in order to not only avoid envisaged interference from other communication applications but also to improve its own systems performance. However, to use an external band-rejection filter along with the antenna will lead to design complexity, large size, and high cost.

The antenna designers have addressed this by using other simple and less complex avenues to replace the external filters, these approaches include, embedding parasitic elements, EBG, and inserting different slit shapes over the radiating element or ground plane of the antenna systems. This will not only create a notched band feature, but will also keep the antenna footprint unaltered. Therefore, within this design, an L-shaped slot was introduced over a proper location on the patch, as shown in Figure 7.1b. The antenna together with the DGP and L-slot still show a smooth wide frequency range from 30 to 45 GHz, but a notched band around 40 GHz was generated, as indicated in Figure 7.5.

The antenna power gains and efficiencies of the antenna with DGP and the antenna including DGP and L-slot are shown in Figure 7.6. It is obvious that the antenna without the rejected band (antenna with DGP) exhibits smooth power gain values ranging from around 3.95 dBi up to 5.8 dBi over the entire frequency range. However, the antenna with the DGP and the L-slot show a stable power gain value from 3.99 dBi to around 6 dBi, except at 40 GHz where the gain decreases sharply to −5.6 dBi where the filtering notch was produced.

The radiation efficiencies of the antenna version with DGP varies from 86% to around 90% over the whole desired wide bandwidth, while the antenna version with the DGP and combined L-slot shows radiation efficiencies varying from 83% to 84% at the lower frequencies (30 GHz up to 33 GHz), whilst showing a smooth curve at around 92% at the

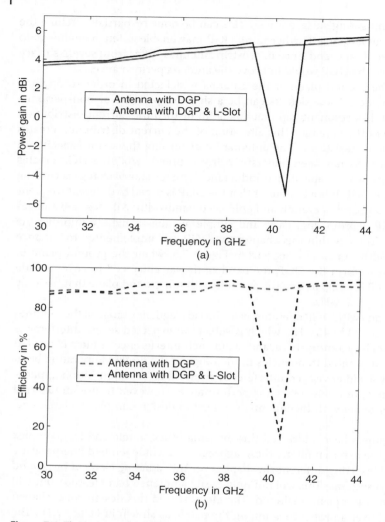

Figure 7.6 The power gains (a) and efficiencies (b) of the proposed antennas with only DGP, and with DGP and L-slot.

higher frequency bands, except at 40 GHz, where the efficiency largely drops to a value of around 18%; as expected the efficiencies are in agreement with the obtained gain.

7.5 Millimeter Wave Antenna Design with Loaded Capacitor

Although the version of the antenna (with DGP and L-slot) maintains the same wide frequency range from 30 to 45 GHz with an inherent filtering notch at the 40 GHz frequency for interference rejection, the design is rigid in nature; once the antenna is fabricated, it is not feasible to alter/shift the produced rejected band. The generated filtering band should be to tuned/switched to cover multiple bands.

The tunable notched-bands have come up with several advantages over their conventional counterparts of fixed notched bands. The antenna with the tuned rejected band frequencies can be exploited to easily avoid interference over multiple frequency bands. This will significantly contribute to size reduction and toward minimizing the cost of the hardware. Their desirable attributes are spurring the proliferation of their application, and will likely play a key role in 5G systems.

In recent years, novel communication architectures have been being developed (software defined radio, MIMO and cognitive radio) as part of the evolution towards 5G, which will alleviate the stress on legacy wireless networks and open new and effective prospects for drastic improvements in network bandwidth and efficiency. Therefore, in order to implement such architectures, antennas that are able to operate in wide band operation are required. However, the defined frequency band for wide band systems may cause interference to the existing narrow wireless communication systems. Therefore, wide band antennas with fixed rejected band can be tuned to operate over a larger range of frequencies while reconfiguring the notched band over a desired range, which effectively enhances the antenna performance. In addition, it can divide the whole wide band into a few sub-bands. This will create more flexibility for practical applications. Thus, within this chapter the created notched band of 40 GHz was tuned by positioning a lumped capacitor over an appropriate location on the L-slot, as shown in Figure 7.1c. The generated notched-band can be widely and continuously tuned to cover the range of 39, 37, 35, and 33 GHz, by varying the capacitance of the attached capacitor over the values of 0.5, 1.5, 5 and 7 pf, respectively, as shown in Figure 7.7.

The power gains and efficiencies of the antenna with tunable notches are indicated in Figure 7.8. In the scenario of the antenna version with tunable rejected bands, the power gain ranges from 3.9 dBi to around 6.2 dBi over all the frequency continuous range, except at four targeted notched bands, namely 39, 37, 35, and 33 GHz (capacitance values 0.5, 1.5, 5 and 7 pf), where the gain values of the four rejected bands were considerably dropped to −2.5, −2.8, −2.6, and −2.7 dBi respectively. Also, radiation efficiencies from 83% to 92% were achieved when the capacitance was varied from 0.5 pf

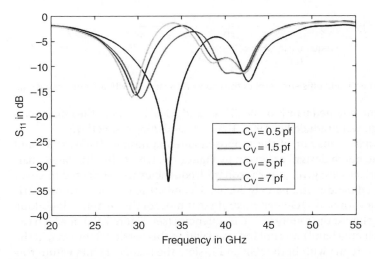

Figure 7.7 The S_{11} variations of the antenna with attached capacitor.

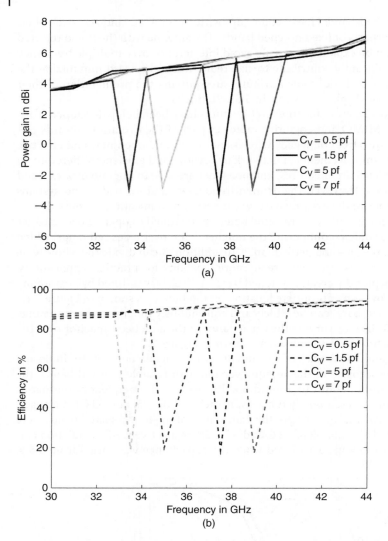

Figure 7.8 The power gains (a) and efficiencies (b) of the proposed antennas with attached capacitor.

up to 7 pf , except at the notched bands of 39, 37, 35, and 33 GHz, where the efficiencies of the proposed design were reduced to 19% , 18%, 20% and 20%, respectively.

Moreover, the resonant behavior of the proposed antenna designs (DGP, DGP and L-slot, and the tunable notch design) are given by Figure 7.9, where the simulated analysis of the current surfaces are given; four resonant frequencies were selected to cover the aggregated obtained bandwidth, namely 30, 33, 37, and 40 GHz. The current distributions of the antenna with only DGP demonstrate that most of the currents flow along the microstrip feeding line at the four targeted frequencies (see Figure 7.9a) in which the filtering feature (no rejected band) was not included within this version. However, in the case of the proposed antenna with both DGP and L-slot , the main currents within this version appear or are induced on the feeding line, except at the frequency of 40 GHz,

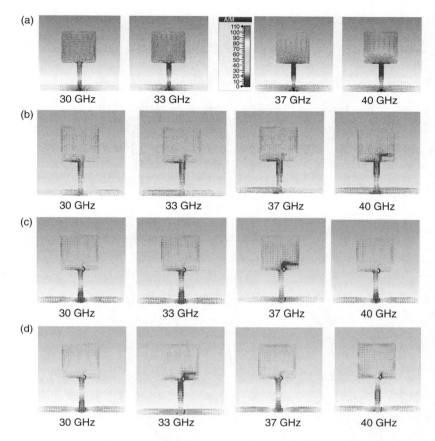

Figure 7.9 The current surfaces for: (a) the unslotted antenna, (b) the slotted antenna, (c) the 1.5 pf loaded antenna, and (d) the 7 pf loaded antenna at 30, 33, 37, and 40 GHz.

where a strong current flows along the strip line and the L-slot, as the desired north was generated as shown in Figure 7.9b; this leads to the conclusion that the embedded L-slot acts as a resonator to be used as an effective and simple band-rejection technique. In the scenario of the tunable notch antenna (antenna with attached capacitor), while the antenna was loaded with 1.5 and 7pf as illustrated in Figures 7.9c and 7.9d, the current distributions were mostly concentrated around the feeding strip and at the boundaries of the L-slot. These currents canceling each other out prevent the antenna from radiating at 37 and 33 GHz where the rejected bands were produced. This means the antenna impedance will abruptly be changed at the desired rejected bands, which in turn introduces the notched band into the proposed monopole antenna structure.

The 3D radiation patterns of the antennas and their fundamental properties in terms of total efficiency and directivity are indicated in Figure 7.10. Four frequency bands at 30, 32, 36, and 42 GHz were selected to cover the entire wide frequency range of the proposed design. As can been seen from Figure 7.10, the proposed antenna achieves desirable radiation performance over the four targeted bands of interest. It is clear that the antenna behavior is the same as an omnidirectional antenna in the frequency range of interest.

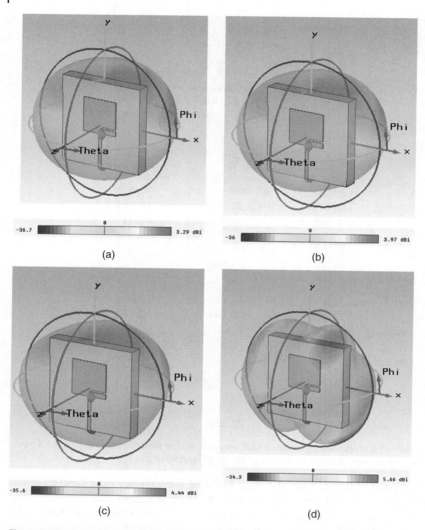

Figure 7.10 The current surfaces for: (a) the unslotted antenna, (b) the slotted antenna, (c) the 1.5 pf loaded antenna, and (d) the 7 pf loaded antenna at 30, 33, 37, and 40 GHz.

7.6 Conclusion

Millimeter wave broadband monopole antennas, which are suitable for future 5G mobile network applications, have been presented. The proposed antennas occupy a compact volume of 2.5×2.5 mm^2 that is printed over an RF4 substrate with size of $5 \times 5 \times 0.8$ mm^3. The first antenna version without the DGP exhibits a narrow bandwidth of around 37.5 GHz. The second version that exploits the defected ground plane approach has results in a wide frequency range of operation from 30 to 45 GHz. In addition, to reduce the foreseen interference between 5G systems and other applications, an L-shaped slot was inserted in a proper location over the radiator to generate a notched band at around 40 GHz. Finally, to tune the created filtering band, a lumped capacitor was attached over the L-shaped slot and its capacitance values varied from 0.5 to 7 pf. This led to a downward shift in the notch from 40 GHz up to 33 GHz, while keeping the same wide bandwidth (30 to 45 GHz) constant. The proposed antennas demonstrate promising performance in terms of return loss, power gain, current surfaces, and efficiency, which ultimately characterizes them as attractive candidate solutions for future 5G wireless systems.

Acknowledgments

This work is carried out under the grant of the Fundacão para a Ciência e a Tecnologia (FCT – Portugal), with the reference number: SFRH/BPD/95110/2013. This work is supported by the European Union's Horizon 2020 Research and Innovation program under grant agreement H2020-MSCA-ITN-2016-SECRET-722424. Our gratitude is also extended to the following funding body Ocean12-H2020-ECSEL-2017-1-783127.

Bibliography

M. Abdollahvand, G. Dadashzadeh, and D. Mostafa. Compact dual band-notched printed monopole antenna for uwb application. *IEEE Antennas and Wireless Propagation Letters*, 9:1148–1151, 2010.

Muhammad Abid, Jalil Kazim, and Owais. Ultra-wideband circular printed monopole antenna for cognitive radio applications. *INTERNATIONAL JOURNAL OF MICROWAVE AND OPTICAL TECHNOLOGY*, 10(3): 184–189, 2015.

S. W. Akram, K. Shambavi, and Z. C. Alex. Design of printed strip monopole antenna for uwb applications. In *2015 2nd International Conference on Electronics and Communication Systems (ICECS)*, pages 823–826, 2015.

Lakbir Belrhiti, Fatima Riouch, Jaouad Terhzaz, Abdelwahed Tribak, and Angel Mediavilla Sanchez. *A Compact Planar Monopole Antenna with a T-Shaped Coupling Feed for LTE/GSM/UMTS Operation in the Mobile Phone*, volume 380 of *Lecture Notes in Electrical Engineering*. Springer International Publishing, 2015.

D. J. Bisharat, S. Liao, and Q. Xue. High gain and low cost differentially fed circularly polarized planar aperture antenna for broadband millimeter-wave applications. *IEEE Transactions on Antennas and Propagation*, 64(1):33–42, 2016.

K. R. Boyle and P. G. Steeneken. A five-band reconfigurable pifa for mobile phones. *IEEE Transactions on Antennas and Propagation*, 55(11):3300–3309, 2007.

Gary Breed. An Introduction to Defected Ground Structures in Microstrip Circuits. tutorial, High Frequency Electronics Copyright©2008 Summit Technical Media, LLC, 2008.

Kuo-Sheng Chin, Ho-Ting Chang, Jia-An Liu, Hsien-Chin Chiu, J. S. Fu, and Shuh-Han Chao. 28-ghz patch antenna arrays with pcb and ltcc substrates. In *Proceedings of 2011 Cross Strait Quad-Regional Radio Science and Wireless Technology Conference*, volume 1, pages 355–358, 2011.

P. N. Choubey, W. Hong, Z. C. Hao, P. Chen, T. V. Duong, and J. Mei. A wideband dual-mode siw cavity-backed triangular-complimentary-split-ring-slot (tcsrs) antenna. *IEEE Transactions on Antennas and Propagation*, 64(6):2541–2545, 2016.

CST-Computer Simulation Technology AG. Antenna simulation, 2014.

A. Dadgarpour, B. Zarghooni, B. S. Virdee, and T. A. Denidni. Single end-fire antenna for dual-beam and broad beamwidth operation at 60 ghz by artificially modifying the permittivity of the antenna substrate. *IEEE Transactions on Antennas and Propagation*, 64(9):4068–4073, 2016.

I. T. E Elfergani, A. S. H., J. RODRIGUEZ, Chan H. SEE, and R. A. Abd-Alhameed. Wideband tunable pifa antenna with loaded slot structure for mobile handset and lte applications. *RADIOENGINEERING*, 23(1):1–11, 2014.

I. T. E Elfergani, A. S. H., J. Rodriguez, and R. A. Abd-Alhameed. A compact dual-band balanced slot antenna for lte applications. In *PIERS Proceedings, Prague, 2015*, 2015a.

I. T. E. Elfergani, A. S. Hussaini, J. Rodriguez, and R. A. Abd-Alhameed. Dual-band printed folded dipole balanced antenna for 700/2600mhz lte bands. In *2016 10th European Conference on Antennas and Propagation (EuCAP)*, pages 1–5, 2016.

Issa Elfergani, Abubakar Sadiq Hussaini, Jonathan Rodriguez, and R.A. Abd-Alhameed. *Antenna Fundamentals for Legacy Mobile Applications and Beyond*. Springer International Publishing, Cham, Switzerland, 1 edition, 2017.

I.T.E. Elfergani, R.A Abd-Alhameed, C.H. See, S.M.R. Jones, and P.S. Excell. A compact design of tunable band-notched ultra-wide band antenna. *Microwave and Optical Technology Letters*, 54(7):1642–1644, 2012.

I.T.E. Elfergani, A. S Hussaini, C See, R.A. Abd-Alhameed Abd-Alhameed, N.J McEwan, S. Zhu, J. Rodriguez, and R. W. Clarke. Printed monopole antenna with tunable band-notched characteristic for use in mobile and ultra-wide band applications. *International Journal of RF and Microwave Computer-Aided Engineering*, 25(5):403–412, 2015b.

Mourad Elhabchi, Mohamed N. Srifi, and Rajae Touahn. A tri-band-notched uwb planar monopole antenna using dgs and semi arc-shaped slot for wimax/wlan/x-band rejection. *Progress In Electromagnetics Research Letters*, 70:7–14, 2017.

M. Elkashlan, T. Q. Duong, and H. H. Chen. Millimeter-wave communications for 5g: fundamentals: Part i [guest editorial]. *IEEE Communications Magazine*, 52(9):52–54, 2014.

S. R. Emadian and J. Ahmadi-Shokouh. Very small dual band-notched rectangular slot antenna with enhanced impedance bandwidth. *IEEE Transactions on Antennas and Propagation*, 63(10):4529–4534, 2015.

X. Gong, L. Tong, Y. Tian, and B. Gao. Design of a microstrip-fed hexagonal shape uwb antenna with triple notched bands. In *Progress In Electromagnetics Research C*, volume 62, pages 77–87, 2016.

Abhik Gorai, Anirban Karmakar, Manimala Pal, and Rowdra Ghatak. Multiple fractal-shaped slots-based uwb antenna with triple-band notch functionality. *Journal of Electromagnetic Waves and Applications*, 27(18):2407–2415, 2013.

M. H. Jamaluddin, R. Gillard, R. Sauleau, and et al. *A dielectric resonator antenna (DRA) reecarray*, chapter 8. John Wiley & Sons, New Jersey, 2005, 2009.

C.-W.P. Huang, A.Z. Elsherbeni, J.J. Chen, and C.E. Smith. Fdtd characterization of meander line antennas for rf and wireless communications. *Journal of Electromagnetic Waves and Applications*, 13(12):1649–1651, 1999.

J.N. Cruz, R.C.S. Freire, A.J.R. Serres L.C.M. Moura, A.P. Costa, and P.H.F. Silva. Parametric study of printed monopole antenna bioinspired on the inga marginata leaves for uwb applications. *J. Microw. Optoelectron. Electromagn. Appl.*, 16(1):312–321, 2017.

Anirban Karmakar, Rowdra Ghatak, Utsab Banerjee, and D.R. Poddar. An uwb antenna using modified hilbert curve slot for dual band notch characteristics. *Journal of Electromagnetic Waves and Applications*, 27(13): 1620–1631, 2013.

M. Khalily, R. Tafazolli, T. A. Rahman, and M. R. Kamarudin. Design of phased arrays of series-fed patch antennas with reduced number of the controllers for 28-ghz mm-wave applications. *IEEE Antennas and Wireless Propagation Letters*, 15:1305–1308, 2016.

Muhammad Kabir Khan, Muhammad Irshad Khan, Iftikhar Ahmad, and Mohammad Saleem. Design of a printed monopole antenna with ridged ground for ultra wideband applications. In *2016 Progress In Electromagnetic Research Symposium (PIERS)*, pages 4394–4396, 2016.

K.P. Ray, S.S. Thakur, and A.A. Deshmkh. Compact slotted printed monopole uwb antenna. In *International Conference on Communication Technology 2013*, pages 16–18, 2013.

Pengcheng Li, Jianxin Liang, and Xiaodong Chen. Study of printed elliptical/circular slot antennas for ultrawideband applications. *IEEE Transactions on Antennas and Propagation*, 54(6):1670–1675, 2006.

C. C. Lin, P. Jin, and R. W. Ziolkowski. Single, dual and tri-band-notched ultrawideband (uwb) antennas using capacitively loaded loop (cll) resonators. *IEEE Transactions on Antennas and Propagation*, 60(1):102–109, 2012.

H. Nakano, H. Tagami, A. Yoshizawa, and J. Yamauchi. Shortening ratios of modified dipole antennas. *IEEE Transactions on Antennas and Propagation*, 32(4):385–386, 1984.

M. Naser-Moghadasi, R. A. Sadeghzadeh, T. Sedghi, T. Aribi, and B. S. Virdee. Uwb cpw-fed fractal patch antenna with band-notched function employing folded t-shaped element. *IEEE Antennas and Wireless Propagation Letters*, 12:504–507, 2013.

L. Q. Orellana and K. Solbach. Study of monopole radiators for planar circuit integration. In *2008 38th European Microwave Conference*, pages 1300–1303, 2008.

V. K. Palukuru, M. Komulainen, M. Berg, H. Jantunen, and E. Salonen. Frequency-tunable planar monopole antenna for mobile terminals. In *The Second European Conference on Antennas and Propagation, EuCAP 2007*, pages 1–5, 2007.

Shan-Cheng Pan and Kin-Lu Wong. Dual-frequency triangular microstrip antenna with a shorting pin. *IEEE Transactions on Antennas and Propagation*, 45(12):1889–1891, 1997.

S. J. Park, D. H. Shin, and S. O. Park. Low side-lobe substrate-integrated-waveguide antenna array using broadband unequal feeding network for millimeter-wave handset device. *IEEE Transactions on Antennas and Propagation*, 64(3):923–932, 2016.

G. S. Reddy, A. Kamma, S. K. Mishra, and J. Mukherjee. Compact bluetooth/uwb dual-band planar antenna with quadruple band-notch characteristics. *IEEE Antennas and Wireless Propagation Letters*, 13: 872–875, 2014.

FIRST REPORT and ORDER. Revision of part 15 of the commission's rules regarding ultra-wideband transmission systems. Technical Report FCC 02-48, Federal Communications Commission, Washington, D.C. 20554, April 2002.

Waleed Tariq Sethi, Hamsakutty Vettikalladi, Habib Fathallah, and Mohamed Himdi. Hexaband printed monopole antenna for wireless applications. *Microwave and Optical Technology Letters*, 59:2816–2822, 2017.

R. M. Shubair, A. M. AlShamsi, K. Khalaf, and A. Kiourti. Novel miniature wearable microstrip antennas for ism-band biomedical telemetry. In *2015 Loughborough Antennas Propagation Conference (LAPC)*, pages 1–4, 2015.

Straight Path Communications Inc. A straight path towards 5g. White paper, StraightPath, 2015.

Y. Sung. Triple band-notched uwb planar monopole antenna using a modified h-shaped resonator. *IEEE Transactions on Antennas and Propagation*, 61(2): 953–957, 2013.

K. F. Tong, K. Li2, and T. Matsui. Performance of millimeter-wave coplanar patch antennas on low-k materials. In *Progress In Electromagnetics Research Symposium 2005, Hangzhou, China,2005*, volume 1, pages 1–2, 2005.

H. Ullah, F. A. Tahir, and M. U. Khan. A honeycomb-shaped planar monopole antenna for broadband millimeter-wave applications. In *2017 11th European Conference on Antennas and Propagation (EUCAP)*, pages 3094–3097, 2017.

D. Wang, H. Wong, K. B. Ng, and C. H. Chan. Wideband shorted higher-order mode millimeter-wave patch antenna. In *Proceedings of the 2012 IEEE International Symposium on Antennas and Propagation*, pages 1–2, 2012.

H. Wong, K. B. Ng, C. H. Chan, and K. M. Luk. Printed antennas for millimeter-wave applications. In *2013 International Workshop on Antenna Technology (iWAT)*, pages 411–414, 2013.

Y. Hong, J. Tak, J. Baek, B. Myeong, and J. Choi. Design of a multiband antenna for lte/gsm/umts band operation. *International Journal of Antennas and Propagation*, pages 1–9, 2014.

C.-W. Zhang, Y.-Z. Yin, P.-A. Liu, and J.-J. Xie. Compact dual band-notched uwb antenna with hexagonal slotted ground plane. *Journal of Electromagnetic Waves and Applications*, 27(2):215–223, 2013.

8

Wireless Signal Encapsulation in a Seamless Fiber–Millimeter Wave System

Pham Tien Dat*,[1], Atsushi Kanno[1], Naokatsu Yamamoto[1], and Testuya Kawanishi[1,2]

[1] *National Institute of Information and Communication Technology, 184-8795, Tokyo, Japan*
[2] *Waseda University, 169-8050, Tokyo, Japan*

8.1 Introduction

In the current fourth generation and future fifth generation mobile networks, deployment of small cells is an important and effective method of providing high data throughput to end users, especially in "hot-spot" areas. Small cells can be deployed in the areas where the communication traffic reaches a high level to complement existing macro-cell networks. The number of small cells can be increased significantly; thus, simplifying antenna sites would be very important to reduce the system cost, power consumption, and management complexity. The cloud radio access network (C-RAN) has been proposed to provide a centralized network for small cell wireless systems [China Mobile Research Institute - 2011; Li et al. - 2012]. In this system, complicated signal processing functions can be located at a central station (CS), to help simplify antenna sites, and enable more advanced and efficient network coordination and management. However, a mobile fronthaul network is necessary to connect the CS with antenna sites. In the current networks, oversampled digital base band streams are transmitted from the CS to antenna sites using interface protocols such as the common public radio interface [CPRI Eri - 2011] or the open base station architecture initiative protocol [OBS AI - 2006]. However, the required data rate of the fronthaul system for the transmission of oversampled signals can be very high. In addition, latency and jitter problems will be the main concerns because media access control and physical layer functions are separated at the CS and antenna sites. Mobile fronthaul systems based on radio-over-fiber (RoF) technology can reduce the required data rate and transmission latency, simplify antenna sites, enable co-existence of multiple radios, and support delay sensitive and critical applications [Liu et al. - 2013]. In this sense, a C-RAN network based on an RoF fronthaul system is very suitable for small cell and ultra-dense small cell networks. However, the installation of fiber cables to each antenna site is difficult, especially in dense urban areas, due to the associated costs and challenges. The use of fiber cable is also infeasible in many cases because of its lack of flexibility. In that case, the transmission of radio signals over a converged fiber and radio link in

---

* Corresponding Author : Pham Tien Dat; ptdat@nict.go.jp

Optical and Wireless Convergence for 5G Networks, First Edition.
Edited by Abdelgader M. Abdalla, Jonathan Rodriguez, Issa Elfergani, and Antonio Teixeira.
© 2020 John Wiley & Sons Ltd. Published 2020 by John Wiley & Sons Ltd.

the high-frequency bands, such as in the millimeter wave (mmWave) band, can be an attractive alternative for an RoF fronthaul system. In this system, the mmWave link can work as an extension to a fiber cable for mobile signal transmission. To realize a simple fiber–mmWave system for low-latency and energy-efficient transmission of mobile signals, a seamless convergence fiber-optic and mmWave links using photonic technologies for the generation, transmission, and up-conversion of radio signals is very attractive. Different wireless signals can be encapsulated and transmitted over the system at the same time. The transmission of radio signals over the seamless fiber–mmWave system is called a radio-on-radio-over-fiber system (RoRoF) [Dat et al. - 2015]. The system has attracted a great interest and several studies have been conducted on the transmission of wireless signals over the converged systems [Nkansah et al. - 2007; James et al. - 2010; Llorente et al. - 2011; Beltran et al. - 2011; Zhu et al. - 2013]. In this chapter, we present the operating principle of wireless signal encapsulation and transmission over the seamless fiber–mmWave system in both downlink (DL) and uplink (UL) directions. We discuss the use of different methods and the corresponding transmission impairments on the signal performance. We present examples of the signal transmission over the system and estimate the possible transmission distance of the mmWave links. The remainder of this chapter is organized as follows. Section 8.2 presents the concept and principle of signal encapsulation. Section 8.3 presents examples of the experimental demonstrations on signal transmission over the seamless systems in the DL and UL directions. Finally, Section 8.4 concludes the chapter.

8.2 Principle of Signal Encapsulation

8.2.1 Downlink System

Figure 8.1a shows the conceptual diagram of the DL transmission of wireless signals over the RoRoF system. In the DL direction, an optical mmWave signal consisting of two optical sidebands is first generated in the CS. The frequency difference between the optical sidebands is equal to the frequency of the mmWave signal to be transmitted over free space in the wireless part of the system. There are several methods for generating an optical mmWave signal, including heterodyning lightwave signals from independent laser diodes [Pang et al. - 2011], optical frequency combs [Dat et al. - 2014], or using optical modulation technology [Kanno et al. - 2012]. Each method has own advantages; however, we should consider the frequency and phase stability of the generated mmWave carrier signal. In addition, the radio signal transmission over the air must adhere to regulations. At the radio frequency of 30–275 GHz, the required tolerance of the center frequency should be less than ±150 ppm, which corresponds to frequency fluctuation of approximately ±14 MHz at the center frequency of 92.5 GHz [Kanno et al. - 2014]. A frequency fluctuation of approximately ±80 MHz, ±5 MHz, and ±2 MHz was observed by heterodyning a 15 Hz fiber laser (FL) with a 100 kHz tunable laser, a 1 kHz FL, and a 15 Hz FL, respectively [Kanno et al. - 2014]. By using optical modulation technology, a frequency fluctuation of less than 100 kHz could be observed. In addition, the phase noise of the generated mmWave signal also has an influence on the signal transmission when coherent detection is used at the receiver. In this sense, the use of the optical modulation technology for generating an optical mmWave signal would be suitable for the system. Through using this method, the generated signal can be expressed as

$$E(t) = A_+\cos\left[\left(\omega_0 + \left(\frac{\omega_{\mathrm{mmWave}}}{2}\right)\right)t + \varphi_0 + (\varphi_{\mathrm{LO}}(t)/2)\right]$$
$$+ A_-\cos\left[\left(\omega_0 - \left(\frac{\omega_{\mathrm{mmWave}}}{2}\right)\right)t + \varphi_0 - (\varphi_{\mathrm{LO}}(t)/2)\right] \tag{8.1}$$

where A_+ and A_- denote the amplitudes of the upper and lower sidebands, respectively, and ω_0 and φ_0 are the angular frequency and phase of the optical signal input to the generator from a laser diode, respectively. ω_{mmWave} is the angular frequency of the generated mmWave signal, and $\varphi_{\mathrm{LO}}(t)$ is the phase of the driving electrical signal to the generator.

The generated mmWave optical signal is modulated by wireless signals at an optical modulator. For the data modulation, both dual- and single-wavelength modulation methods can be used [Dat et al. - 2016a]. In the dual-wavelength modulation method, both optical sidebands of the generated optical mmWave signal are modulated by the wireless signals. In the single-wavelength modulation scheme, only one of the sidebands is modulated by the wireless signals, and the other sideband is kept unmodulated. The optical sidebands are then recombined to form an RoF signal for up-converting to the mmWave signal at the receiver. The former scheme is simple; however, the effects of fiber dispersion on the transmitted wireless signals are larger because the received signals are formed from many different signal components, which are transmitted over the fiber links at different velocities. However, the phase error induced from differential transmission delays between the optical sidebands can be minimized. For the later modulation scheme, the effect of fiber dispersion can be minimized. However, the effect of optical phase noise is larger. For the transmission of narrow-bandwidth wireless signals over a short to medium fiber link (up to approximately 20 km), the dual-wavelength modulation scheme can be a suitable solution. By contrast, for the transmission of large bandwidth signals and/or over a long fiber link (over 30 km), single-wavelength modulation would be the better choice [Dat et al. - 2016a]. Here we consider the case of dual-wavelength modulation. We assume that $S(t) = \sum_{k=1}^{N} S_k t$ is the total signal to be transmitted from the CS to the remote antenna head (RRH), where N is the number of wireless signals and $S_k(t)$ is the kth wireless signal; k varies from 1 to N. When the optical modulator is biased at a quadrature point, the output signal can be written as [Alves and Cartaxo - 2012]

$$E_1(t) = (\mathrm{IL})E(t)\cos[(\pi/2V_\pi)S(t) - \pi/4]$$
$$= (\mathrm{IL}/\sqrt{2})E(t)[\sin[(\pi/2V_\pi)S(t)] + \cos[(\pi/2V_\pi)S(t)]] \tag{8.2}$$

where IL is the insertion loss and $2V_\pi$ is the switching voltage of the optical modulator. Using the Taylor series for the $\cos(x)$ and $\sin(x)$ functions, we can rewrite 8.2 as

$$E_1(t) = (\mathrm{IL}/\sqrt{2})E(t)[1 + (\pi/2V_\pi)S(t) - 1/2!(\pi/2V_\pi)^2 S(t)^2 + 1/3!(\pi/2V_\pi)^3 S(t)^3]. \tag{8.3}$$

The modulated signal is then fed into a single-mode fiber (SMF) and transmitted to a remote antenna unit (RAU) where the signal is converted from optical to the mmWave band directly. The conversion can be easily and effectively performed by a high-bandwidth photodiode. The signal at the output of the photodiode can be represented by

$$E_1(t) = (\mu \cdot \text{IL}^2 \cdot A_+ \cdot A_-/2)\cos[(\omega_{\text{mmWave}})t + (\varphi_{LO}(t))]$$
$$\times [1 + (\pi/2V_\pi)S(t) - 1/2!(\pi/2V_\pi)^2 S(t)^2 + 1/3!(\pi/2V_\pi)^3 S(t)^3] \qquad (8.4)$$

where μ is the photodiode efficiency. The optical carrier ω_0 and the low-frequency components are not generated by the photodiode. The signal, expressed in the form of 8.4, is emitted directly into free space by an antenna, transmitted over the air to the RRH, and is received by another antenna. The received signal at the output of the receiving antenna can be expressed as

$$I_{RX} = \frac{I(t) \cdot G_{TX} \cdot G_{RX}}{Loss} \times r \times e^{-j\theta} \qquad (8.5)$$

where G_{TX} *and* G_{RX} are the gains of the antennas at the transmitter and receiver, respectively, r and θ are the random amplitude and phase due to the Rician fading, respectively, and *Loss* is the free-space loss calculated by the Friss equation [Friis - 1946]. The received signal is then down-converted from the mmWave band to the original wireless signals in the microwave band using a coherent or incoherent detection method. In the coherent method, an electrical mixer, which is driven by a local oscillator (LO) signal, is used. This method can help increase the receiver sensitivity and the dynamic range of the system; however, the effect of electrical phase noise is very large for narrow bandwidth signal transmission. To stabilize the phase noise effect, a phase lock of the LO source with those at the transmitter used to generate the optical mmWave signal should be implemented [Dat et al. - 2014]. However, it is difficult to implement such a phase lock in a practical system. A self-homodyne receiver can be an alternative solution for coherent detection. In this scheme, the received mmWave signal is divided into two parts. One of the signals is used as the usual received signal, while the other is amplified and used as an LO signal to drive an electrical mixer for the signal down conversion. This method can reduce the phase noise effect and the complexity of the receiver compared to a coherent detection receiver and can improve the receiver sensitivity and dynamic range compared to an incoherent detection [Dat et al. - 2015; Dat et al. - 2016b]. In simple systems, the incoherent detection method using a square-law envelope detector can be used to simplify the structure of the RRHs, and the recovered signal can be expressed as

$$r_{DL}(t) \propto (I_{DL-r}(t))^2$$
$$= K.[1 + (\pi/2V_\pi)S(t) - 1/2!(\pi/2V_\pi)^2 S(t)^2 + 1/3!(\pi/2V_\pi)^3 S(t)^3]^4 + n(t)$$
$$= k + k\alpha_1 \cdot S(t) + k \cdot \sum_{n=2}^{n=12} \alpha_k.S^k(t) + n(t) \qquad (8.6)$$

where α_k, $k = 1, \ldots, 12$ is the constant coefficient and $n(t)$ is the system noise. Equation 8.6 shows that the desired signal $S(t)$ can be recovered after being transmitted over the system. However, the received signal also consists of other products of intermodulation and noises. Among the high-order intermodulation products, the second and third products greatly affect the performance of wireless signals owing to their high power and close frequency to the desired signals. To reduce the intermodulation effects, wireless signals with the appropriate frequencies should be transmitted. Optimizing the number of wireless signals for transmission over the system is also important to minimize the intermodulation effects. Finally, the transmitted wireless signals can be recovered using appropriate electrical filters.

8.2.2 Uplink System

A signal encapsulation on a combined mmWave and RoF system can also be realized for the UL transmission of mobile signals over a fronthaul system, as shown in Figure 8.1b. Different wireless signals can be combined before being up-converted to the mmWave band using an electrical mixer. An LO signal is needed for the signal up-conversion. The up-converted signal is transmitted over free space to an RAU where a fiber cable link is available for further transmission to the CS. For the signal transmission over the fiber link, the signal in the mmWave band can be applied directly to the optical link using a high-speed optical modulator. However, the effect of fiber dispersion will be very high. In addition, the system cost will also increase due to the use of high-bandwidth optical modulators. To simplify the optical system, the received mmWave signals should be down-converted to the originally transmitted wireless signals in the microwave bands before being fed into the fiber system. For the signal down-conversion, similar to the DL system, a coherent/incoherent detection and a self-homodyne receiver can be used. The down-converted signal is converted to an optical format and transmitted to the CS using a common electrical to optical (E/O) converter. At the CS, the optical signal is converted back to the electrical format, and finally the wireless signals are recovered using appropriate electrical filters. The characteristics of the fiber links such as radio frequency signal gain and noise figure have a large impact on the quality of the wireless signal transmission and the transmission range of the mmWave link [Orange et al. - 2012]. In addition, in the UL direction, low-cost lightwave sources at the antenna sites can suddenly turn on after a power recovery. The UL radio signals are only transmitted in given periods of time. Thus, there is a possibility that optical burst signals are transmitted over the UL system. However, the response of the optical amplifiers is not flat to the input optical burst signals. The optical surge induced from the transience of the optical amplifiers can damage the photodetectors at the receivers, and degrade the signal performance. To avoid damage to photodetectors

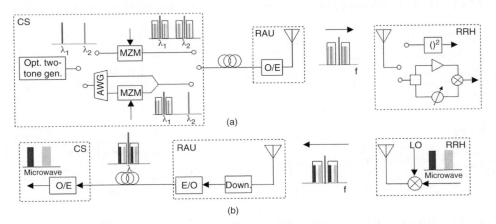

Figure 8.1 Concept of wireless signal transmission over the RoRoF systems: (a) DL. (b) UL.

and to improve the signal performance, a burst-mode optical amplifier should be used [Dat et al. - 2015].

8.3 Examples of Signal Encapsulation

In this section, we present examples of the performance of wireless signal encapsulation and transmission over a seamless fiber–mmWave system in both downlink and uplink directions. Standard compliant very-high-throughput WLAN 802.11ac and high-speed LTE-A signals at different frequencies are combined and transmitted over the system, and the performance in term of error vector magnitude (EVM) is evaluated.

8.3.1 Downlink Transmission

Figure 8.2a shows the setup for the DL transmission of wireless signals over the seamless system. Figure 8.2b shows the detail of an optical mmWave signal generator. Figures 8.2c and d show the setup for the system characterization and the simultaneous transmission of the DL LTE-A and IEEE 802.11ac signals over the system, respectively. To generate a stable optical mmWave signal, we use a high-extinction ratio dual-parallel Mach–Zehnder interferometer modulator (DPMZM) [Kawanishi et al. - 2007]. As described in Figure 8.2b, an electrical signal is an input to an electrode of the main MZM to generate even-order sidebands. The generated optical signal is passed through an optical band elimination filter (OBEF) to suppress the carrier component. A coherent two-tone optical signal with a frequency separation of quadrupling those of the fed signal is achieved. An optical amplifier (EDFA) is used to boost the optical power, and an optical bandpass filter (OBPF) is used to reduce the amplified spontaneous emission noise. For the signal transmission over the system, in this experiment, we use a dual-wavelength modulation scheme. In this scheme, the two optical sidebands at the output of the optical mmWave generator are modulated by wireless signals at an MZM. The modulated optical signal is amplified by an EDFA, input into an SMF (10 km in the experiment), and transmitted to an optical receiver at the RAU. The signal is converted to a radio-on-radio (RoR) signal at 94.1 GHz by a high-speed photodetector. The signal is amplified by a W-band power

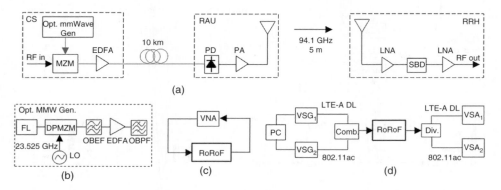

Figure 8.2 Experimental setup for transmission of LTE-A and 802.11ac signals over the RoRoF system in the DL.

Table 8.1 System parameters and requirements.

Parameters	Values	Parameters	Values
Input power to PD	8 dBm	Antenna gain	23 dBi
PD 3dB bandwidth	90 GHz	PD responsivity	0.5 A/W
LN MZM Vpi	2 V	LN MZM insertion loss	5 dB
W-band LNA gain	20 dB	SBD responsivity	2000 V/W
Signal requirements			
WLAN signal EVM [IEEE Std 802.11TM-2012 - 2012]		16-QAM, 3/4	−19 dB
QPSK, 3/4	−13 dB	64-QAM, 5/6	−27 dB
64-QAM, 3/4	−25 dB	256-QAM, 3/4	−30 dB
LTE-A signal EVM [3rd (2016)]		16-QAM	12%
QPSK	17.5%	64-QAM	8%

amplifier (PA) before being emitted into free space by a horn antenna (23 dBi gain). After being transmitted over a 5 m link in free space, the signal is received by another horn antenna at the RRH, amplified by a low-noise amplifier (LNA), and down-converted to the original signals by an envelope detector (zero-biased Schottky barrier diode: SBD). The recovered signal is amplified by an LNA before being analyzed. For the system characterization, as shown in Figure 8.2c, the input and output of the system are connected to a vector network analyzer (VNA). For the wireless signal transmission, shown in Figure 8.2d, the WLAN and LTE-A signals are generated from a laptop using commercially available signal studios, downloaded to vector signal generators (VSGs), and combined by a power combiner before driving the MZM. At the receiver, the recovered signals are divided by a power divider and inputted to vector signal analyzers (VSAs). Finally, the signals are analyzed using Keysight 89600 series VSA software. Table 8.1 lists the main system parameters and signal requirements.

We first evaluate important characteristics of the system, including the phase noise and the spurious free-dynamic range (SFDR). Figure 8.3a shows the single-sideband phase noise of the system using the incoherent detection. For comparison, the phase noise of the system employing a coherent detection at the RRH is also presented. In the coherent detection system, the phase noise becomes too unstable to transmit wireless signals when the LOs are freely running. Using a phase lock can help stabilize the phase noise; however, the system becomes complicated, especially the RRH. The phase noise in the incoherent detection is similar to that of the coherent detection using phase-locked LOs. SFDR is another important parameter that needs to be considered when transmitting analog radio signals, and Figure 8.3b shows the measurement result. A total SFDR of approximately 58 dB $Hz^{2/3}$ is achieved, which indicates a sufficiently high dynamic range for DL wireless signal transmission [Al-Raweshidy and Komaki - 2002]. The use of the SBD with sufficiently good phase noise and dynamic range characteristics significantly simplifies the system and remote sites, and help minimize the system cost and power consumption. Therefore, it is suitable for small cell based heterogeneous cellular systems. For a higher dynamic range system, a self-homodyne receiver should be used for the signal down-conversion.

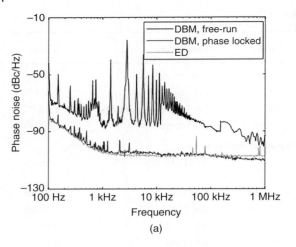

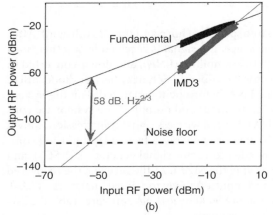

Figure 8.3 DL system characteristics: (a) Phase noise; (b) SFDR.

We then evaluate the performance of the IEEE 802.11ac and LTE-A signals after being transmitted over the system. Figure 8.4a shows an example of the received spectrum of the combined LTE-A and WLAN 802.11ac signals. A spectrum with clear separation of the transmitted signals is observed. The third-order intermodulation can be neglected, whereas second-order intermodulation is quite small. A clear zoom view of the signal spectrum is shown in Figure 8.4b. Figure 8.5 shows the constellation diagrams of 20 MHz 256-QAM and 80 MHz 64-QAM WLAN 802.11ac signals. We can observe clear constellations, even for the high-order modulation signals. The measurement results for WLAN 802.11ac signals under different transmit power values are shown in Figures 8.6a–c for the 20, 40, and 80 MHz signals, respectively. Compared with the requirements listed in Table 8.1, signals using the 20 MHz 256-QAM and 40 MHz 256-QAM can be successfully transmitted over the system. For the 80 MHz signal, the performance is relatively degraded because of the distortion and aliasing effects as the bandwidth of the generated signal is the same as those of the signal generator. However, the performance is still better than the requirement for a 64-QAM signal. For generation and transmission of higher throughput signals using an 80 + 80 MHz or a 160 MHz

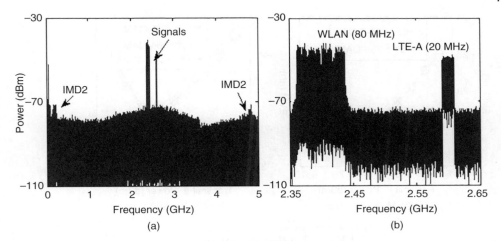

Figure 8.4 Received spectrum of the combined LTE-A and 802.11ac signals.

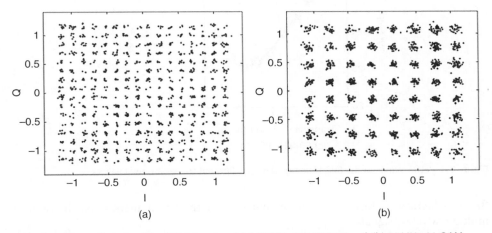

Figure 8.5 Constellations of the 802.11ac signal: (a) 20 MHz 256-QAM signal; (b) 80 MHz 64-QAM signal.

bandwidth, a signal generator with large internal bandwidth should be used. Figure 8.7 shows the performance of the 802.11ac signal when transmitted simultaneously with the LTE-A signal over the system. Satisfactory performance can be maintained for the 802.11ac signal even when the LTE-A signal is transmitted with relatively high power. Figures 8.8a and b show the EVM performance of the LTE-A signal for different transmit powers, and different transmit powers of the co-transmitting WLAN 802.11ac signal, respectively. All signals carried by different carrier components of the LTE-A signal are successfully transmitted with satisfactory EVM performance. Although the signals are affected by the noise and nonlinear distortion associated with the system, the transmit power dynamic range is sufficiently large. Increasing the transmit power of the co-transmitting WLAN signal affects the performance of the LTE-A signal due to nonlinear distortion. However, the EVM increase is not very significant, confirming

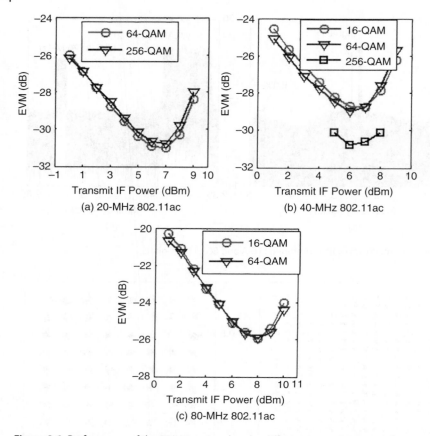

Figure 8.6 Performance of the 802.11ac signal under different transmit power values.

that satisfactory performance can be achieved for the simultaneous transmission of multiple wireless signals.

8.3.2 Uplink Transmission

The setup for the transmission of wireless signals over the converged mmWave and RoF system in the UL direction is shown in Figure 8.9a. Figures 8.9b and c show the setup for the system characterization and simultaneous transmission of the LTE-A (UL) and IEEE 802.11ac signals over the system. First, the input wireless signal is up-converted to a mmWave signal at 94.1 GHz using a W-band mixer at the RRH. The up-converted signal is amplified by a W-band PA to reach a power level of approximately 5 dBm before being emitted into free space by a horn antenna. The signal is transmitted over a 5 m free-space link and received by another horn antenna at the RAU. The received signal is amplified by an LNA before being coupled into a SBD to down-convert to the original wireless signals. It is then amplified by another LNA before feeding to an RoF transmit module for conversion to an optical signal. The optical signal is transmitted over a 20 km SMF to the receiver located at the CS. To emulate different practical application cases,

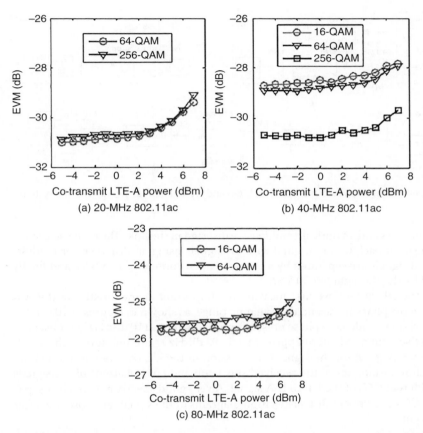

Figure 8.7 Performance of the 802.11ac signal for different transmit powers of the co-transmit LTE-A signal.

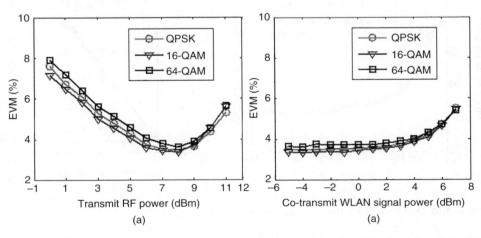

Figure 8.8 Performance of the LTE-A signal for: (a) different transmit powers and (b) transmit powers of the co-transmitted WLAN signal.

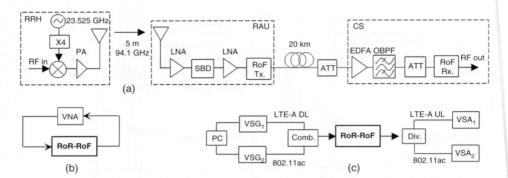

Figure 8.9 Experimental setup for transmission of LTE-A and WLAN 802.11ac signals over the system in the UL.

variable attenuators and an optical amplifier are inserted in the link. The received optical signal is converted back to the original wireless signals using an RoF receiver module. The recovered signals are separated by a power divider, connected to VSAs, and finally are analyzed by the Keysight 89600 VSA.

Similar to the DL system, we first measure the important characteristics of the system, including the phase noise and the dynamic range, as shown in Figures 8.10a and b, respectively. A single-sideband phase noise of approximately -110 dBc Hz^{-1} is observed at a 1 MHz offset and an SFDR of approximately 93 dB Hz$^{2/3}$ are measured. These values are sufficiently good for the signal transmission in the UL transmission direction. Figure 8.11 shows examples of the received spectrum and constellations of the signals carried by different CCs of the UL LTE-A signal. A very good spectrum with clear separation of the CCs and very clear constellation diagrams of the signals are observed after the transmission.

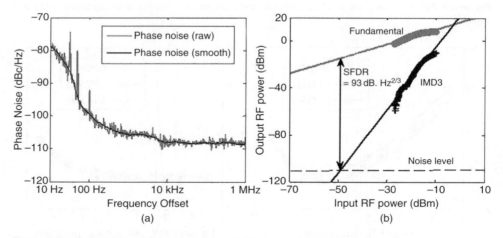

Figure 8.10 Characteristics of the UL system. (a) Phase noise. (b) SFDR.

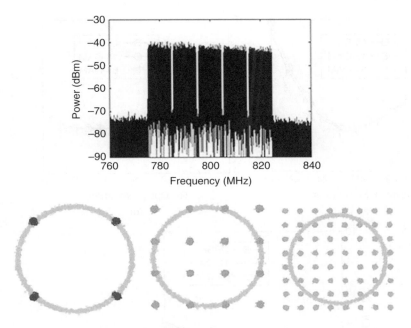

Figure 8.11 Received spectrum and constellation diagrams of the UL LTE-A signals.

The performance of the UL LTE-A signal under different transmit power values, received optical power at the CS, and different transmit power values of the WLAN 802.11ac signal, are shown in Figures 8.12a–c, respectively. We should note that for the UL transmission, the transmit power range in which the performance is better than the requirement is very important because it determines the range of the received power at the RRHs when mobile terminals move around. Figure 8.12a shows that this range for a 64 QAM signal is approximately 20 dB, which is sufficiently high for a UL transmission. Similarly, the dynamic range of the received optical power at the receiver is essential because it reflects the attenuation range in the optical transport networks due to various fiber lengths and the insertion of a splitter in passive optical networks (PONs). According to the PON standard, this range should be larger than 15 dB to satisfy the UL transmission requirement [ITU - 2010]. Figure 8.12b shows that a range of approximately 22 dB is achieved for a 64-QAM LTE-A signal transmission. We should note that, in the experiment, the maximum received power is limited to approximately 6 dBm to prevent damage to the optical receiver. Using another RoF module with higher input power can help to further increase the range of the received optical power, which shows the possibility of applying PONs for the optical link in the UL transmission. Figure 8.12c shows the effect of simultaneous transmission of multiple signals on the performance of the LTE-A signal. As shown in the figure, increasing the transmit power of the co-transmitting WLAN signal affects the signal performance. However, the influence is relatively small, and the performance is still much better than

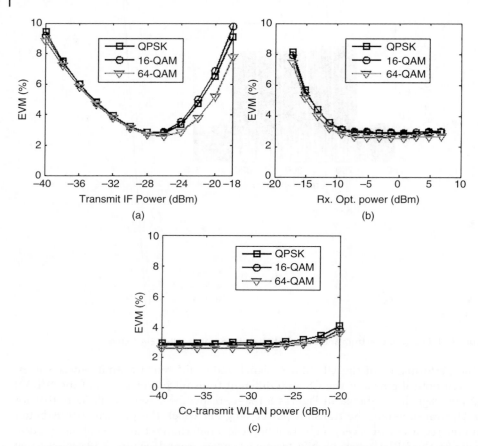

Figure 8.12 Performance of the UL LTE-A signal as a function of: (a) transmit power, (b) received optical power, (c) co-transmitted 802.11ac signal power.

the requirements. Figure 8.13 shows the performance of the WLAN 802.11ac signal. By setting the transmit power in the optimal range, as shown in Figure 8.13a, we can successfully transmit a 256-QAM signal over the UL system. The received optical power range for the 256-QAM signal is approximately 19 dB, as shown in Figure 8.13b, which is higher than the requirement for a PON system. The effect of nonlinear distortion due to a simultaneous transmission with the LTE-A signal is shown in Figure 8.13c, which indicates some effects on the performance of the 802.11ac signals when the transmit power of the LTE-A signal is increased. However, by transmitting appropriate power levels for both LTE-A and 802.11ac signals, the simultaneous transmission over the UL system is possible even for signals using high-order modulation of the 256 QAM.

8.3.3 MmWave Link Distance

The transmission distance of the radio-wave link is an important parameter in many practical applications. In the experiments, the mmWave link is set at 5 m because of the

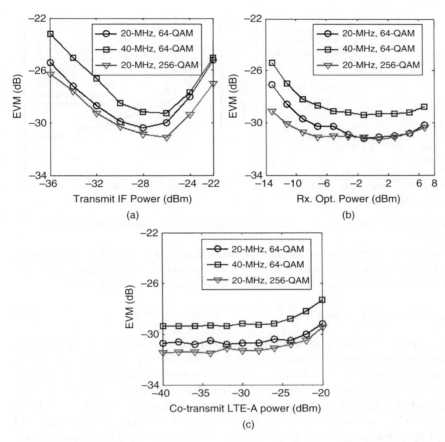

Figure 8.13 Performance of the 802.11ac signal under: (a) transmit power, (b) received optical power, (c) co-transmitted LTE-A signal.

space limitation in the experiment room. However, the distance can be further extended provided that the transmission performance still satisfies the requirements. We investigate the performance of the 20 MHz 802.11ac and 3-CC LTE-A signals for different transmitting powers of the mmWave signals in the DL system, as shown in Figure 8.14. The measurement is performed by changing the transmit power of the mmWave signal at the input of the transmit antenna. From the results, the allowed minimum received power, and thus the transmission range can be estimated for different wireless signals. For instance, a minimum power of approximately −34 and −41 dBm must be received to demodulate the 20 MHz 256-QAM 802.11ac and 64-QAM LTE-A signals successfully, respectively. These received power values correspond to a maximum transmission distance of approximately 7 and 14 m, respectively, using the setup in the experiments. For low-order modulation signals, the transmission distance is longer owing to the less stringent requirement for the received EVM. The transmission distance can be further extended to a sufficiently long distance to meet practical applications using high-gain antennas and high-output PAs.

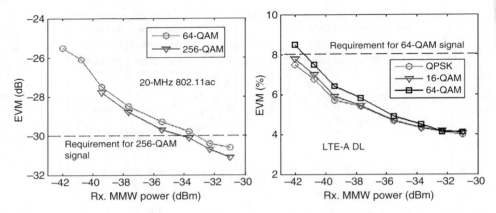

Figure 8.14 Performance of the 802.11ac and DL LTE-A signals versus received mmWave power.

We estimated the transmission distance for different antenna gains and transmit powers on the basis of the minimum allowed received power shown in Figure 8.14. The received power after transmission over the mmWave link can be calculated by

$$P_{RX} = P_{TX} + G_{TX} - 20 \times \log \left(\frac{4 \cdot \pi \cdot d \cdot f}{c} \right) - (L + \gamma_R) \times d + G_{RX} - M, \qquad (8.7)$$

where G_{TX} and G_{RX} are the gains of transmitter and receiver antennas, respectively. L and γ_R are the atmospheric and rain attenuation coefficients, respectively, P_{TX} is the transmit power, d and f are the distance and frequency of the mmWave link, respectively, and M is the link margin. The transmission distance depends on the designed link margin and the amount of attenuation due to rainfall. It is related to the availability of the system, which can be determined using statistical rainfall data over a long period in a specific operation area [Dat et al. - 2015]. Availability is an important parameter for a mobile backhaul/fronthaul system, and a ratio of approximately 99.999% is required [Orange et al. - 2012]. For example, a rainfall rate of more than 120 mm h^{-1}, which corresponds to a 36.9 dB km^{-1} attenuation, is approximately 0.0001% over a long-term observation in Japan [Hirata et al. - 2006]. With a required availability of 99.999%, the system must be designed to overcome an attenuation of at least 36.9 dB km^{-1}. By inputting the allowed minimum received power shown in Figure 8.14 and the determined rain attenuation to 8.7, we can estimate the transmission distance under different antenna gains and transmit powers, as shown in Figure 8.15. In this calculation, a link margin of 10 dB is reserved for compensation for other effects, such as fading and fog attenuation. An mmWave link distance of approximately 1 km can be achieved for transmission of the 64-QAM LTE-A signal over the DL direction if a high power of 30 dBm power is transmitted and high-gain antennas are used. The link distance for the transmission of the 256-QAM 802.ac signal is shorter because of the higher received power requirement, as shown in Figure 8.14.

Similar to the DL system, the performance of the LTE-A and 802.11ac signals for different transmit powers of the mmWave signal in the UL system is shown in Figure 8.16. Using the figure, we can determine the minimum received the power of the mmWave

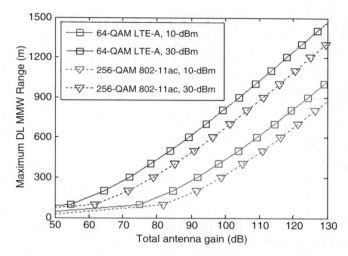

Figure 8.15 Maximum mmWave range for the LTE-A and 802.11ac signals.

Figure 8.16 Performance of: (a) UL LTE-A, (b) 802.11ac signals under different received mmWave signal powers.

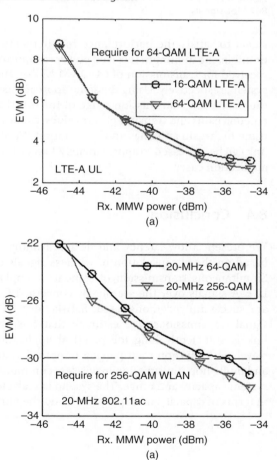

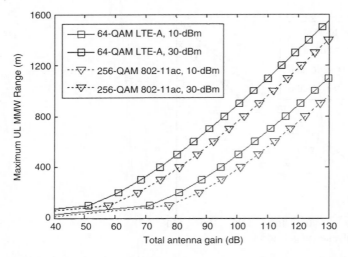

Figure 8.17 Maximum mmWave range of the UL transmission of 64-QAM LTE-A and 256-QAM 802.11ac signals.

signal to satisfy the performance requirements for the different wireless signals. For example, the minimum received powers of approximately −44.5 and −38 dBm should be received for transmission of 64-QAM LTE-A and 256-QAM 802.11ac signals over the system, respectively. Using the same approach as that of the DL system, we can estimate the maximum transmission range of the mmWave link of the UL system for different antenna gains and transmit power values, as shown in Figure 8.17. The figure shows that using high-gain antennas and high-output PA of 30 dBm, the distance of the mmWave link can be increased to approximately 1 km, which is sufficiently long for many practical application cases.

8.4 Conclusion

We present a low-latency and flexible fronthaul for heterogeneous cellular networks based on the encapsulation of wireless signals on a seamless fiber–mmWave system. Simultaneous transmission of standard compliant IEEE 802.11ac and 3GPP DL/UL LTE-A signals over the seamlessly converged RoF and mmWave systems at 94.1 GHz are successfully demonstrated. Satisfactory EVM performance is confirmed for all signal transmissions. We estimate that the transmission distance of the mmWave link is sufficiently long for practical applications. Advanced technologies, such as wavelength division multiplexing in optical systems, arrayed antenna, beam forming, and multi-hop transmission in wireless communication can be exploited to increase the system capacity and range. The system is scalable to meet future deployment demands in terms of capacity, multiple radios, and the number of wireless services.

Bibliography

Evolved Universal Terrestrial Radio Access (E-UTRA); Physical channels and modulation. 3rdGPP, June 2016. E-UTRA Technical specification (TS) Release 13, [Online]. Available: https://portal.3gpp.org/desktopmodules/Specifications/SpecificationDetails.aspx? specificationId=2425.

Hamed Al-Raweshidy and Shozo Komaki. *Radio Over Fiber Technologies for Mobile Communications Networks*. Artech House, 2002.

T. Alves and A. Cartaxo. Transmission of multiband ofdm-uwb signals along lr-pons employing a mach-zehnder modulator biased at the quasi-minimum power transmission point. *Journal of Lightwave Technology*, 30(11): 1587–1594, 2012.

M. Beltran, J. B. Jensen, X. Yu, R. Llorente, R. Rodes, M. Ortsiefer, C. Neumeyr, and I. T. Monroy. Performance of a 60-ghz dcm-ofdm and bpsk-impulse ultra-wideband system with radio-over-fiber and wireless transmission employing a directly-modulated vcsel. *IEEE Journal on Selected Areas in Communications*, 29(6):1295–1303, 2011.

China Mobile Research Institute. C-ran: The road towards green ran. White paper, China Mobile, 2011.

P. T. Dat, A. Kanno, K. Inagaki, and T. Kawanishi. High-capacity wireless backhaul network using seamless convergence of radio-over-fiber and 90-ghz millimeter-wave. *Journal of Lightwave Technology*, 32(20):3910–3923, 2014.

P. T. Dat, A. Kanno, and T. Kawanishi. Radio-on-radio-over-fiber: efficient fronthauling for small cells and moving cells. *IEEE Wireless Communications*, 22(5):67–75, 2015.

P. T. Dat, A. Kanno, N. Yamamoto, and T. Kawanishi. Low-latency fiber-millimeter-wave system for future mobile fronthauling. In *Proc. SPIE 9772, Broadband Access Communication Technologies X, OPTO Photonic West, 97720D (12 February 2016)*, pages 1–12, 2016a.

P. T. Dat, A. Kanno, N. Yamamoto, and T. Kawanishi. Full-duplex transmission of lte-a carrier aggregation signal over a bidirectional seamless fiber-millimeter-wave system. *Journal of Lightwave Technology*, 34(2):691–700, 2016b.

P. T. Dat, A. Kanno, and T. Kawanishi. Performance of uplink packetized lte-a signal transmission on a cascaded radio-on-radio and radio-over-fiber system. *IEICE Transactions on Electronics*, E98.C(8): 840–848, 2015.

Common Public Radio Interface (CPRI): Interface Specification. Ericsson AB and Huawei Technologies Co. Ltd and NEC Corporation and Alcatel Lucent and and Nokia Siemens Networks GmbH & Co. KG, September 2011. The CPRI Specification version 5.0, [Online]. Available: http://www.cpri.info/downloads/CPRI__&c.caron;5_0_2011-09-21 .pdf.

H. T. Friis. A note on a simple transmission formula. *Proceedings of the IRE*, 34(5):254–256, 1946.

A. Hirata, T. Kosugi, H. Takahashi, R. Yamaguchi, F. Nakajima, T. Furuta, H. Ito, H. Sugahara, Y. Sato, and T. Nagatsuma. 120-ghz-band millimeter-wave photonic wireless link for 10-gb/s data transmission. *IEEE Transactions on Microwave Theory and Techniques*, 54(5):1937–1944, 2006.

IEEE Std 802.11TM-2012. Iso/iec/ieee international standard - information technology–telecommunications and information exchange between systems local and metropolitan area networks–specific requirements part 11: Wireless lan medium access control (mac) and physical layer (phy) specifications. *ISO/IEC/IEEE 8802-11:2012(E) (Revison of ISO/IEC/IEEE 8802-11-2005 and Amendments)*, pages 1–2798, 2012.

G.987.2 : 10-Gigabit-capable passive optical networks (XG-PON): Physical media dependent (PMD) layer specification. ITU-T, April 2010. Recommendation G.987.2 , [Online]. Available: https://www.itu.int/rec/T-REC-G.987.2-201001-S/en.

J. James, P. Shen, A. Nkansah, X. Liang, and N. J. Gomes. Nonlinearity and noise effects in multi-level signal millimeter-wave over fiber transmission using single and dual wavelength modulation. *IEEE Transactions on Microwave Theory and Techniques*, 58(11):3189–3198, 2010.

A. Kanno, P. T. Dat, T. Kuri, I. Hosako, T. Kawanishi, Y. Yoshida, Y. Yasumura, and K. Kitayama. Coherent radio-over-fiber and millimeter-wave radio seamless transmission system for resilient access networks. *IEEE Photonics Journal*, 4(6):2196–2204, 2012.

A. Kanno, P. T. Dat, T. Kuri, I. Hosako, T. Kawanishi, Y. Yoshida, and K. Kitayama. Evaluation of frequency fluctuation in fiber-wireless link with direct iq down-converter. In *2014 The European Conference on Optical Communication (ECOC)*, pages 1–3, 2014.

T. Kawanishi, T. Sakamoto, and M. Izutsu. High-speed control of lightwave amplitude, phase, and frequency by use of electrooptic effect. *IEEE Journal of Selected Topics in Quantum Electronics*, 13(1):79–91, 2007.

J. Li, D. Chen, Y. Wang, and J. Wu. Performance evaluation of cloud-ran system with carrier frequency offset. In *2012 IEEE Globecom Workshops*, pages 222–226, 2012.

C. Liu, L. Zhang, M. Zhu, J. Wang, L. Cheng, and G. Chang. A novel multi-service small-cell cloud radio access network for mobile backhaul and computing based on radio-over-fiber technologies. *Journal of Lightwave Technology*, 31 (17):2869–2875, 2013.

R. Llorente, S. Walker, I. T. Monroy, M. Beltrán, M. Morant, T. Quinlan, and J. B. Jensen. Triple-play and 60-ghz radio-over-fiber techniques for next-generation optical access networks. In *2011 16th European Conference on Networks and Optical Communications*, pages 16–19, 2011.

A. Nkansah, A. Das, N. J. Gomes, and P. Shen. Multilevel modulated signal transmission over serial single-mode and multimode fiber links using vertical-cavity surface-emitting lasers for millimeter-wave wireless communications. *IEEE Transactions on Microwave Theory and Techniques*, 55(6):1219–1228, 2007.

Open Base Station Architecture Initiative (OBSAI). OBSAI, 2006. OPEN BASE STATION ARCHITECTURE INITIATIVE V2.0, [Online]. Available: http://www.obsai.com/specs/OBSAI_System_Spec_V2.0.pdf.

Orange, Alcatel Lucent, Nokia Siemens Networks, NEC, Huawei, Cisco, and Everything Everywhere. Small cell backhaul requirements. White paper, NGMN Alliance, 2012.

X. Pang, A. Caballero, A. Dogadaev, V. Arlunno, R. Borkowski, J. S. Pedersen, L. Deng, F. Karinou, F. Roubeau, D. Zibar, X. Yu, and I. T. Monroy. 100 gbit/s hybrid optical fiber-wireless link in the w-band (75–110 ghz). *Opt. Express*, 19(25): 24944–24949, Dec 2011.

P. T. Dat, A. Kanno, and T. Kawanishi. High-speed and low-latency front-haul system for heterogeneous wireless networks using seamless fiber-millimeter-wave. In *2015 IEEE International Conference on Communications (ICC)*, pages 994–999, 2015.

M. Zhu, L. Zhang, J. Wang, L. Cheng, C. Liu, and G. Chang. Radio-over-fiber access architecture for integrated broadband wireless services. *Journal of Lightwave Technology*, 31(23):3614–3620, 2013.

9

5G Optical Sensing Technologies

Seedahmed S. Mahmoud*[1], Bernhard Koziol[2], and Jusak Jusak[3]

[1] *Department of Biomedical Engineering, College of Engineering, Shantou University, 243 Daxue Road, Shantou, Guangdong, China*
[2] *Swinburne University of Technology, Victoria Hawthorn, Australia*
[3] *Department of Computer Engineering, Institute of Business and Informatics Stikom Surabaya, 60298, East Java, Surabaya, Jl. Raya Kedung Baruk 98, Indonesia*

9.1   Introduction

Demand for high speed data wireless communication is increasing due to the proliferation of multimedia services and applications in the last couple years. The explosion of this traffic need is also a result of the exponential growth in mobile internet as well as smart devices products that lure customers with numerous features and internet dedicated applications such as video, game and social media. Although the current standard can support the maximum data rate to enable video streaming running steadily, there are some emerging applications like ultra high definition (UHD) video and games as well as remote health-care utilizing HD cameras that require a higher data rate to meet with a certain standard quality. We are now facing a new challenge to enable a speed higher than that provided by fourth generation (4G) cellular communications. Hence the birth of fifth generation (5G) wireless communication networks. The 5G is foreseen to offer approximately a thousand fold capacity and throughput improvement and to achieve reduced latency, very low energy consumption, high scalability and connectivity, and improved security [Andrews et al. - 2014].

The 5G networks are expected to support billions of connected devices and sensors that will generate and exchange data in the order of zettabytes (ZB). Unlike previous generations of cellular networks, the 5G network is envisioned to enable access to massive machine-type communications such as smartphones, smartwatches, appliances, sensors, and remote monitoring devices that are all connected to the internet to form the so-called "Internet of Things" (IoT). There will be a paradigm shift from the current connected people to the connected things concept whereby millions of connected mobile phones and computers will be complemented by billions of devices and sensors that need an internet connection. In this context, the standard body of the International Telecommunication Union (ITU) released three types of services to be supported by 5G including: mobile broadband services, reliable and low latency communications,

* Corresponding Author: Seedahmed S. Mahmoud seedahmed.sharif@gmail.com

Optical and Wireless Convergence for 5G Networks, First Edition.
Edited by Abdelgader M. Abdalla, Jonathan Rodriguez, Issa Elfergani, and Antonio Teixeira.
© 2020 John Wiley & Sons Ltd. Published 2020 by John Wiley & Sons Ltd.

machine-type communications, as mentioned in [ITU-R M.2083-02 Union - 2015]. This ever increasing number of various connected devices pushes the current cellular network architecture to be able to cater for the data rate and area capacity that satisfy the quality of experience expected by users. Therefore, to accommodate this high data rate requirement due to the growing number of devices, researchers in industry and academia have proposed some favorable solutions, including network densification, millimeter wave communications and massive multiple-input-multiple-output (MIMO) among other methods.

Figure 9.1 depicts an imaginary example of a heterogeneous network for a healthcare system that connects a number of nodes with several medical sensors indirectly to a macrocell base station, a core network and the internet sequentially. Both spatial and frequency densification creates more layers of cells than the traditional macrocell network architecture, as in the 4G system and before. As seen in Figure 9.1, portable electrical devices in a healthcare system scenario equipped with several medical sensors are attached to the patients, called an Internet of Medical Things (IMedT) [Jusak and Puspasari - 2015; Jusak et al. - 2016; Jusak and Mahmoud - 2018]. These sensors undertake measurement by recording input signals of all vital signs sensors such as body temperature, blood pressure, pulse rate, and respiration rate sensors, and encode them into the digital form for transmission. In case of fiber optic sensors, measurement is employed by sensing either intensity or phase change in one or more lights beams. Furthermore, the recorded data in the form of analog or digital waveforms is then shortly transmitted to the healthcare provider through the communication networks. Recent development of fiber optic sensors technology has embodied its potential applications and services to serve real time measurement and

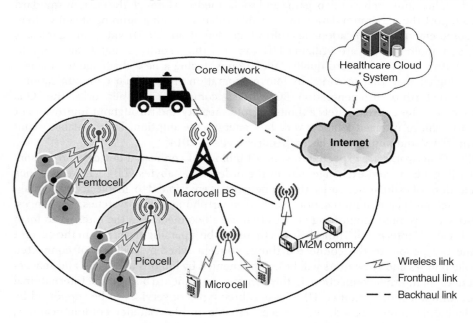

Figure 9.1 Illustration of a multi-tier heterogeneous network in a 5G communication network for a healthcare system.

data acquisition. In particular, this type of sensor exhibits a number of advantages to comply with most 5G requirements, such as high data rates, low energy consumption, light weight, immunity to electromagnetic interference, high sensitivity, low latency, and operated in both point and distribution configuration.

In Figure 9.1, the different sizes of cells consist of femtocells, picocells, and microcells that cover a certain area with various radio access technologies linked to the sensors appropriately forming a multi-tier network to a macrocell. Therefore, these heterogeneous cells create densification in this area. In this scenario, a macrocell base station may be equipped with large antenna arrays utilizing the massive MIMO technology, which consists of hundreds of antennas to increase the capacity, throughput, and spatial degrees of freedom. This structure of antenna arrays plays an important role in providing access to the surrounding smaller cells, which possibly incorporate multiple radio access networks protocols such as 2G, 3G, or 4G networks. Adoption of this densification network architecture is expected to become one of the prominent models in future 5G communication networks [Akyildiz et al. - 2016].

However, it is very likely that amalgamation of the various radio access technologies has to pay the price for the raising of security issues, including increasing threats in sensor nodes or the user equipment side, access networks, core networks, and even in the cloud storage, varying from gaining access to secure services up to data integrity attacks. A number of factors that may contribute to the security hazard arising from the new networs architecture include: (i) utilization of various types of fiber optic sensors available in the market and their interconnecting devices, (ii) limited bandwidth of wireless communication restricts the accomplishment of some security features for instance authentication, integrity, availability, confidentiality, and non-repudiation [Fang et al. - 2017], (iii) integration of several types of radio access networks initiates various malicious threats to the wireless information transmission, (iv) an enormous number of and various types of connected devices with different operating systems complicate the authentication procedure, and (v) implementation of internet protocol (IP) based mobile communication in 5G networks. Therefore, as a consequence of deployment planning for the new architecture, as well as motivation by the above security challenges, there should be parallel actions among researchers in academia and industry to provide secure and reliable access for mobile communication in the new generation 5G networks.

This chapter discusses and reviews recent developments in fiber optic sensors used in technology network security, and their potential applications and services in the context of 5G network development. Important parameter indicators will be discussed to serve as a broad view to the extension of fiber optic sensor requirements for future 5G networks. Different types of fiber optic sensors such as phase modulated sensors, intensity modulated sensors and scattering based sensors will be presented with their application to intrusion detection systems.

Additionally, enabling the communication of sensing devices through 5G networks requires security mechanisms capable of providing efficient and reliable functionality of applications and to ensure resources and critical communications are protected. In fact, security will become one of the biggest issues in 5G network deployment incorporating such IoT devices. Therefore, throughout this chapter we will cover optical fiber based sensors and pay particular attention to optical fiber network physical security solutions.

9.2 Optical Fibre Communication Network: Intrusion Methods

The use of optical fibers as the main backbone of most communication systems including the 5G network can efficiently and cost effectively help in transferring large amounts of information from point to point. Modern fiber optic communication networks deploy optical fiber over millions of kilometers worldwide, carrying important and confidential information of government, military and financial sectors. Although, it was initially thought that optical fiber transmission would be inherently secure, it is now known that it is very easy to tap information out of an optical fiber with negligible interference to the optical signal.

Optical fiber cables are designed to protect the delicate glass fibers within from the stresses and strains of manual installation. A typical optical fiber cable can have a few or hundreds of individual optical fibers [FOS - June 2018; AFL - June 2018; Optical Communications - June 2018], see Figure 9.2.

In order to tap into optical fibers; several methods can be employed:

1. Bend a single optical fiber and bleed the light onto a photodetector, see Figure 9.3. There are commercial units available that can demonstrate this effect [Everett - 2007; Ruppe - 2018].

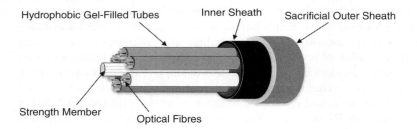

Figure 9.2 A typical optical fiber cable for communication applications.

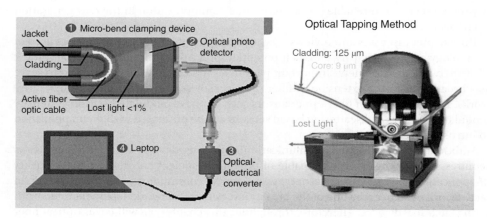

Figure 9.3 Optical tapping method using a microbend device [Murray - 2010] and a commercially available device [Neidlinger - 2014].

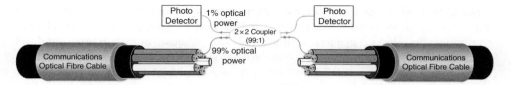

Figure 9.4 Fusion spliced low loss optical coupler in a communications link.

2. Evanescent coupling. This method exploits the cladding modes that are lost in the first meters of laser light entering an optical fiber. This method can only work when the laser is accessible or in distribution racks where connectors are used.
3. Cutting a groove into an optical fiber close to the core of the target optical fiber to intercept evanescent field light and redirecting it into a optical fiber. This method requires high skill so as not to sever the target fiber, especially in field environments) [Iqbal et al. - 2011].
4. Creation of scattering centers within the optical fiber through the use of high powered UV lasers. A high powered laser can impart a large amount of energy into a small volume and change the glass structure (similar to writing a fiber Brag grating) [Iqbal et al. - 2011].
5. Optical splicing of a low ratio coupler, Figure 9.4, to bleed out only 1% of the light, although this method requires the optical fiber to be severed and also requires knowledge from the telecommunications carrier for this to occur [Everett - 2007; Thorlabs - June 2018].

9.3 Physical Protection of Optical Fiber Communication Cables

There are a number of techniques that have been deployed to secure fiber optic communication links from tampering and information theft. Some examples of these methods are:

1. Use of a fiber optic sensor as a physical intrusion detection system [Mahmoud and Katsifolis - 2009, 2010; Mahmoud et al. - 2012].
2. Encryption methods of the transmitted information over the fiber optic communication link [Coomaraswamy et al. - 1991].
3. Utilization of the optical time domain reflectometer (OTDR) to detect reflection and attenuation along the fiber. OTDRs are ineffective at detecting dynamic or transient disturbance to a fiber cable.
4. Monitoring optical power level at discrete points within a fiber optic communication network.
5. Physical security systems where fiber communication cable can be manufactured with a thicker coating and a harder protective jacket on the cables and also housing fiber cables in conduits, metal security ducts, or encasing them in concrete.

This chapter will mainly focus on the first option where spare fibers in the fiber optic communication cables can be used as a fiber optic sensor to detect tampering and intrusion.

Distributed fiber optic sensors have been used in many commercial and defense applications. These sensors have been used to protect assets such as optical fiber communication links, airports, commercial and defense infrastructures, and oil and pipeline systems. A number of underlying sensing technologies can be implemented when designing distributed fiber optic sensors, which include Mach–Zehnder interferometers (MZIs) [Kersey - 1996; Katsifolis and McIntosh - 2009], Michelson interferometers [Kersey - 1996], fiber-Bragg grating arrays [Wu et al. - 2011], Sagnac loops [Zhu and He - 2014] and phase optical time domain reflectometry (Φ-OTDR) [Juarez et al. - 2005; Juarez and Taylor - 2007]. The advantages of using fiber optic sensors in intrusion detection systems over conventional technologies are well recognized and include their immunity to electromagnetic interference, high sensitivity, no power required in the field, intrinsic safety in volatile environments, and high reliability and cost effectiveness over large distances. Optical fiber sensors can be used to measure a myriad of physical parameters [Udd - 1995; Instruments - 2011]; this section however focuses on dynamic strain measurements as a means of detecting illegal tapping on optical fiber communication systems.

Optical fiber sensors can be categorized as either location or zone based systems. Figure 9.5 presents a classification scheme for optical fiber based intrusion detection systems (IDSs) that are used to physically secure communication networks. Location based systems are able to associate an event on the sensor to a location in the temporal domain and thence location can be calculated to meter scale or smaller. Location based systems will locate anywhere along the sensor length. There is however a subset of location based systems, quasi-distributed; this implies that the location of an event on the sensor can be known to within a region of several meters and are often used in fiber Bragg grating based sensors. The range of location based systems is up to 40 km with

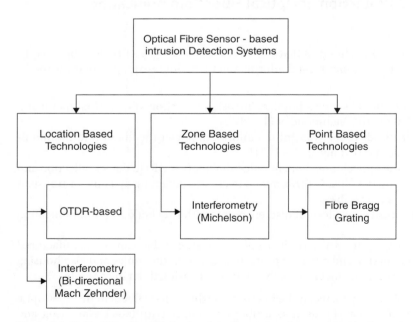

Figure 9.5 Optical fiber based intrusion detection system for a communication network.

some sensors exceeding 40 km [Pradhan and Sahu - 2015]. Location based sensors offer a precise location and are more expensive than zone based systems. Location accuracy depends on the signal processing method. Robust signal processing produces pin-point location accuracy along the sensing length.

Zone based systems on the other hand can also detect intrusion events; however, these are unable to return an accurate location, only an indication that somewhere along the senor activity is occurring. Therefore zone based sensor lengths are often in the range of 100 m to allow for ease of localization; therefore these systems are best utilized in intra or inter building communication links and are more commonly used as such as they are less expensive than distributed sensors.

Point based sensors are sensitive within the sensing element and are best situated in specific locations where access to communication cables is very easy and not necessarily secure (i.e. field buried pits or optical fiber patch racks used in data exchanges). Point sensors are simple to install and can provide excellent sensing capabilities when installed correctly.

9.3.1 Location-Based Optical Fibre Sensors

There are a number of location based optical fiber sensors that can be used to protect 5G communication networks. In this chapter three types of optical fiber sensor (OFS) technologies will be presented. These technologies are: OTDR based sensor, MFI and fiber Bragg grating (FBG) sensor.

9.3.1.1 OTDR Based Sensor

OTDR technology has been used extensively in the telecommunication industry to evaluate the performance of optical fiber links and to find the location of faults such as areas of high attenuation or reflections or complete fiber breaks for immediate repair [Juarez - 2005]. Optical fibers are not homogeneous host scattering centers that also attenuate optical power (i.e. single mode optical fibers have an attenuation of approximately 0.2 dB km^{-1} at 1550 nm). This technology is able to not only detect but also to detect these abnormalities with the following:

$$\text{Location} = \frac{ct}{2n} \tag{9.1}$$

where c is the speed of light in a vacuum, t is the elapsed time and n is the refractive index of the optical fiber. The intensity of detected light is measured and is proportional to the current created from the detection transducer, hence a spatial domain light level (attenuation) plot can be created, from which optical fiber communication health information can be analyzed.

OTDR systems use homodyne optical detection circuits and require a temporal domain sample average to measure accurately the attenuation profile of an optical fiber. When a tap has been attached to an optical fiber it can take up to several minutes to detect the attenuation that the tap has induced on the communications line, dependent on the length of the communication fiber and the parameter settings. Location can also be improved by varying the pulse width; however, care is required to select an appropriate pulse width and averaging time for a given length of optical fiber (i.e. short duration pulses for short lengths and vice versa for long sensor lengths).

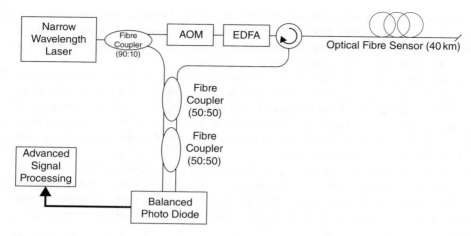

Figure 9.6 Generalized Φ-OTDR configuration for optical fiber intrusion detection.

Coherent OTDR (COTDR) improves the sensitivity of OTDR sensing; it uses a highly coherent light source coupled with optical heterodyne detection. Each scattering point essentially creates an interferometer and this is highly sensitive to dynamic changes in strain. Figure 9.6 shows a general configuration for COTDR.

Polarization induced fading is an issue in Φ-OTDR technology [Zhou et al. - 2013], which is under investigation to be solved [Zhang et al. - 2016]. The ability to use only one end of an optical fiber is also an advantage in this system; however, despite their expense, sensors using this technology are available on the [open market Technology - 2018; Solutions - 2018; Senstar - 2018].

9.3.1.2 Mach–Zehnder Interferometry

Interferometry sensors utilize the phase properties of electromagnetic signals: in summary light splits into two pathways, travels in different directions (or fibers in this case) and are later recombined to measure the phase change $\Delta\varphi$ of light, which in this case a change in light intensity over time [Liang et al. - 2010; Allwood et al. - 2016]. Figure 9.7 shows the MF configuration.

The time domain signal received by the two detectors (Detector 1 and Detector 2) due to disturbance in the communication fiber can be expressed as

$$I_1(t) = I_1\{1 + K_1 \cos[\Delta\varphi(t - t_1) + \varphi_0]\} \tag{9.2}$$

$$I_2(t) = I_2\{1 + K_2 \cos[\Delta\varphi(t - t_2 = t_3) + \varphi_0]\} \tag{9.3}$$

where t_1, t_2 and t_3 are the propagating time through L_1, L_2 and L_3, respectively. I_1 and I_2 are determined by the input intensity of the interferometers and the total loss of the

Figure 9.7 A bi-directional MZ optical fiber sensor configuration.

fiber. K_1 and K_2 represent fringe visibility, which can be approximated to 1 with the inclusion of adequate polarization maintaining techniques that mitigate the effect of polarization induced fading within L_1 and L_2. φ_0 is the initial phase caused by the difference between the two arms of the MZI. There is a time delay difference $\tau = t_2 + t_3 - t_1$ between $I_1(t)$ and $I_2(t)$. Hence, the location of disturbance along the communication fiber can be obtained from the time delay τ [Zyczkowski et al. - 2004]. In the case of a counter-propagating MZ system, the location of an event can be measured using a number of signal processing techniques that estimate the time lag between the two counter-propagating signals. Cross-correlation is one of the methods that can be used to estimate this lag.

Interferometry can also be used to detect changes in strain on the optical fiber and can therefore be used to measure and locate intrusions on communication optical fibers. Also, a unidirectional MZI can be used as a zone-based sensor to secure a short range communication link.

The physical implementation of the sensor as a communication network based intrusion detection system is achieved by housing the two sensing fibers as well as the insensitive lead-out fiber $L_{\text{lead-out}}$ within a single data fiber cable, as shown in Figure 9.7 and can be installed in communication optical fiber cables that contain with three or more fiber cores.

9.3.2 Point-Based OFSs

Point-based OFSs such as FBGs can also be used to protect communication networks.

9.3.2.1 FBGs

FBGs have been used widely within the optical fiber communications industry; they are often used for multiplexing/de-multiplexing communication channels within an optical fiber [Allwood et al. - 2016]. FBGs are created by a written periodic change in refractive index of the fiber glass core; this allows a wavelength range of light to counter-propagate the incident light path, the reflected wavelength can be determined by:

$$\lambda_b = 2n\Lambda \tag{9.4}$$

where λ_b is the reflected wavelength, n is the effective refractive index of the optical fiber core and Λ represents the grating periodicity. FBGs have also been used as strain sensors for infrastructure such as dam walls, bridges, and buildings for pre-emptive maintenance control [Instruments - 2016; Othonos and Kalli - 1999; Kashyap - 2009].

FBGs have the ability to be multiplexed either in time or frequency domains [34]. Time domain multiplexing allows a large number of FBGs to be used within a single continual optical fiber, where the full bandwidth of the instrument is available to each sensor and all sensors can have optical characteristics, simplifying system design. However, each sensor requires a spacing of at least 2 m to allow for resolution between each sensor. The time resolution for a sensor is also dependent on the number of monitored sensors [Allwood et al. - 2016].

Wavelength division multiplexing (WDM) allows FBG sensors to be located closer together as each FGB has a unique central frequency and can be located within a component for multiple point sensing. WDM, however, is limited to the number of monitored sensors as the bandwidth of the light source is divided between each sensor; also,

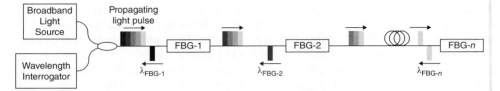

Figure 9.8 A point sensor using FBGs; each FBG is tuned to reflect a unique wavelength range.

cross-talk can exist between FBGs if there is insufficient wavelength spacing between each FBG. Maintenance of this sensor configuration is difficult as each FGB has unique optical characteristics.

FBG sensor design must ensure that the correct multiplexing method is utilized. Even though FBGs are usually used as point sensors, they have been used as quasi-distributed sensors for perimeter intrusion detection [Instruments - 2016; Wu et al. - 2011]; this can easily be applied to data security, especially in buried cables where temperature change is slow. A series of FBGs can be connected along an optical fiber to create a quasi-distributed sensor (Figure 9.8), each FBG reflecting a slightly different wavelength (at least 4 nm), and care must be taken that there is no overlap in wavelength between FBGs [Instruments - 2016; Othonos and Kalli - 1999; Wu et al. - 2011].

An advantage of this system is that this sensor will be able to measure strain up to the furthest illuminated FBG and another advantage is only one end is required to be attached to the light source.

9.3.3 Zone-Based OFSs

A zone-based OFS that can be used to protect communication networks is a Michelson interferometer.

9.3.3.1 Michelson Interferometer

The optical configuration of a Michelson interferometer is shown in Figure 9.9. In this configuration a laser beam contained within an optical fiber splits into two paths via a 50:50 coupler. The beams propagate the length of the fiber, which are the sensor arms. The light within each sensor arm is reflected off Faraday rotatory mirrors at the end of their respective paths and return to the beamsplitter. The light beams from the two paths are summed, where their respective phases interfere within the coupler and the resultant light is detected by a photodetector [Wild and Hinckley - 2008]. The amplitude of the detected light is determined by the total phase difference between the combined

Figure 9.9 A zone based Michelson intrusion detection system.

beams. The output intensity at the detector due to the reflected beams from the mirrors is given by

$$I = \frac{E_0^{\,2}}{2}[1 - \cos(2k(L_1 - L_2))] \qquad (9.5)$$

where $k = 2\pi/\lambda$ is the wavenumber and λ is the wavelength of the laser. L_1 and $L_{(1=2)}$ are the path lengths. In optical fiber based sensing both arms detect different signals from the same event as they propagate through different fibers. Therefore, both arm are sensitive to changes in strain that will impart a change in phase.

Michelson interferometry can be used as an optical fiber based communication link intrusion detector without the benefit of locating the exact location of intrusion along the communication link. The sensor can therefore be used as a zone sensor for a short section of cable, either within a building, between buildings, or between fixed lengths for longer lengths, although more infrastructure would be required to monitor longer lengths [Wild and Hinckley - 2008].

A zone-based Michelson intrusion detection system requires common photonic components, it can also be wavelength multiplexed within existing passive optical fiber telecommunication systems to get the benefit of the existing communication link. Figure 9.9 shows the optical fiber configuration of the Michelson intrusion detection system.

Faraday rotatory mirrors (FRM) are required at the end of each sensing arm to maintain a high fringe visibility, these mirrors rotate the polarization state of the reflected light by $\pi/2$ with respect to the incident light. This arrangement cancels the effect of random changes in birefringence within the optical fibers of the sensing arms, resulting in high SNR values when a strain is applied.

A zone-based Michelson intrusion detection system for communication link security is a low cost solution. It has high sensitivity where intrusion disturbance to the sensor cable creates minute phase differences between the two interfering light beams.

9.4   Design Considerations and Performance Characteristics

To design an intrusion detection system for a communication network there are a number of design considerations that should be addressed. Also the robustness of any intrusion detection system is measured by three performance parameters. In this section, these performance parameters and design considerations will be discussed.

9.4.1   Performance Parameters

The success of any intrusion detection system depends on three important performance parameters: the probability of detection (POD), the nuisance alarm rate (NAR), and the false alarm rate [Mahmoud and Katsifolis - 2009, 2010; Mahmoud et al. - 2012]. The POD provides an indication of a system's ability to detect an intrusion within the protected area. A nuisance alarm is any alarm that is generated by an event that is not of interest. Nuisance alarms are typically generated by environmental conditions such as rain, wind, snow, wildlife, and vegetation, as well as man-made sources such as traffic crossings, industrial noise and other ambient noise sources. While increasing the sensitivity of a system increases its POD, it also increases its sensitivity to nuisance events. Basic event detection algorithms with little event discrimination capability that are applied to a wide range of intrusion events can suffer from this POD to nuisance trade-off in the form of increased NARs or decreased POD. Therefore, advanced signal processing algorithms that can improve the POD and eliminate nuisance alarms are crucial in communication network intrusion detection systems [Mahmoud and Katsifolis - 2009, 2010; Mahmoud et al. - 2012].

Vulnerability to defeat is another measure of the effectiveness of fiber optic sensors and system design. Since there is no single sensor that can reliably detect all types of intrusion yet still have an acceptably low NAR, the potential for defeat can be reduced by designing overlapping sensor coverage using multiple units of complementary technologies. Each of these three performance characteristics will vary according to the technology selected and the unique site conditions.

9.4.2   The Need for Robust Signal Processing Methods

The use of fiber optic sensors in telecommunication network security is prone to nuisance alarms. In this application, there exists a performance trade-off between the POD and NAR [Tarr and Leach - 1998]. The most important challenge of distributed fiber optic intrusion detection systems is to minimize the NAR without compromising the system sensitivity or probability of detection for a wide range of operating environments.

In any sensing system, a nuisance alarm can be defined as an alarm caused by an event that is not of interest for that sensing system [Garcia - 2008]. For outdoor systems such as a fiber optic communication IDS between buildings, this includes the elimination of nuisance alarms caused by non-intrusion events, such as nearby construction, vehicular traffic, and other environmentally related non-intrusion events. Nuisance alarms can adversely affect the performance of intrusion detection systems, as well as the confidence of the system operator. The minimization of the nuisance alarm rate of intrusion detection systems, and indeed of any sensing system, is therefore critical for its successful performance and confidence of operation.

An important part of nuisance alarm handling involves being able to recognize and discriminate between nuisance events and valid intrusion events. Employing advanced signal recognition and discrimination techniques is a very effective way of minimizing the effects of nuisance alarms without compromising system sensitivity or the probability of detection. A number of different signal processing techniques can be used to achieve this and can range from simple filtering techniques, to adaptive filtering techniques, to a number of time-frequency analyses [Madsen et al. - 2007; Egorov et al. - 1998; Griffin and Connelly - 2004]. The crux of all event recognition and discrimination techniques is the signal classification process, which involves extracting and identifying unique features in event signals [Lee and Stolfo - 2000]. The event signals may represent isolated individual events (for example intrusion, rain, wind or traffic), or a number of events occurring simultaneously (for example, an intrusion event during environmental noise or man-made construction). In this latter case of simultaneously occurring events, an effective technique for extracting the event of interest from the event of non-interest was developed in [Mahmoud and Katsifolis - 2009, 2010].

A typical intrusion detection and classification system consists of a preprocessing stage that extracts unique features from the detected event, and a classifier that assigns

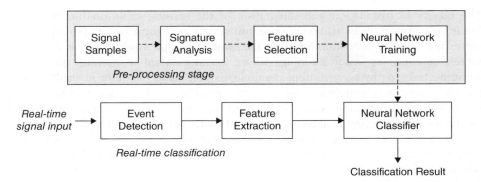

Figure 9.10 Example of an event classification system for optical fiber intrusion detection system [Mahmoud et al. - 2012].

the computed features to a particular class of intrusion or nuisance, as shown in Figure 9.10 [Mahmoud et al. - 2012]. Feature samples of nuisance and intrusion events are used to train a classifier offline in the case of the neural network example, and when training is complete the system will classify new instances based on what was learnt in the training phase. Accurate nuisance and intrusion event classification requires both features that are highly discriminative with respect to the classes of interest and a classifier that can form arbitrary boundaries in the feature space. There still huge demand for robust classification algorithms, especially for the COTDR technology.

9.4.3 System Installation and Technology Suitability

For a communication network intrusion detection system, the implementation of the sensor can be part of the existing optical fiber data cable. The dark spare fiber can be configured in either MZ, Michelson or COTDR configurations. Figure 9.11 shows the fiber optic intrusion detection system for a telecommunication network.

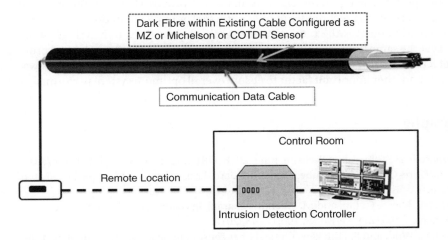

Figure 9.11 Fibre optic intrusion detection system for a telecommunication network.

The quality of installation and integration of the sensing cable on the network cable is very important in achieving optimum system performance and is very often underestimated. The telecommunication intrusion detection system can be installed between buildings or inside buildings where it can be used to monitor an entire communications pathway, including manhole covers, cable conduits in vaults, and multi-ducts between manhole locations. The implementation of the telecommunication network intrusion detection system in such noisy or hostile environments, however, presents some interesting challenges that need to be overcome in order to achieve acceptable performance. In all outdoor perimeter intrusion detection systems including communication networks, there exists a performance trade-off between the POD and NAR [Lee and Stolfo - 2000].

The location based telecommunication network intrusion detection system provides preventative, real-time and location of third party interference, and illegal tapping of data. The location accuracy can vary from ±20 m to few meters, depending on the sensor technology used. When the intrusion detection system raises an alert, the network operator can make informed decisions on shutting down or re-routing the network prior to a potential security breach. The system should also provide an advance alarm/alert of many naturally occurring events, often before critical infrastructure is damaged or data compromised such as land slides and flooding. With a plethora of optical fiber sensing technologies available in today's market there is a solution for many situations and circumstances to enhance data protection in a 5G future.

9.5 Conclusions

This chapter has described the future network architecture in 5G communication systems and the recent development of fiber optic sensors to provide not just only secure but also reliable access for the new mobile communication systems. The chapter mainly focused on the fiber optic sensors for intrusion detection systems and technology to cover security issues that may arise in the new network architecture. Design considerations and performance parameters were discussed briefly. However, the need for advance signal processing techniques in fiber optic sensor implementation remain a big issue, particularly urgent is the requirement for robust algorithms to handle nuisance alarms that at the same time are able to maintain high system sensitivity and signal probability detection. Finally, proper integration of fiber optic sensors in the networks was highlighted as another important factor to ensure overall optimum system performance.

Bibliography

AFL. Standard loose tube cable. Website, June. 2018. URL https://www.aflglobal.com/au/ Products/Fibre-Cable/Loose-Tube-Cable/Standard-Loose-Tube-Cable.aspx. last visited: June 2018.

I. F. Akyildiz, S. Nie, S. C. Lin, and M. Chandrasekaran. 5g roadmap: 10 key enabling technologies. *Computer Networks*, 106:17–48, 2016.

G. Allwood, G. Wild, and S. Hinckley. Optical fiber sensors in physical intrusion detection systems: A review. *IEEE Sensors Journal*, 16(14):5497–5509, 2016.

J. G. Andrews, S. Buzzi, W. Choi, S. V. Hanly, A. Lozano, A. C. K. Soong, and J. C. Zhang. What will 5g be? *IEEE Journal on Selected Areas in Communications*, 32(6):1065–1082, 2014.

G. Coomaraswamy, S. P. R. Kumar, and M. E. Marhic. Fiber-optic lan/wan systems to support confidential communication. *Computers & Security*, 8(10): 765–776, 1991.

S. A. Egorov, A. N. Mamaev, I. G. Likhachiev, and Y. A. Ershov. Advance signal processing method for interferometric fibre-optic sensors with straightforward spectral detection. In *Proceedings of SPIE 3201, Sensors and Controls for Advanced Manufacturing*, pages 44–48, Pittsburgh, PA, jan 1998.

B. Everett. Tapping into fibre optic cables. *Network Security*, 5(5):13–16, 2007.

D. Fang, Y. Qian, and R. Q. Hu. Security for 5g mobile wireless networks. *IEEE Access*, 6:4850–4874, 2017.

FOS. Sacrificial sheath loose tube cable. Website, June. 2018. URL https://www.fibreoptic.com.au/product/sacrificial-sheath-loose-tube-cable/. last visited: June 2018.

M. L. Garcia. *The design and evaluation of physical protection systems*. Butterworth-Heinemann, Burlington, 2 edition, 2008. ISBN 9780750683524.

B. Griffin and M. J. Connelly. Digital signal processing of interferometric fibre-optic sensors. In *Proceedings of the IEEE Light-wave Technologies in Instrumentation and Measurement Conference*, pages 153–156, Palisades, NY, December 2004.

National Instruments. Overview of fiber optic sensing technologies. Website, 6 2011. URL http://www.ni.com/white-paper/12953/en/. last visited: June 2018.

National Instruments. Fundamentals of fibre bragg grating (fbg) optical sensing. Website, 1 2016. URL http://www.ni.com/white-paper/11821/en/. last visited: June 2018.

M. Z. Iqbal, H. Fathallah, and N. Belhadj. Optical fiber tapping: Methods and precautions. In *High Capacity Optical Networks and Enabling Technologies (HONET)*, pages 164–168, Riyadh, Saudi Arabia, December 2011.

J. Juarez. *Distributed Fiber Optic Intrusion Sensor System for Monitoring Long Perimeters*. PhD thesis, Graduate Studies of Texas A&M University, 8 2005.

J. C. Juarez and H. F. Taylor. Field test of a distributed fiber-optic intrusion sensor system for long perimeters. *Applied Optics*, 46(11):1968–1971, 2007.

J. C. Juarez, E. W. Maier, K. N. Choi, and H. F. Taylor. Distributed fiber-optic intrusion sensor system. *Journal of Lightwave Technology*, 23(6):2081–2087, 2005.

J. Jusak and S. S. Mahmoud. A novel and low processing time ecg security method suitable for sensor node platforms. *International Journal of Communication Networks and Information Security*, 10(1):213–222, 2018.

J. Jusak and I. Puspasari. Wireless tele-auscultation for phonocardiograph signal recording through the zigbee networks. In *IEEE Asia Pacific Conference on Wireless and Mobile (APWiMob)*, pages 95–100, Bandung, Indonesia, August 2015.

J. Jusak, H. Pratikno, and V. H. Putra. Internet of medical things for cardiac monitoring: paving the way to 5g mobile networks. In *IEEE Int. Conference on Communication, Networks and Satellite (COMNETSAT 2016)*, pages 75–79, Surabaya, Indonesia, December 2016.

R. Kashyap. *Fiber Bragg Gratings*. Academic Press, San Diego, CA, USA, 2 edition, 2009. ISBN 9780123725790.

J. Katsifolis and L. McIntosh. Apparatus and method for using a counter-propagating signal method for locating events. *U.S. Patent*, (7499177), 2009.

A. D. Kersey. A review of recent developments in fiber optic sensor technology. *Optical Fiber Technology*, 36(2):291–317, 1996.

W. Lee and S. Stolfo. A framework for constructing features and models for intrusion detection systems. *ACM Transaction on Information and System Security*, 3(4):227–261, 2000.

S. Liang, C. Zhang, B. Lin, W. Lin, Q. Li, X. Zhong, and L. J. Li. Influences of semiconductor laser on mach-zehnder interferometer based fiber-optic distributed disturbance sensor. *Chinese Physics B*, 19(12), 2010.

C. K. Madsen, T. Bae, and T. Snider. Intruder signature analysis from a phase-sensitive distributed fibre-optic perimeter sensor. In *Proceedings of SPIE 6770, Fiber Optic Sensors and Applications*, pages 67700K–1–67700K–8, Boston, MA, oct 2007.

S. Mahmoud and J. Katsifolis. A real-time event classification system for a fibre-optic perimeter intrusion detection system. In *Proc. SPIE 7503, 20th International Conference on Optical Fibre Sensors*, page 75031P-1–75031P-4, Edinburgh, UK, October 2009.

S. Mahmoud and J. Katsifolis. Performance investigation of real-time fiber optic perimeter intrusion detection systems using event classification. In *IEEE International Carnahan Conference on Security Technology*, page 387–389, San Jose, CA, December 2010.

S. S. Mahmoud, Y. Visagathilagar, and J. Katsifolis. Real-time distributed fiber optic sensor for security systems: Performance, event classification and nuisance mitigation. *Photonic Sensors*, 2(3):225–236, 2012.

K. D. Murray. Fiber optics easier to wiretap than wire. Website, 11 2010. URL https:// spybusters.blogspot.com/2010/11/fiber-optics-easier-to-wiretap-than.html. last visited: June 2018.

S. Neidlinger. Security in optical networks- useless or necessary? Website, 7 2014. URL https://www.slideshare.net/ADVAOpticalNetworking/security-optical-networks. last visited: June 2018.

Corning Optical Communications. High fibre count optical fibres. Website, June. 2018. URL https://ecatalog.corning.com/optical-communications/AU/en/. last visited: June 2018.

A. Othonos and K. Kalli. *Fiber Bragg Gratings: Fundamentals and Applications in Telecommunications and Sensing*. Artech House, Boston, MA, USA, 1999. ISBN 9780890063446.

H. S. Pradhan and P. K. Sahu. A survey on the performances of distributed fiber-optic sensors. In *Proceedings of the International Conference on Microwave, Optical and Communication Engineering (ICMOCE)*, pages 243–246, Bhubaneswar, India, December 2015.

Zhu Q. and Ye W. Distributed fiber-optic sensing using double-loop sagnac interferometer. *9th IEEE Conference on Industrial Electronics and Applications*, page 499–503, 2014.

J Ruppe. Fiber optic tapping - tapping setup. Website, 2018. URL https://www.joshruppe .com/fiber-optic-tapping-tapping-setup. last visited: June 2018.

Senstar. Fiberpatrol® pipeline. Website, 2018. URL https://senstar.com/products/ fiberpatrol-pipeline/. viewed: 3 March 2018.

Fotech Solutions. Smart security management. Website, 2018. URL http://www.fotech .com. viewed: 3 March 2018.

S. Tarr and G. Leach. The dependence of detection system performance on fence construction and detector location. In *IEEE International Carnahan Conference on Security Technology*, pages 196–200, Alexandra, VA, October 1998.

Future Fibre Technology. Website, 2018. URL http://www.fftsecurity.com. viewed: 3 March 2018.

Thorlabs. Single mode fused fiber optic couplers/taps, 1550*nm*. Website, June. 2018. URL https://www.thorlabs.com/newgrouppage9.cfm?objectgroup_id=9152. last visited: June 2018.

E. Udd. An overview of fiber-optic sensors. *Review of Scientific Instruments*, 66 (8):4015–4030, 1995.

International Telecommunication Union. Imt vision - framework and overall objectives of the future development of imt for 2020 and beyond. *International Telecommunication Union*, 9:1–19, 2015.

G. Wild and S. Hinckley. Acousto-ultrasonic optical fiber sensors: Overview and state-of-the-art. *IEEE Sensors Journal*, 8(7):1184–1193, 2008.

H. J. Wu, Y. J. Rao, C. Tang, Y. Wu, and Y. Gong. A novel fbg-based security fence enabling to detect extremely weak intrusion signals from nonequivalent sensor nodes. *Sensors and Actuators A: Physical*, 167(2):548–555, 2011.

X. Zhang, J. Zeng, Y. Shan, Z. Sun, W. Qiao, and Y. Zhang. Polarization-relevance noise compensation for an ϕ-otdr based optical communication network maintenance system. In *Proceedings of the 15th International Conference on Optical Communications and Networks (ICOCN)*, pages 243–246, Hangzhou, China, 9 2016.

J. Zhou, Z. Pan, Q. Ye, H. Cai, R. Qu, and Z. Fang. Characteristics and explanations of interference fading of a ϕ-otdr with a multi-frequency source. *Journal of Lightwave Technology*, 31(17): 2947–2954, 2013.

M. Zyczkowski, M. Szustakowski, N. Palka, and M Kondrat. Proc. spie 5611, unmanned/unattended sensors and sensor networks. In *Proc. SPIE 5611, Unmanned/Unattended Sensors and Sensor Networks*, pages 71–78, London, UK, November 2004.

10

The Tactile Internet over 5G FiWi Architectures

Amin Ebrahimzadeh, Mahfuzulhoq Chowdhury, and Martin Maier*

Optical Zeitgeist Laboratory, Institut National de la Recherche Scientifique (INRS), Montréal, Canada

10.1   Introduction

Today's telecommunication networks enable people and devices to exchange a tremendous amount of audiovisual and data content. With the advent of commercially available haptic/tactile sensory and display devices and conventional triple-play (i.e. audio, video, and data) content communication now extends to encompass the real-time exchange of haptic information (i.e. touch and actuation) for the remote control of physical and/or virtual objects through the internet. This paves the way towards realizing the so-called *Tactile Internet* (TI) [Simsek et al. - 2016], whereby human–machine interaction will convert today's content delivery networks into skillset/labor delivery networks [Aijaz et al. - 2017]. The TI holds great promise to have a profound socio-economic impact on a broad array of applications in our everyday life, ranging from industry automation and transport systems to healthcare, telesurgery, and education. In most of these industry verticals, very low latency and ultra-high reliability are key for realizing immersive applications such as robotic teleoperation.

In real-time cyber-physical systems (CPSs) including virtual and augmented reality, an extremely low round-trip delay of approximately 1–10 ms is required. Current cellular and WLAN systems miss this target by at least one order of magnitude. According to Aptilo Networks, despite the ongoing competition between LTE and WiFi, the two technologies are, in fact, complementary, as the latency being roughly ten times less in WiFi leads to a higher user experience, whereas LTE provides long-range outdoor coverage (see http://www.anpdm.com). End-to-end latency measurements in an LTE network in a dense urban environment for a low-mobility scenario with a proprietary application running on an android smartphone showed an average end-to-end delay of roughly 47 ms and 54 ms for low and high cell load scenarios [Schulz et al. - 2017]. During peak hours in the afternoon, this figure increased to 85 ms. This demonstrates that public cellular network performance in terms of latency is far from the 1–10 ms requirement of the TI. In addition, rather than focusing only on the average delay, one should

---

* Corresponding Author: Martin Maier maier@emt.inrs.ca

Optical and Wireless Convergence for 5G Networks, First Edition.
Edited by Abdelgader M. Abdalla, Jonathan Rodriguez, Issa Elfergani, and Antonio Teixeira.
© 2020 John Wiley & Sons Ltd. Published 2020 by John Wiley & Sons Ltd.

take into consideration the exact delay experienced by a packet. That is, even if an average round-trip delay of 1–10 ms is achieved, it is not guaranteed that the experienced delay by an arbitrary packet does not exceed a given upper limit.

A round-trip delay of 1–10 ms can transfer today's mobile broadband experience into the new world of the TI. In addition to voice and data communications, current cellular networks enable real-time access to richer content and enable early applications of machine-to-machine (M2M) or machine type communication (MTC). After machines become connected, the next step is the remote control of them. This, in turn, will create a totally new paradigm of control communications to steer objects in our surroundings. A round-trip latency of 1–10 ms along with carrier-grade availability will enable the TI to steer/control real and/or virtual objects. However, we note that the TI will amplify the differences between machines and humans. Besides lowering latency and jitter, another key challenge little discussed in existing TI surveys is how we can make sure that the potential of the TI can be unleashed for a race with (rather than against) machines. The ultimate goal of the TI should be the production of new goods and services by means of empowering rather than automating machines that complement humans rather than substitute for them.

In order to better understand the TI, it is helpful to compare it to the Internet of Things (IoT) and 5G mobile networks and elaborate on their commonalities and subtle differences. Figure 10.1a shows the revolutionary leap of the TI according to a recent ITU-T Technology Watch Report [Maier et al. - 2016]. The high availability and security, low response times, and carrier-grade reliability of the TI will add a new dimension to human–machine interaction by enabling tactile and haptic sensations. In addition, future 5G networks will have to be able to cope with a rapid growth in mobile data traffic as well as the large amount of data from the smart devices that will power the IoT. Toward this end, the 5G vision anticipates one thousand fold gains in area capacity, 10 Gbps peak data rates, and connections for billions of devices. The main challenge of 5G wireless access and core network architectures is to realize machine centric applications, which are not addressed by current cellular networks. Some of these envisioned 5G applications require very low latency and ultra-high reliability with guaranteed availability. Therefore, in addition to very low latency, 5G should enable connectivity, with a higher reliability than that of current radio access networks. Unlike the previous four generations, 5G aims to be highly integrative, which, in turn, will lead to an increasing integration of cellular and WiFi technologies and standards. Another important aspect of the 5G vision is decentralization by evolving the cell centric architecture into a device centric one and exploiting edge intelligence at the device side.

Obviously, the discussion above indicates that there is a notable amount of overlap between IoT, 5G, and the TI, although each one of them has its own unique features. For illustration, Figure 10.1b presents an overview of the mentioned commonalities and differences through the three lenses of IoT, 5G, and the TI. The main differences can be best expressed in terms of underlying communications paradigms and enabling devices. In particular, IoT mainly relies on M2M communications between smart devices (e.g., sensors and actuators). In co-existence with emerging MTC, 5G will maintain its traditional human-to-human (H2H) communication paradigm for conventional triple-play services (i.e. voice, video, and data) with a growing focus on the integration with other wireless technologies (most notably WiFi) and decentralization.

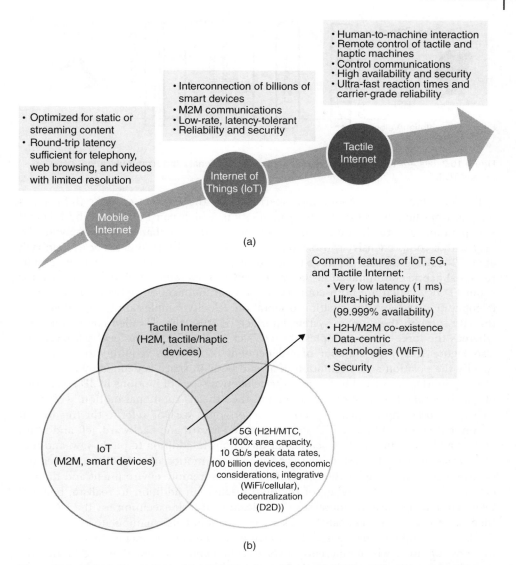

Figure 10.1 (a) Evolutionary leap of the TI; (b) the three lenses of IoT, 5G, and the TI: commonalities and differences [Maier et al. - 2016].

Conversely, the TI will be centered around human-to-machine (H2M) communications focusing on tactile/haptic devices. Moreover, despite their differences, IoT, 5G, and the TI are expected to share the following important design objectives:

- Ultra low latency on the order of 1–10 ms.
- Ultra-high reliability with availability of 99.999%.
- H2H/M2M/H2M coexistence.
- Data centric technologies (e.g., WiFi).
- Security.

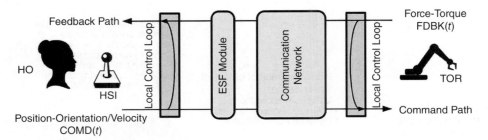

Figure 10.2 Bilateral teleoperation system based on bidirectional haptic communications between HO and TOR.

The TI, which is also commonly referred to as the *Internet of Skills* (IoS), relies on real-time human–robot/machine interaction, which can be realized by bilateral teleoperation, where the quality of experience (QoE) is enhanced by supplying the user with a comprehensive sensory (i.e. audio, video, and tactile) feedback [Steinbach et al. - 2012]. Figure 10.2 depicts a typical bilateral teleoperation system based on bidirectional haptic communications between a human operator (HO) and a teleoperator robot (TOR), which are both connected via a communication network. In a typical teleoperation system[1], the position-orientation/velocity samples are transmitted from the HO through the human–system interface (HSI) in the so-called command path, whereas the force-torque samples are fed back to the HO in the feedback path (see also Figure 10.2]. The number of independent coordinates required to completely specify the position and orientation of a rigid body in space is defined by its degrees of freedom (DoF). A rigid body can experience two types of motions in 3D space, one of which is called translational motion and the other is rotational motion, which are caused by force and torque, respectively. Accordingly, six DoF refer to the freedom of movement of a rigid body in 3D space to move forward and backward, left and right, and up and down in three perpendicular directions in addition to pitch, yaw, and roll. By interfacing with the HSI, the HO commands the motion of the TOR in the remote environment. This couples the HO closely with the remote environment and thereby creates a more realistic feeling of remote presence. In addition to realism, bilateral teleoperation significantly increases the efficiency of task execution for the HO, who then can not only see and hear the task environment but also manipulate it.

Bilateral teleoperation is centered around human perceptions, including, in particular, the sense of touch, which inherently involves action and reaction. Thus, the TI involves the inherent human-in-the-loop (HITL) nature of haptic interaction, thus allowing a human centric design approach via advanced perceptual coding techniques in order to substantially reduce the haptic packet rate. Unlike robots, humans can perceive the change of stimulus only if it exceeds a given threshold. This allows the so-called *deadband coding* technique to effectively reduce the haptic packet rate without a noticeable reduction in transparency. More specifically, with deadband coding, a sample is transmitted only if its relative change with respect to the preceding transmitted sample exceeds a given deadband parameter (usually given in percent). Note that in order to ensure stability and tracking performance, a bilateral teleoperation system requires

---

1 e.g., OMEGA, PHANTOM Premium, VIRTUOSE, FALCON, PHANTOM Omni, SIDEWINDER, da VINCI surgical system.

proper control schemes, which are deployed at local control loops at both HO and TOR sides, as depicted in Figure 10.2.

To meet the aforementioned requirements of very low latency and ultra-high reliability, the performance gains obtained from enhancing coverage centric 4G LTE-Advanced (LTE-A) heterogeneous networks (HetNets) with capacity centric FiWi access networks based on low-cost, data centric ethernet next-generation passive optical network (PON) and Gigabit-class WLAN technologies have been explored recently. Towards realizing immersive and transparent real-time teleoperation, we envision our FiWi network architecture enhanced with multi-access edge computing (MEC) capabilities that leverage edge-intelligence, as depicted in Figure 10.3. The fiber backhaul consists of a time or wavelength division multiplexing (TDM/WDM) IEEE 802.3ah/av 1/10 Gbps ethernet PON (EPON) with a typical fiber range of 20 km between the central optical line terminal (OLT) and remote optical network units (ONUs). The EPON may comprise multiple stages, each stage separated by a wavelength-broadcasting splitter/combiner or a wavelength multiplexer/demultiplexer. There are three different subsets of ONUs. An ONU may serve fixed (wired) subscribers. Alternatively, it may connect to a cellular network base station (BS) or an IEEE 802.11n/ac/s WLAN mesh portal point (MPP), giving rise to a collocated ONU-BS or ONU-MPP, respectively. Depending on his/her trajectory, a mobile user (MU) may communicate through the cellular network and/or WLAN mesh front-end, which consists of ONU-MPPs, intermediate mesh points (MPs), and mesh access points (MAPs). Given the typical WiFi-only operation of state-of-the-art robots [Maier et al. - 2016], HOs and TORs are assumed to communicate only via WLAN, as opposed to MUs using their dual-mode 4G/WiFi smartphones. Teleoperation is done

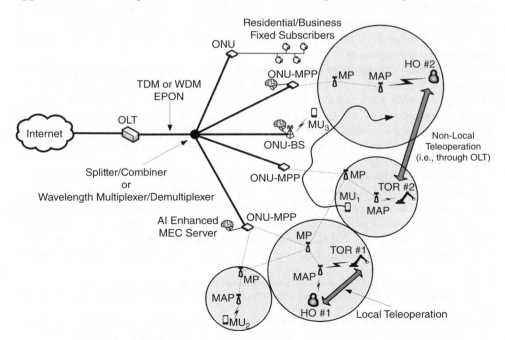

Figure 10.3 Envisioned TI networking infrastructure: FiWi enhanced LTE-A HetNets with AI based MEC capabilities.

either locally or non-locally, depending on the proximity of the involved HO and TOR, as illustrated in Figure 10.3. In local teleoperation, the HO and corresponding TOR are associated with the same MAP and exchange their command and feedback samples through this MAP without traversing the fiber backhaul. Conversely, if HO and TOR are associated with different MAPs, non-local teleoperation is generally done by communicating via the backhaul EPON and central OLT. In addition, we equip selected ONU-BSs/MPPs with artificial intelligence (AI) based MEC servers and place them at the optical–wireless interface (see Figure 10.3] in order to improve the QoE of HOs by forecasting delayed haptic samples in the feedback path (to be discussed in technically greater detail shortly).

Another distinct aspect of the TI is the fact that it should amplify the differences between machines and humans. By building on the areas where machines are strong and humans are weak, TI human-to-robot (H2R) communication leverages on their "ooperative" and "collaborative" autonomy such that humans and robots complement each other. This approach is known as *human-agent-robot teamwork (HART)* [Johnson et al. - 2012; Bradshaw et al. - 2012], whose specific design goal is to keep humans in rather than out of the loop. Historically, HART extends the so-called *humans-are-better-at/machines-are-better-at* (HABA-MABA) approach, which divides up work between humans and machines, whereas HART focuses on how humans and machines could work together. Considering the underlying human–machine interaction of the TI, the main challenge is to orchestrate how tasks can be best shared by both in concert. Collaboration and communication among HART members is important to cope with dynamic changes in the task environment, thereby improving the task execution latency. Note that the interdependent activities of HART may lead to increased complexity and resource consumption. To facilitate cost-effective HART, further research in the area of centralized/decentralized service coordination and resource management is mandatory.

The successful development of HART centric H2R applications necessitates efficient task allocation among robots, taking the different capabilities of robots and specific task requirements, such as robot distance to task location, task execution deadline, and energy consumption of robots into account [Khamis et al. - 2015]. Moreover, the lack of proper task allocation policies and the limited computing, battery, and storage resources of robots may result in an increased task execution delay and excessive energy consumption of selected robots.

In response to the aforementioned challenges, mobile devices/robots increasingly seek assistance from collaborative nodes (i.e. mobile cloud computing[2], device-to-device communications) for executing their computation tasks, a trend also known as cyber-foraging [Patil et al. - 2016] or collaborative computing[3] [Langford et al. - 2013; Zhang et al. - 2013]. Importantly, collaborative computing allows resource-constrained robots to migrate their computation tasks to the collaborative cloud nodes

---

2 Mobile cloud computing (MCC) is an emerging technology in which cloud services (i.e. computing, storage) are utilized to speed up the running of mobile computation and data-intensive applications [Zhang et al. - 2015].

3 Collaborative computing is a task execution framework in which a mobile device exploits the capabilities offered by remote/local cloud servers and nearby devices to execute its own task [Langford et al. - 2013].

(remote cloud or decentralized cloudlet[4]) for execution by means of computation offloading[5]. By enabling collaborative computing among mobile devices and cloud servers, the executable task load on each mobile device is reduced and the lifetime of the mobile device is extended. However, the communication between cloud server and mobile device incurs an additional latency while offloading the computation task. Thus, collaborative computing needs to tackle several challenges for TI H2R application execution such as minimizing task offloading latency, ensuring adequate bandwidth, and a non-disruptive network connection for computation offloading, among others.

Collaborative computing based H2R communications in advanced FiWi based TI infrastructures may offer significant benefits in terms of improved task execution time, cost reduction, and scalability. More interestingly, co-working with robots will favor geographical clusters of local production (inshoring) and will require human expertise in the coordination of the human–robot symbiosis with the objective of inventing new jobs humans can hardly imagine or did not even know they wanted done. FiWi enabled TI H2R communications may be a stepping stone to merging mobile internet, IoT, and advanced robotics with automation of knowledge work and cloud technologies, which together represent the five technologies with the highest estimated potential economic impact in 2025 [Maier et al. - 2016].

The remainder of this chapter is organized as follows. After reviewing and classifying the state of the art, existing challenges, and prior work on the TI, we focus on our proposed FiWi enhanced LTE-A HetNets, on which emerging 5G systems are envisioned to rely. Section 10.2 reviews recent progress and open challenges in realizing the TI vision. We discuss related work in Section 10.3. In Section 10.4, we set forth to present in-depth technical insights into realizing HITL centric teleoperation TI over FiWi enhanced networks, including trace-based haptic traffic modeling, perceptual deadband coding, haptic sample forecasting, and trace-driven simulations. We then address the robot task allocation problem as a use case of the TI in Section 10.5, where we report on our progress toward HART centric collaborative computing and multi-robot task allocation over FiWi enhanced networks. Finally, Section 10.6 concludes the chapter.

10.2 The TI: State of the Art and Open Challenges

To obtain a low round-trip latency, the authors in [Simsek et al. - 2016] emphasize that even at the speed of light (i.e. dedicated optical/wireless path without any intermediate queue), a round-trip propagation delay of 1 ms requires a computing/processing server or cloudlet (i.e. decentralized proxy cloud server with processing and storage capabilities) within 150 km. This computing/processing server at the edge of the mobile radio access network is a central part of the *mobile-edge cloud* concept. It is expected that advanced caching techniques and user-oriented traffic management in addition to AI at

---

4 Cloudlet is a relatively powerful server offering cloud services (computation, processing, and storage) at the network edge in close proximity to mobile users [Zhang et al. - 2015].
5 Computation offloading aims to alleviate resource constraints of mobile devices by migrating part or all of the computation tasks of a running application to powerful servers [Zhang et al. - 2015].

the edge of access networks will improve not only network performance but also QoE in H2R interaction. To realize the vision of TI H2R communications, recently in [Maier et al. - 2016], the authors elaborated on the role of several key enabling technologies such as FiWi enhanced LTE-A HetNets, cloudlets, and cloud robotics. A cloud robotic system architecture that leverages the combination of an ad hoc cloud with M2M communications among participating robots and an infrastructure cloud with machine-to-cloud communications between robots and the remote cloud/decentralized cloudlet play a significant role in extending the capabilities of robots for executing H2R applications.

TI traffic is expected to require the underlying communication networks to undergo profound modifications, both from architectural and medium access control (MAC) viewpoints. As new communication services evolve, more accurate models have to be devised to predict the system performance. Note, however, that despite the growing interest in TI, there is still limited understanding of the characteristics of real TI traffic. For the sake of simplicity and analytical tractability, Poisson traffic has been assumed for modeling H2H and M2M communications. In a recent study, the authors of [Wong et al. - 2017a] considered Poisson and Pareto traffic models to characterize TI traffic. Nevertheless, these assumptions were not verified by real-world haptic teleoperation experiments.

More recently, the authors of [Beyranvand et al. - 2017] proposed the concept of so-called *FiWi enhanced LTE-A HetNets*, which were shown to achieve the 5G and TI key requirements of very low latency and ultra-high reliability by unifying coverage centric 4G mobile networks and capacity centric FiWi broadband access networks based on data centric ethernet technologies. By means of probabilistic analysis and verifying simulations based on recent and comprehensive smartphone traces the authors of [Beyranvand et al. - 2017] showed that an average end-to-end latency of 1 ms can be achieved for a wide range of traffic loads and that mobile users can be provided with highly fault-tolerant FiWi connectivity for reliable low-latency fiber backhaul sharing and WiFi offloading. Note, however, that only conventional H2H communications was considered in [Beyranvand et al. - 2017] without any coexistent H2R or M2M communications. In [Condoluci et al. - 2017], the authors have proposed a soft resource reservation mechanism in cellular networks to achieve low-delay teleoperation over mobile access networks. In doing so, the uplink grant from one transmission is softly reserved for the following transmissions, thus reducing the latency while maintaining the spectral efficiency. Nevertheless, the achieved round-trip delay was roughly 50 ms, thus missing the 1–10 ms target for highly dynamic applications and environments.

The multi-attribute and multi-dimensional nature of haptic communications motivated several research studies on haptic data reduction and compression. Haptic data reduction and compression can be classified into two distinct categories: (i) statistical and (ii) perceptual. Techniques for the former category make use of the statistical properties of haptic signals to achieve efficient data compression, whereas the latter category involves methods that use the perceptual limitation of humans. Specifically, the haptic perceptibility is generally analyzed by using Weber's law of the so-called just noticeable difference (JND), which typically varies between 5% and 15% [Burdea and Brooks - 1996]. This research direction has given rise to a vast variety of haptic data reduction methods, referred to as deadband coding schemes, where the JND is determined by a deadband parameter d. With deadband coding, an update sample is transmitted only if the proportional change with respect to the preceding sample exceeds a given deadband parameter value.

The approaches that address the stringent delay requirements of teleoperation can be classified into three main categories. The first category comprises haptic data reduction techniques that address given delay limits by reducing the packet rate, thus alleviating the network burden [Baran et al. - 2016]. Second, a large variety of control mechanisms have been examined with the objective of guaranteeing stability in the presence of communication uncertainties (interested readers are referred to [Steinbach et al. - 2012] and references therein). By trading off stability and transparency, approaches under the second category were shown to be able to guarantee system stability for delays of up to 200 ms [Li and Kawashima - 2016]. Third, the expected stringent requirements of teleoperation, such as ultra-low latency, user experience continuity, and high reliability, drives the need for highly localized services in close proximity to the HO. Towards this end, edge computing is a new paradigm, in which computing and storage resources – variously referred to as cloudlets, micro datacenters, or fog nodes – are placed at the internet's edge in proximity to wireless end devices in order to achieve low end-to-end latency, low jitter, and scalability. A similar concept, originally known as mobile edge computing (MEC), has been standardized by ETSI for 5G networks. Note that since September 2016, ETSI has dropped the "mobile" out of MEC and renamed it *multi-access edge computing (MEC)* in order to broaden its applicability to heterogeneous networks, including WiFi and fixed access technologies (e.g., fiber) [Taleb et al. - 2017].

To enable HART centric TI applications based on the H2R communications paradigm, human task allocation among robots has attracted significant attention among researchers. The task allocation among robots needs to tackle several challenges including minimization of task execution latency, energy consumption of robots, and failure avoidance [Khamis et al. - 2015]. Recent studies on computation task offloading onto collaborative nodes come in three flavors: (i) system or infrastructure based (e.g., central cloud [Osunmakinde and Vikash - 2014; cloudlet Liu et al. - 2016]), (ii) method based (e.g., application partitioning and code migration to infrastructure-based cloud [Zhang et al. - 2015]), and (iii) optimization based (e.g., minimizing mobile devices energy consumption [Osunmakinde and Vikash - 2014] and task response time [Liu et al. - 2016]) offloading. The aforementioned task offloading studies neither presented any resource allocation scheme for both broadband and task offloading traffic nor considered mobility. Different scenarios need to be investigated, where both collaborative cloud nodes may satisfy or fail to satisfy given task offloading requirements. Furthermore, considering the mobility of end-users and taking the idea of task offloading a step further, task migration has emerged as a promising option to reduce the H2R task execution latency [Gkatzikis and Koutsopoulos - 2013]. As a rule of thumb, task migration is beneficial for TI applications only if the anticipated task execution time at the new server is smaller than that at the current server. Migration provides a more fine-tuned means for balancing the load throughout the system, since migration may take place at any time during the lifetime of a task. There exists an inherent trade-off in selecting the optimal task migration strategy in collaborative computing based H2R task execution such as optimizing the clients' QoE, which for a particular task translates into the minimization of execution time along with reduction of collaborative computing operating costs (e.g., reduced energy consumption by turning off underutilized servers). Hence, the question of how and where a task should migrate to is one of the important research issues for HART centric TI applications. To answer this question, several aspects need to be taken into account, e.g., information about the state of the current host and the tentative destination server, both the number of tasks running on each server but also their interaction, and task migration latency.

10.3 Related Work

The discussion above clearly demonstrates that TI lies at the nexus of robotics, cloud computing, and communications. In this section, we compile the research work related to TI and after classifying them into three separate yet interdependent categories, we discuss each in greater detail. The main branches of our classification are low-latency communications and networking architectures, advanced robotic-enabled teleoperation, MEC strategies. Further, we elaborate on various multi-robot task allocation mechanisms.

Toward realizing TI applications, ensuring the stringent end-to-end packet delay of 1–10 ms as well as the ultra-high reliability of 99.999% place high demands on radio resource allocation and MAC strategies in 5G infrastructures. In [Pilz et al. - 2016], the first TI demonstration of a wireless single-hop communication system was shown to achieve a round-trip delay below 1 ms. A suitable radio resource allocation scheme for haptic communications in LTE-A RAN was presented in [Aijaz - 2016; Khorov et al. - 2017]. In [She and Yang - 2016; She et al. - 2016], an energy-efficient resource allocation design approach for TI over time division duplex cellular networks was investigated. The authors of [Wong et al. - 2016] showed that an average end-to-end packet delay of 200 µs is achievable for TI over TWDM PON with predictive dynamic bandwidth allocation. More recently, feasibility of IEEE 802.11 hybrid coordination function controlled channel access to support TI applications under different system settings was investigated in [Feng et al. - 2017a,b]. The authors of [Wong et al. - 2017a] combined TWDM technologies on a passive optical LAN with a predictive resource allocation algorithm for TI applications.

It is well known that communications induced artifacts such as packet delay may jeopardize the stability and transparency performance of teleoperation systems with geographically separated HOs and TORs. Leveraging on the limitations of human perception, the authors of [Hinterseer et al. - 2008] presented the concept of perceptual deadband coding to reduce the haptic packet rate in teleoperation systems. The authors of [Xu et al. - 2016a] conducted a comprehensive survey on the research conducted on model-mediated control schemes, which are proven beneficial to increase the stability as well as the transparency of time-delayed teleoperation systems.

Given the importance of scaling up research in network integration and convergence in support of MEC enabled 5G, the feasibility and performance gains obtained from MEC over ethernet-based FiWi networks, accounting for both network architecture and enhanced resource management, was examined in [Rimal et al. - 2017a,b,c]. In [Ateya et al. - 2017], a multi-level hierarchical cloud architecture for FiWi based cellular systems for TI was proposed. Further, the authors of [Wong et al. - 2017b] addressed the cloudlet placement problem in FiWi networks from the network planning point of view.

The authors of [Trigui et al. - 2014] proposed a task allocation scheme that considers only the robot distance to a given task location for selection. In [de Mendonça et al. - 2016], a decentralized framework was proposed where tasks of higher priority were allocated before lower-priority tasks, while task re-allocation and failure avoidance were not studied. Neither local (humans and robots residing under the same BSs or MAPs) nor non-local (humans and robots residing under different BSs or MAPs) H2R task allocation has been examined in sufficient detail previously. To improve H2R task execution latency, robots may offload their computation tasks onto collaborative cloud computing servers/nodes for processing. A task allocation strategy that combines suitable host robot selection and computation task offloading onto collaborative nodes was presented in [Chowdhury and Maier - 2017]. A hierarchical auction-based resource allocation for cloud robotics systems in ad hoc networks was presented in [Wang et al. - 2017]. An effective strategy to realize the required security and privacy in an ultra-dense green 5G radio access network for TI was presented in [Szymanski - 2017].

10.4 HITL Centric Teleoperation over AI Enhanced FiWi Networks

As a use case of TI applications that allows for remote control and immersion, we study HITL centric local and non-local teleoperation over FiWi enhanced LTE-A HetNets with AI embedded MEC capabilities. Unlike the mobile internet and IoT, the TI will facilitate *haptic communications* [Aijaz et al. - 2017] by providing the medium for transporting haptic senses (i.e. touch and actuation) in real-time in addition to non-haptic data, video, and audio traffic. In contrast to audiovisual senses, the sense of touch occurs bilaterally in haptic communications, i.e. it is sensed by imposing a motion on an environment and feeling the environment by a distortion or reaction force. Haptic information is composed of two distinct types of feedback: *kinesthetic* feedback (providing information about force, torque, position, velocity, etc.), and *tactile* feedback (providing information about surface texture, friction, etc.). The TI will enable networked control systems, where master and slave domains are connected and highly dynamic processes are controlled. The key difference between haptic and non-haptic control is that the haptic feedback is exchanged through a global control loop with stringent latency constraints, whereas non-haptic feedback is only audio-visual and there is no notion of a closed control loop.

Let us first perform a statistical analysis using packet traces gathered from real teleoperation experiments carried out by the authors of [Xu et al. - 2016b]. The experiments were performed over roughly 16 s with different deadband parameter values set to $d = 20\%, 15\%, 10\%, 5\%, 0$. In the teleoperation experiments, two Phantom Omni devices were used as master (i.e. HO) and slave (i.e. TOR) devices to create a one DoF teleoperation scenario. The slave was controlled by a proportional derivative controller to ensure stability at the slave side (see [Xu et al. - 2016b] for details). Using a basic sampling rate of 1 kHz, deadband coding with deadband parameter d was applied to reduce the packet rate. The communication channel between HO and TOR was emulated by using a queue architecture to generate constant or time-varying delays. The velocity signal at the HO side was sampled before being transmitted to the TOR, which in turn fed the force signal back to the HO. When using deadband coding at teleoperation wireless nodes, an update sample is transmitted only if the proportional change with respect to the preceding sample exceeds a given deadband parameter value d. Therefore, in order to retrieve the transmitted force feedback samples $s_i^{(m)}$ associated with time stamp t_i, it is necessary to perform an interpolation of the received update samples $s_j^{\prime(m)}$ associated with time stamp t_j'. Toward this end, we use the following linear interpolation:

$$
s_i^{(m)} = \begin{cases} s_j^{\prime(m)} & \text{if } t_i = t_j' \\ \dfrac{s_{j+1}^{\prime(m)} - s_j^{\prime(m)}}{t_{j+1}' - t_j'}(t_i - t_j') + s_j^{\prime(m)} & \text{otherwise.} \end{cases} \tag{10.1}
$$

Let n_s and $\bar{s}$ denote the sample size and sample mean, respectively, whereby the latter one is estimated by $1/n_s \sum_{i=1}^{n_s} s_i$. The sample autocorrelation function, $\rho(h)$, which indicates the correlation between samples lagged by h time units is obtained as follows [Priestley - 1983]:

$$
\hat{\rho}(h) = \frac{\hat{\gamma}(h)}{\hat{\gamma}(0)}, \tag{10.2}
$$

where $\hat{\gamma}(h)$ denotes the sample autocovariance function given by

$$
\hat{\gamma}(h) = \frac{1}{n_s} \sum_{i=1}^{n_s - h} (s_{i+h} - \bar{s})(s_i - \bar{s}). \tag{10.3}
$$

Let N_{DoF} denote the number of DoFs in the teleoperation system in use. Typically, N_{DoF} haptic samples coming from the application layer are encapsulated in a single segment with a prepended RTP/UDP/IP header. Each DoF haptic sample typically consists of 8 bytes, thus translating into a total of $8N_{\text{DoF}}$ bytes of payload. The haptic samples are packetized using real-time transport protocol (RTP), user datagram protocol (UDP), and internet protocol (IP) with a typical header size of 12, 8, and 20 bytes, respectively. Hence, the size of the resultant packet at the MAC layer service access point is equal to $8N_{\text{DoF}} + 40$ bytes.

We first consider the velocity signal in the command path. Let t_n, $n = 1, 2, \ldots$, denote the time instance when packet n is transmitted. The packet interarrival time in the command path is defined as the difference between transmission time instances associated with packets n and $n - 1$, i.e. $I_n = t_n - t_{n-1}$, $n = 2, 3, \ldots$. Our preliminary evaluations suggested that the packet interarrival times may exhibit an exponential or a Pareto distribution. We therefore fit the best exponential probability density function (PDF), generalized Pareto (GP), and truncated Pareto (TP) PDF to that of the experimental packet traces. Our method of fitting distributions to the traces comprises the following two steps. First, the maximum likelihood estimation (MLE) method is used to estimate the parameters of the desired distribution. Second, the results obtained from the first step are visually verified by using the complementary cumulative distribution function (CCDF) (i.e. $\overline{F}(x) = P(X > x)$). Table 10.1 summarizes the PDF of our considered distributions in addition to the parameters to be estimated.

In the following, we use the MLE method to estimate the parameters of each distribution listed in Table 10.1. For the truncated Pareto distribution, we use the method presented in [Aban et al. - 2006]. We observe that the GP distribution fits the traces

Table 10.1 Considered distributions and their identifying parameters.

Distribution	PDF
Exponential	$f_X(x) = \frac{1}{m}e^{-\frac{x}{m}}, x \geq 0$
Generalized Pareto	$f_X(x) = \frac{1}{\sigma}\left(1 + k\frac{x - T_p^{\text{haptic}}}{\sigma}\right)^{-1-\frac{1}{k}}, x \geq T_p^{\text{haptic}}$
Truncated Pareto	$f_X(x) = \frac{\alpha\gamma^\alpha x^{-\alpha-1}}{1-(\frac{\gamma}{v})^\alpha}, 0 < \gamma \leq x \leq v < \infty$

reasonably well, as compared to the exponential and TP distributions. For illustration, Figure 10.4a depicts the CCDF of the empirical data in the command path for $d = 10\%$ and fitted distributions. We observe from the figure that the GP distribution fits the empirical data better than the exponential and TP distributions. Moreover, the GP distribution accurately captures the heavy-tailed characteristics of packet interarrival times in the command path. It is worth mentioning that the same observation was made for different values of d, which are not shown here to save space.

The same procedure was applied to the feedback path as well. By fitting the exponential, TP, and GP distributions, we observed that the GP distribution fits the packet interarrival times reasonably well. Due to space constraints, we present here only the CCDF for $d = 10\%$ in Figure 10.4b, which verifies that the GP model accurately captures the heavy-tailed feature of the packet arrival in the feedback path.

Next, we estimate the mean packet arrival rate at the MAC queue of the HO. The packet arrival rate is equal to $1/\mathbb{E}[I]$, where $\mathbb{E}[I]$ is the average packet interarrival time. Assuming a GP distribution with parameters σ, k, and T_p^{haptic}, we have $\mathbb{E}[I] = T_p^{\text{haptic}} + \frac{\sigma}{1-k}$. Let $\lambda_H^{c,d}$ and $\lambda_H^{f,d}$ denote the mean packet arrival rate of haptic traffic in the command path and feedback path with deadband parameter d, respectively. Table 10.2 contains the average arrival rates of haptic traffic in the command path and feedback path for different values of deadband parameter d. Note that the packet rates without using deadband coding (i.e. $d = 0$) correspond to the base sampling rate of 1000 packets per second. Increasing deadband parameter d generally reduces the packet rate, although in Table 10.2 we observe an outlier in the command path in the transition from $d = 15\%$ to $d = 20\%$, where the packet rate increases. This unexpected increase is due to the fact that the available packet traces are gathered from a rather limited number of experiments.

In the following, we explore the key role of AI-embedded MEC servers in our FiWi architecture in greater detail. For enhanced TI reliability performance, we present our proposed edge sample forecast (ESF) module, which is inserted at the edge of our communication network in close proximity to the HO, as shown in Figure 10.3. Our proposed ESF module leverages MEC servers with embedded AI capabilities that are placed at the optical–wireless interface of FiWi enhanced LTE-A HetNets to compensate for delayed haptic samples in the feedback path by means of multiple-sample-ahead-of-time forecasting. In doing so, the response time of the HO can be kept small, resulting in a tighter togetherness with, and thereby an improved safety in, the remote TOR environment. More specifically, we use a type of parameterized artificial neural network (ANN) known as a multi-layer perceptron (MLP), which is capable of approximating any linear/non-linear function to an arbitrary degree of accuracy. The weights of the ANN are calculated by the corresponding

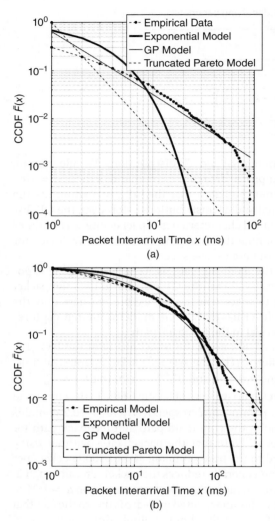

Figure 10.4 CCDF of empirical data and fitted distributions in (a) the command path and (b) the feedback path for $d = 10\%$.

Table 10.2 Packet rates for different deadband parameter d in command and feedback paths.

	Packet rate (packets per second)				
	$d = 20\%$	$d = 15\%$	$d = 10\%$	$d = 5\%$	$d = 0$
$\lambda_H^{c,d}$	427.63	246.24	410.22	473.75	1000
$\lambda_H^{f,d}$	13.78	27.47	33.49	135.88	1000

MEC server and are subsequently transmitted to the HO in close proximity. The ESF module is responsible for generating at any time t a forecast sample θ^* for time instant $t_0 = t - T_{\text{deadline}}$, where T_{deadline} is the waiting threshold until which the HO can wait to receive the actual force sample θ corresponding to time t_0. Leveraging on the existence of an autocorrelation in the past observations of the force feedback signal, the ESF module (rather than waiting for the delayed samples) immediately generates an estimation of the actual sample at a given time and then delivers it to the HO. This, in turn, increases the response time of the teleoperation system, thus resulting in an increased coupling and togetherness between the HO and the remote environment.

Next, we examine the performance of teleoperation over FiWi networks with AI enhanced MEC servers. In our haptic trace-driven simulation, we apply the same default FiWi enhanced LTE-A HetNets parameter settings as in [Beyranvand et al. - 2017]. We assume that MUs as well as HOs and TORs are mostly within WiFi coverage of the ONU-APs. We consider four ONU-APs, each associated with two MUs, one HO, and one TOR, whereby two MUs communicate with each other via the ONU-AP and the remaining two MUs communicate with uniformly randomly selected MUs associated with a different ONU-AP via the backhaul EPON. Similarly, two of the total of four HO-TOR pairs communicate with each other via the same ONU-AP (i.e. local teleoperation), whereas the other two HOs communicate with two uniformly randomly selected TORs associated with a different ONU-AP via the EPON (i.e. non-local teleoperation). In addition, we consider four ONUs serving fixed (wired) subscribers that communicate with each other. The fixed subscribers and MUs together generate background Poisson traffic at an average packet rate of λ_{BKGD} (given in packets per second).

Figure 10.5a depicts the average end-to-end delay simulation results (shown with 95% confidence interval) for HOs in the local teleoperation scenario with and without deadband coding equally applied in command and feedback paths, i.e. $d_c = d_f$. By exploiting deadband coding the average end-to-end delay can be kept below 1 ms for background traffic loads of up to roughly $\lambda_{\text{BKGD}} = 102$ packets per second, as shown in Figure 10.5a. Clearly, our presented haptic trace driven simulation results show that the QoE requirements of HITL centric local and non-local teleoperation can be met by achieving low delay and jitter performance on the order of 1–10 ms. However, in highly dynamic environments with latency requirements of 1 ms and below (e.g., industrial control systems) humans are not capable of interacting with machines swiftly enough. This calls for complementary AI based forecasting solutions, as explained next.

Recall from above that selected ONU-BSs/MPPs are equipped with AI enhanced MEC servers, which serve as agents in our HART centric task coordination. The MEC servers rely on the computational capabilities of cloudlets placed at the optical-wireless interface that are enhanced with AI capabilities to create, train, and use ANNs. We use the AI-enhanced MEC servers for performing multi-sample-ahead-of-time forecasting of delayed feedback force samples coming from a given TOR. Feedback samples are considered delayed if they do not arrive at their refresh time instants, which occur every 1 ms due to the fixed haptic sampling rate of 1 kHz. By delivering the forecast samples to the HO rather than waiting for the delayed ones, the MEC servers enable HOs to perceive the remote environment in real time at a 1 ms granularity and thus achieve tighter togetherness and improved safety control therein. As quality of control (QoC) metric we compute the mean absolute percentage error (MAPE) between the real force samples sent by the TOR and the forecast force samples delivered to the HO. Figure 10.5b shows

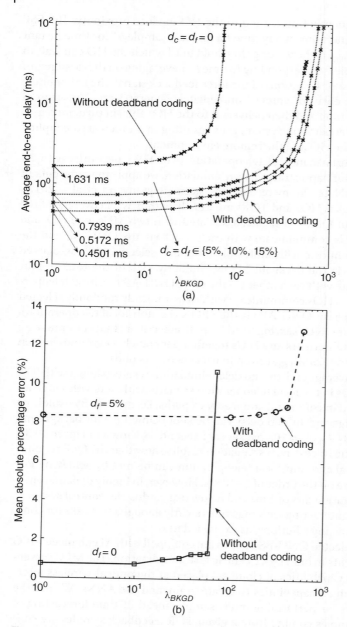

Figure 10.5 (a) Average end-to-end delay of HOs versus background traffic λ_{BKGD} for local teleoperation with and without deadband coding. (b) Performance of sample forecasting by AI enhanced MEC servers in terms of mean absolute percentage error (MAPE) versus background traffic λ_{BKGD}.

the achievable MAPE versus λ_{BKGD} with and without deadband coding. We observe that our forecasting scheme is particularly effective without deadband coding by reducing MAPE below 2%. This is due to the fact that without deadband coding, all force samples are actually fed back by the TOR, which can be used by our MEC servers to achieve a higher forecast accuracy. Importantly, note that MAPE in both cases without deadband coding (<2%) and with deadband coding (>8%) stays below the typical JND threshold of 10–20% of humans, who could not achieve such an accurate recovery of delayed haptic samples in real-time without augmentation by intelligent servers.

10.5 HART Centric Task Allocation over Multi-Robot FiWi Based TI Infrastructures

Taking the respective areas where robots are strong and humans are weak into account, FiWi enabled H2R communications aim to leverage on their *cooperative* and *collaborative* autonomy such that humans and robots may complement each other, thus allowing for a HART centric design approach [Chowdhury and Maier - 2017]. Note, however, that highly reliable and secure networking infrastructures along with intelligent task allocation and service coordination strategies are of utmost importance to meet their stringent QoS requirements. To ensure proper coordination in both local and non-local HART centric design of task allocation, integrated FiWi multi-robot networking infrastructures play a crucial role, whereby a human user delegates his/her task request to a robot through a nearby agent, which is placed at the optical-wireless interface of integrated FiWi multi-robot networks. The agent in turn coordinates the tasks to the selected robots that are responsible for executing the task. Clearly. for the efficient utilization of robotic resources, proper task allocation among robots is crucial by taking into consideration the different capabilities of robots as well as the specific task requirements such as task execution deadline and energy consumption of robots.

Toward this end, we develop an efficient H2R task allocation strategy that includes both host robot and collaborative node selection in integrated FiWi multi-robot networks. We propose to use not only the central cloud and local cloudlets as collaborative nodes but also available neighboring robots for computation sub-task offloading. To achieve energy savings of the robots and accomplish H2R tasks within their required time, the objective is to select the proper policy for H2R task execution by evaluating the performance of a non-collaborative task execution scheme, in which the selected host robot executes the full H2R task, and the collaborative/joint H2R task execution, in which the selected host robot performs only the sensing subtask while the selected collaborative node executes the computation sub-task via computation offloading.

The generic FiWi multi-robot network architecture for coordinating the local and non-local allocation of the H2R task is shown in Figure 10.3. Tasks consist of both sensing and computation sub-parts, whereby humans, robots, and agents actively participate in the task allocation process. We exploit the remote central cloud, cloudlets (colocated with ONU-MPPs), and neighboring robots as collaborative nodes for computation sub-task offloading. We assume that the central cloud servers are connected to the OLT via dedicated fiber links. In the following, after elaborating on our proposed unified resource management scheme, we explain the computation offloading operation followed by the proposed task allocation algorithm.

First, we aim to resolve contention in the wireless as well as the optical fiber sub-networks. Our proposed unified resource management scheme leverages a two-layer time division multiple access (TDMA)-based operation in both optical and wireless sub-networks, as illustrated in Figure 10.6 [Chowdhury and Maier - 2017]. During the initial task allocation phase, the agent located at the ONU-MPP exchanges three control messages (RTS, CTS, and ACK) with its associated robots. The proposed resource management scheme exploits IEEE 802.11ac WLAN frames in the wireless front-end, whereas medium access in the optical fiber backhaul is controlled by using the IEEE 802.3av multipoint control protocol.

Second, we elaborate on our computation offloading phase of the proposed HART centric scheme. Note that the transmission opportunity for computation offloading is kept separate from conventional broadband transmissions to permit both broadband and computation offloading operations within a polling cycle. After receiving the computation offloading request from a given host robot, the ONU-APP selects where to offload the computation sub-task onto the subject to given offloading requirements and sends an extended REPORT message to the OLT, which embeds the computation offloading request. As soon as the host robots are notified about their computation offloading time slot information by the ONU-MPP, they transmit the computation sub-task data frame to the ONU-MPP via its assigned offloading time slot. After receiving the computation sub-task input data frame from the task offloading host robot, the ONU-MPP forwards them to the selected collaborative node (central cloud/cloudlet/neighboring robot) for further processing. Once the ONU-MPP receives the results of the computation sub-task from the cloudlet/neighboring robot, it immediately sends them to the task offloading host robot. For central cloud offloading, the ONU-MPP sends the computation sub-task input data frame to the OLT. Then, after receiving the computation sub-task input data, the OLT transfers them to the central cloud. The OLT receives the computation sub-task results from the central cloud after processing and sends them to the ONU-MPP. Once the ONU-MPP receives the computation sub-task results from the OLT, it immediately forwards them to the corresponding host robot.

Third, we elaborate on our proposed task allocation algorithm, which accounts for two different task execution schemes: (i) a non-collaborative scheme, where the suitable host robot executes the full task and (ii) a collaborative scheme, where the suitable host robot and collaborative node (central cloud, cloudlet, or neighboring robot) conduct the sensing and computation sub-task, respectively. Our proposed task allocation algorithm, which performs both suitable host robot and collaborative node selection, comprises the following four steps:

1. First, an MU sends his/her H2R task request message to the agent node during her assigned upstream transmission time slot containing the following information: task location, task type, remaining energy threshold, and task deadline.
2. When the agent at the ONU-MPP receives the task request message from the MU, it selects those robots in its wireless coverage area that satisfy the given availability, the energy threshold to conduct the task, and the task execution deadline requirements. Toward this end, the agent broadcasts a task announcement message to all nearby robots. After receiving the task announcement message, the available robots send task reply messages to the agent containing the following information: remaining energy, location, moving and processing speed, and precalculation of task execution

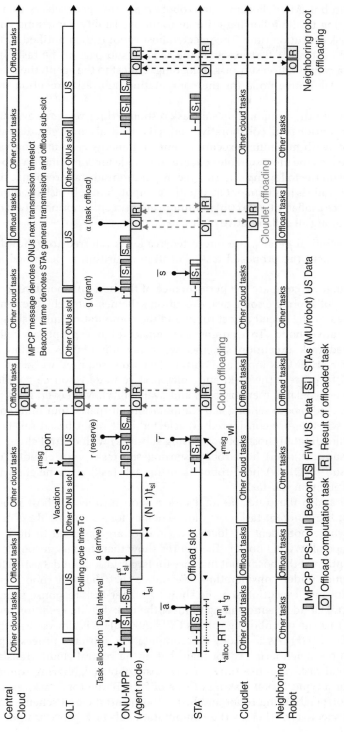

Figure 10.6 Two-layer TDMA based resource management scheme.

time of each robot. After checking each robot's reply, the agent selects a suitable host robot according to the following criteria: robot availability, remaining energy, and minimum task execution time. The selected host robot is then notified by the agent.

3. The selected host robot first executes the sensing sub-part of the task. If the host robot sends a computation sub-task offloading request to the agent, the agent selects a suitable collaborative node for the computation sub-task execution (remaining sub-part).

4. The agent checks the computation sub-task response time, resource availability, and energy consumption of all collaborative nodes (i.e. central cloud, cloudlet, and neighboring robots with minimum execution time and energy consumption value). The agent then selects the most suitable collaborative node for computation sub-task execution based on the following criteria: (i) computation sub-task response time of the collaborative node is less than or equal to the computation sub-task deadline; (ii) sufficient resource availability; and (iii) minimum energy consumption of task offloading host robot among all collaborative nodes.

For further technical details, the interested reader is referred to [Chowdhury and Maier - 2017], where the proposed task allocation algorithm is described in a more formal way.

In the following, we compare the performance of the non-collaborative (i.e. without offloading) and collaborative/joint task execution schemes, whereby the sensing sub-task is conducted by the selected host robot and the computation sub-task is offloaded onto a collaborative node. To examine the impact of our proposed collaborative computing-based task execution scheme, we study different evaluation scenarios based on different H2R task input and output data sizes, required workload (in terms of CPU cycles) to process the task, and collaborative node resource conditions (i.e. processing power, available memory size, and availability). Moreover, for a particular H2R task that includes both sensing and computation sub-parts, four different types of task execution schemes are considered: (i) selected host robot-based full-task execution without offloading, (ii) host robot (sensing sub-task) with central cloud execution (computation sub-task), (iii) host robot (sensing sub-task) with cloudlet execution (computation sub-task), (iv) host robot (sensing sub-task) with neighboring robot execution (computation sub-task).

Figures 10.7a and b illustrate the total task response time and host robot energy consumption of the different task execution schemes for scenario 1 described on top of the figure, where both central cloud and cloudlet are assumed to have the same computation capability/CPU power. The figures show that the task response time and energy consumption of the host robot increase for larger task input data sizes for all task execution schemes under consideration. We notice that the host robot/neighboring robot based joint task execution scheme shows a higher task response time than the host robot central cloud scheme, failing to meet the task deadline requirement. This is due to the fact that the neighboring robot's CPU (500 MHz) is slower than the central cloud CPU (3200 MHz). As a result, the computation sub-task processing delay is much higher in the neighboring robot than that of the central cloud, translating into an increased total task response time for the host robot/neighboring robot scheme. For instance, for a typical total task input size of 240 kB, the total task response time in the host robot/neighboring robot and host robot/central cloud scheme equals 4.56 s and 2.95 s, respectively, whereas the computation sub-task processing delay of the

Scenario 1: $\mu_{cl} = \mu_{ct} = 3200$ MHz, $m_{ct} = 35$ MB, $m_o = 35$ MB, $\mu_o = 500$ MHz, $m_r = 25$ MB, $I_u = (40,60,80,100,120)$KB, $I_r = (16,24,32,40,48)$KB

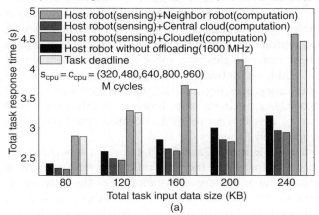

Scenario 1: $\mu_{cl} = \mu_{ct} = 3200$ MHz, $m_{ct} = 35$ MB, $m_o = 35$ MB, $\mu_o = 500$ MHz, $m_r = 25$ MB, $I_u = (40,60,80,100,120)$KB, $I_r = (16,24,32,40,48)$KB

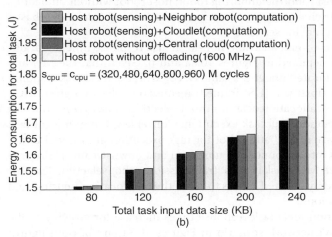

Figure 10.7 Task response time and energy consumption variation of collaborative and non-collaborative task execution schemes versus total task input data size.

neighboring robot and central cloud equals 1.92 s and 0.3 s, respectively. Hence, the computation sub-task offloading delay of the neighboring robot and central cloud are equal to 0.049 s and 0.056 s, respectively. On the downside, however, the energy efficiency gain of the host robot/neighboring robot compared to the host robot/central cloud is negligible, amounting to less than 1%. This is because the difference between the energy consumption of the host robot for the central cloud and neighboring robot to computation sub-task processing delay is very small. The average energy consumption of the host robot (per second) is very low during the neighboring robot (for the host robot/neighboring robot scheme) and central cloud (for the host robot/central cloud scheme) computation sub-task processing, as the host robot is idle during that time. Therefore, the difference between the host robot energy consumption for the host robot/neighboring robot and host robot/central cloud total task execution is also very low. Due to the lower energy consumption of the host robot for central cloud computation sub-task processing in Figure 10.7b the host-robot/central-cloud execution shows a 1% higher energy efficiency gain compared to the host robot/neighboring robot scheme. For instance, for a typical total task input size of 240 kB, the host robot's energy consumption for the host robot/neighboring robot and host robot/central cloud schemes equal 1.72 J and 1.71 J, respectively, whereby the host robot's energy consumption for computation sub-task processing at the neighboring robot and central cloud is equal to 0.00192 J and 0.0003 J, respectively. Note, however, that the host robot's energy consumption in the case of neighboring robot and central cloud offloading equals 0.00418 J and 0.0049 J, respectively.

Furthermore, we observe from both Figures 10.7a and b that the host robot/cloudlet based joint task execution scheme outperforms the host robot/central cloud based scheme in terms of task response time and energy consumption of the host robot. This is mainly due to the fact that the involved cloudlet causes a smaller computation offloading delay than the central cloud. The host robot/cloudlet based scheme shows a 36%, 8%, and 1% increase of task response time efficiency and a 2%, 15%, and 1% higher energy efficiency than the host robot/neighboring robot, host robot without offloading, and host robot/cloud based scheme, respectively. Thus, the host robot/cloudlet based task execution scheme is optimal for scenario 1.

Finally, we evaluate the end-to-end local and non-local task response time under different FiWi traffic loads, as depicted in Figures 10.8a and b. Given that the MU and robot for task execution are located under the same ONU-MPP, the local task response calculation accounts for the upstream (US) frame transmission delay of a given MU's task request transmission, task allocation delay for robot selection, time to reach the task location, sensing, and computation subtask execution. Conversely, the non-local task response time calculation, where MU and robot for task execution are located under different ONU-MPPs, takes into account both upstream and downstream (DS) frame transmission delays of a given MU's task request transmission, robot selection delay for task allocation, time required by selected robots to reach the task location, sensing, and computation sub-task execution.

Note that both local and non-local task response times increase for growing traffic loads in our considered FiWi network scenario in Figures 10.8a and b. Both figures clearly indicate that the host-robot/cloudlet based scheme provides a lower task response time than the alternative host robot/central cloud (2%) and host robot without offloading (10%) schemes. This is because the host robot/central cloud based scheme incurs a higher computation offloading delay than host robot/cloudlet based execution. Moreover, host robot based non-collaborative task execution experiences a much higher task response time than the alternative collaborative schemes. This result is expected, given that the host robot is less powerful than the central cloud and cloudlet. We also note that the end-to-end local task response times of all compared schemes (collaborative and non-collaborative) are lower than the non-local task response time. The reason behind this is that besides the task execution delay, the end-to-end non-local task response time involves both US and DS frame transmission delays for end-to-end task allocation, while the local task response time involves only the one-way US frame transmission delay.

$t_{non-local}$ vs. Traffic load (M = 10, N = 16, t_{wl}^{msg} = .512 µs, μ_{cl} = μ_{ct} = 3200 MHZ, RTT = .2ms, V = 1.512 µs, t_{alloc} = .315 ms, t_{pon}^{msg} = .512 µs, x_{max} = 12.14 µs)

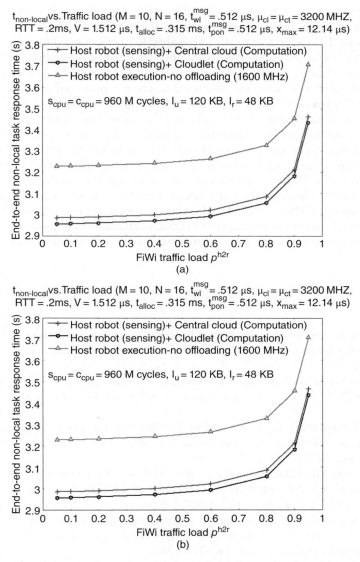

Figure 10.8 End-to-end local and non-local task response time variation versus FiWi traffic load.

10.6 Conclusions

After shedding some light on the commonalities and subtle differences between the IoT and emerging TI, we provided insights into how the TI can create value for society and human well being. Unlike the IoT without any human involvement in its underlying machine-to-machine communications, the TI aims to involve the inherent HITL nature of haptic interaction, thus allowing for a human centric design approach towards creating novel immersive experiences via the internet. Our approach to the TI is based on the combination of the two following original ideas. Besides the design of reliable FiWi enhanced LTE-A HetNets with AI embedded MEC capabilities to meet the QoE requirements of HITL centric local and non-local teleoperation in terms of low latency performance on the order of 1–10 ms, we developed a HART centric multi-robot task allocation strategy, which relies on the proposed unified resource allocation in FiWi based TI infrastructures by means of suitable host robot selection and computation task offloading onto collaborative nodes. We showed that our proposed MLP based forecasting scheme is able to achieve QoC by improving the forecast accuracy of delayed haptic samples in real time, thus reducing MAPE below 2%. Further, the obtained results of both collaborative/joint and non-collaborative task execution schemes showed that for a typical task input size of 240 kB, the collaborative task execution scheme is able to decrease the task response time by up to 8.75% and the energy consumption by up to 14.98% compared to the non-collaborative task execution scheme. The proposed collaborative computing based H2R task allocation and resource management scheme proves instrumental in enabling future low-latency multi-robot TI applications.

Bibliography

Inmaculada B Aban, Mark M Meerschaert, and Anna K Panorska. Parameter Estimation for the Truncated Pareto Distribution. *Journal of the American Statistical Association*, 101(473):270–277, Mar. 2006.

A. Aijaz. Towards 5G-enabled Tactile Internet: Radio resource allocation for haptic communications. In *2016 IEEE Wireless Communications and Networking Conference*, pages 1–6, 2016.

A. Aijaz, M. Dohler, A. H. Aghvami, V. Friderikos, and M. Frodigh. Realizing the Tactile Internet: Haptic Communications over Next Generation 5G Cellular Networks. *IEEE Wireless Communications*, 24(2):82–89, 2017.

A. A. Ateya, A. Vybornova, R. Kirichek, and A. Koucheryavy. Multilevel Cloud based Tactile Internet System. In *2017 19th International Conference on Advanced Communication Technology (ICACT)*, pages 105–110, 2017.

E. A. Baran, A. Kuzu, S. Bogosyan, M. Gokasan, and A. Sabanovic. Comparative Analysis of a Selected DCT-Based Compression Scheme for Haptic Data Transmission. *IEEE Transactions on Industrial Informatics*, 12(3):1146–1155, 2016.

H. Beyranvand, M. Lévesque, M. Maier, J. A. Salehi, C. Verikoukis, and D. Tipper. Toward 5G: FiWi Enhanced LTE-A HetNets With Reliable Low-Latency Fiber Backhaul Sharing and WiFi Offloading. *IEEE/ACM Transactions on Networking*, 25(2):690–707, 2017.

J. M. Bradshaw, V. Dignum, C. M. Jonker, and M. Sierhuis. Human-Agent-Robot Teamwork. In *2012 7th ACM/IEEE International Conference on Human-Robot Interaction (HRI)*, pages 487–487, 2012.

Grigore C Burdea and Frederick P Brooks. *Force and Touch Feedback for Virtual Reality*. Wiley, New York, 1996.

M. Chowdhury and M. Maier. Collaborative Computing for Advanced Tactile Internet Human-to-Robot (H2R) Communications in Integrated FiWi Multirobot Infrastructures. *IEEE Internet of Things Journal*, 4(6):2142–2158, 2017.

M. Condoluci, T. Mahmoodi, E. Steinbach, and M. Dohler. Soft Resource Reservation for Low-Delayed Teleoperation Over Mobile Networks. *IEEE Access*, 5:10445–10455, 2017.

Rafael Mathias de Mendonça, Nadia Nedjah, and Luiza de Macedo Mourelle. Efficient distributed algorithm of dynamic task assignment for swarm robotics. *Neurocomputing*, 172:345–355, 2016.

Y. Feng, C. Jayasundara, A. Nirmalathas, and E. Wong. Hybrid Coordination Function Controlled Channel Access for Latency-Sensitive Tactile Applications. In *GLOBECOM 2017 - 2017 IEEE Global Communications Conference*, pages 1–6, 2017a.

Y. Feng, C. Jayasundara, A. Nirmalathas, and E. Wong. IEEE 802.11 HCCA for Tactile Applications. In *2017 27th International Telecommunication Networks and Applications Conference (ITNAC)*, pages 1–3, 2017b.

L. Gkatzikis and I. Koutsopoulos. Migrate or Not? Exploiting Dynamic Task Migration in Mobile Cloud Computing Systems. *IEEE Wireless Communications*, 20(3):24–32, 2013.

P. Hinterseer, S. Hirche, S. Chaudhuri, E. Steinbach, and M. Buss. Perception-Based Data Reduction and Transmission of Haptic Data in Telepresence and Teleaction Systems. *IEEE Transactions on Signal Processing*, 56(2):588–597, 2008.

M. Johnson, J. M. Bradshaw, P. Feltovich, C. Jonker, B. van Riemsdijk, and M. Sierhuis. Autonomy and Interdependence in Human-Agent-Robot Teams. *IEEE Intelligent Systems*, 27(2):43–51, 2012.

Alaa Khamis, Ahmed Hussein, and Ahmed Elmogy. Multi-robot task allocation: A review of the state-of-the-art. *Cooperative Robots and Sensor Networks, Springer*, 604:31–51, May 2015.

E. Khorov, A. Krasilov, and A. Malyshev. Radio Resource Scheduling for Low-Latency Communications in LTE and Beyond. In *2017 IEEE/ACM 25th International Symposium on Quality of Service (IWQoS)*, pages 1–6, 2017.

T. Langford, Q. Gu, A. Rivera-Longoria, and M. Guirguis. Collaborative Computing On-demand: Harnessing Mobile Devices in Executing On-the-Fly Jobs. In *2013 IEEE 10th International Conference on Mobile Ad-Hoc and Sensor Systems*, pages 342–350, 2013.

Hongbing Li and Kenji Kawashima. Bilateral teleoperation with delayed force feedback using time domain passivity controller. *Robotics and Computer-Integrated Manufacturing*, 37:188–196, 2016.

J. Liu, Y. Mao, J. Zhang, and K. B. Letaief. Delay-Optimal Computation Task Scheduling for Mobile-Edge Computing Systems. In *2016 IEEE International Symposium on Information Theory (ISIT)*, pages 1451–1455, 2016.

M. Maier, M. Chowdhury, B. P. Rimal, and D. P. Van. The Tactile Internet: Vision, Recent Progress, and Open Challenges. *IEEE Communications Magazine*, 54(5):138–145, 2016.

Isaac Osunmakinde and Ramharuk Vikash. Development of a Survivable Cloud Multi-Robot Framework for Heterogeneous Environments. *International Journal of Advanced Robotic Systems*, 11(10):164–186, Oct. 2014.

P. Patil, A. Hakiri, and A. Gokhale. Cyber Foraging and Offloading Framework for Internet of Things. In *2016 IEEE 40th Annual Computer Software and Applications Conference (COMPSAC)*, volume 1, pages 359–368, 2016.

J. Pilz, M. Mehlhose, T. Wirth, D. Wieruch, B. Holfeld, and T. Haustein. A Tactile Internet demonstration: 1ms ultra low delay for wireless communications towards 5G. In *2016 IEEE Conference on Computer Communications Workshops (INFOCOM WKSHPS)*, pages 862–863, 2016.

Maurice Bertram Priestley. *Spectral Analysis and Time Series.* Academic press, 1983.

B. P. Rimal, D. P. Van, and M. Maier. Mobile Edge Computing Empowered Fiber-Wireless Access Networks in the 5G Era. *IEEE Communications Magazine*, 55(2):192–200, 2017a.

B. P. Rimal, D. Pham Van, and M. Maier. Cloudlet Enhanced Fiber-Wireless Access Networks for Mobile-Edge Computing. *IEEE Transactions on Wireless Communications*, 16(6):3601–3618, 2017b.

B. P. Rimal, D. Pham Van, and M. Maier. Mobile-Edge Computing Versus Centralized Cloud Computing Over a Converged FiWi Access Network. *IEEE Transactions on Network and Service Management*, 14(3):498–513, 2017c.

P. Schulz, M. Matthe, H. Klessig, M. Simsek, G. Fettweis, J. Ansari, S. A. Ashraf, B. Almeroth, J. Voigt, I. Riedel, A. Puschmann, A. Mitschele-Thiel, M. Muller, T. Elste, and M. Windisch. Latency Critical IoT Applications in 5G: Perspective on the Design of Radio Interface and Network Architecture. *IEEE Communications Magazine*, 55(2):70–78, 2017.

C. She and C. Yang. Energy Efficient Design for Tactile Internet. In *2016 IEEE/CIC International Conference on Communications in China (ICCC)*, pages 1–6, 2016.

C. She, C. Yang, and T. Q. S. Quek. Cross-Layer Transmission Design for Tactile Internet. In *2016 IEEE Global Communications Conference (GLOBECOM)*, pages 1–6, 2016.

M. Simsek, A. Aijaz, M. Dohler, J. Sachs, and G. Fettweis. 5G-Enabled Tactile Internet. *IEEE Journal on Selected Areas in Communications*, 34(3):460–473, 2016.

E. Steinbach, S. Hirche, M. Ernst, F. Brandi, R. Chaudhari, J. Kammerl, and I. Vittorias. Haptic Communications. *Proceedings of the IEEE*, 100(4): 937–956, 2012.

T. H. Szymanski. Strengthening security and privacy in an ultra-dense green 5G Radio Access Network for the industrial and tactile Internet of Things. In *2017 13th International Wireless Communications and Mobile Computing Conference (IWCMC)*, pages 415–422, 2017.

T. Taleb, K. Samdanis, B. Mada, H. Flinck, S. Dutta, and D. Sabella. On Multi-Access Edge Computing: A Survey of the Emerging 5G Network Edge Cloud Architecture and Orchestration. *IEEE Communications Surveys Tutorials*, 19(3):1657–1681, 2017.

Sahar Trigui, Anis Koubaa, Omar Cheikhrouhou, Habib Youssef, Hachemi Bennaceur, Mohamed-Foued Sriti, and Yasir Javed. A distributed market-based algorithm for the multi-robot assignment problem. *Elsevier Procedia Computer Science*, 32:1108–1114, Jun. 2014.

L. Wang, M. Liu, and M. Q. H. Meng. A Hierarchical Auction-Based Mechanism for Real-Time Resource Allocation in Cloud Robotic Systems. *IEEE Transactions on Cybernetics*, 47(2):473–484, 2017.

E. Wong, M. P. I. Dias, and L. Ruan. Tactile Internet Capable Passive Optical LAN for Healthcare. In *2016 21st OptoElectronics and Communications Conference (OECC) held jointly with 2016 International Conference on Photonics in Switching (PS)*, pages 1–3, 2016.

E. Wong, M. Pubudini Imali Dias, and L. Ruan. Predictive Resource Allocation for Tactile Internet Capable Passive Optical LANs. *IEEE/OSA Journal of Lightwave Technology*, 35(13):2629–2641, 2017a.

E. Wong, S. Mondal, and G. Das. Latency-Aware Optimisation Framework for Cloudlet Placement. In *2017 19th International Conference on Transparent Optical Networks (ICTON)*, pages 1–2, 2017b.

X. Xu, B. Cizmeci, C. Schuwerk, and E. Steinbach. Model-Mediated Teleoperation: Toward Stable and Transparent Teleoperation Systems. *IEEE Access*, 4:425–449, 2016a.

X. Xu, C. Schuwerk, B. Cizmeci, and E. Steinbach. Energy Prediction for Teleoperation Systems That Combine the Time Domain Passivity Approach with Perceptual Deadband-Based Haptic Data Reduction. *IEEE Transactions on Haptics*, 9(4):560–573, 2016b.

H. Zhang, Q. Zhang, and X. Du. Toward Vehicle-Assisted Cloud Computing for Smartphones. *IEEE Transactions on Vehicular Technology*, 64(12):5610–5618, 2015.

W. Zhang, Y. Wen, and D. O. Wu. Energy-Efficient Scheduling Policy for Collaborative Execution in Mobile Cloud Computing. In *2013 Proceedings IEEE INFOCOM*, pages 190–194, 2013.

11

Energy Efficiency in the Cloud Radio Access Network (C-RAN) for 5G Mobile Networks: Opportunities and Challenges

Isiaka Ajewale Alimi*[1], Abdelgader M. Abdalla[1], Akeem Olapade Mufutau[1], Fernando Pereira Guiomar[1], Ifiok Otung[2], Jonathan Rodriguez[1,2], Paulo Pereira Monteiro[1,3], and Antonio Luís Teixeira[1,3]

[1] *Instituto de Telecomunicações, 3810-193, Aveiro, Portugal*
[2] *Satellite Communications, Faculty of Computing, Engineering and Science, University of South Wales, Pontypridd, CF37 1DL, United Kingdom*
[3] *Department of Electronics, Telecommunications and Informatics (DETI), Universidade de Aveiro, 3810-193, Aveiro, Portugal*

11.1 Introduction

The Internet of Things technology has unprecedentedly revolutionized the number of network devices such as mobile phones, actuators, sensors, and radio-frequency identification tags. The supported bandwidth-intensive applications and services by the entities have put huge pressure on mobile operators regarding system requirements like coverage, capacity, and latency [Alimi et al. - 2017a,b].

The fifth generation (5G) and beyond-5G (B5G) networks are expected to be viable solutions for supporting the anticipated huge traffic. Moreover, the key path toward 5G network requirements realization is cell-densification schemes through the deployment of more base stations (BSs). In general, the aim of cellular densification is to meet the required coverage and capacity by the subscribers [Alimi et al. - 2018, 2017c]. However, this leads to a dramatic rise in the power consumption [Chi - 2011]. As illustrated in Figure 11.1, in a typical cellular network, the BS takes about 60% of the total power consumption of cellular systems. This offers a clear perception that the BS energy consumption deserves to be addressed [Hasan et al. - 2011; Karmokar and Anpalagan - 2013]. Moreover, the actual energy consumption statistics of BSs varies, depending on factors such as the cell size, technology, components as well as radiation power. It is noteworthy that the majority of the earlier studies on the radio access network (RAN) are mainly aimed at enhancing the system coverage and capacity while little or no attention is being paid to a means of addressing the significant growth in the energy consumption. Furthermore, the increase in the energy consumption by the networks has led to an unprecedented rise in energy bills and carbon footprint.

For instance, EE has only been taken into account for uplink communication owing to the limited battery power of mobile devices, while EE in the downlink direction

---

* Corresponding Author: Isiaka Ajewale Alimi iaalimi@ua.pt

Optical and Wireless Convergence for 5G Networks, First Edition.
Edited by Abdelgader M. Abdalla, Jonathan Rodriguez, Issa Elfergani, and Antonio Teixeira.
© 2020 John Wiley & Sons Ltd. Published 2020 by John Wiley & Sons Ltd.

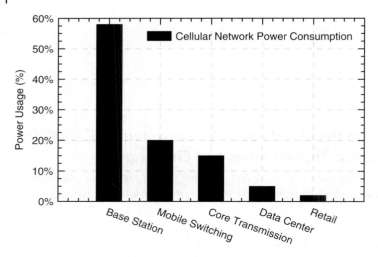

Figure 11.1 Typical wireless cellular network power consumption breakdown [Hasan et al. - 2011; Han et al. - 2011].

is normally ignored [Karmokar and Anpalagan - 2013]. Consequently, the increasing energy demand has brought about extensive research on the topics of *green communications* in both academia and industry [Hasan et al. - 2011; Alsharif et al. - 2013; Karmokar and Anpalagan - 2013].

Currently, significant consideration is being given to energy-related aspects of the network by the mobile service providers and regulatory bodies such as the Third Generation Partnership Project (3GPP) and International Telecommunication Union (ITU). Based on this, the European Commission has been presenting projects such as the European Union's FP7 projects Energy Aware Radio and Network Technologies (EARTH), Towards Real Energy-Efficient Network Design (TREND), and Cognitive Radio and Cooperative Strategies for Power Saving in Multi-standard Wireless Devices (C2POWER) in an effort to address the EE of the systems [Hasan et al. - 2011; Karmokar and Anpalagan - 2013]. In the following subsections, we explain the major reasons for considering the increase in the RAN energy consumption.

11.1.1   Environmental Effects

One of the reasons for this development is the need to minimize the environmental impact of the ICT sector on climate change. Climate change is as a result of an upsurge in the accumulation of carbon dioxide (CO_2) as well as other greenhouse gases (GHGs) such as nitrous oxide, methane, and ozone in the atmosphere. Moreover, GHGs are usually emitted owing to the use of fossil fuels as a main source of electrical energy production and transportation [Humar et al. - 2011; TheClimateGroup - 2008]. In the ICT sector, the CO_2 emissions are primarily due to off-grid sites (or in its absence to grid energy) that are normally powered by diesel generators [Alsharif et al. - 2013]. In addition, it has been predicted that, the ICT sector's emissions will increase from 0.53 billion tonnes (Gt) CO_2 equivalent (CO_2e) in 2002 to 1.43 $GtCO_2$e in 2020 [TheClimateGroup - 2008]. In addition, it has been estimated that the overall telecom footprint will be fairly constant by 2020, with the mobile network contributions being dominant. For instance, as depicted in Figure 11.2, the growth in telecoms emissions has increased from 150 $MtCO_2e$ in 2002 (Figure 11.2a) to 300 $MtCO_2e$ in 2007. Furthermore, it is envisaged to increase to 350 $MtCO_2e$ by 2020 (Figure 11.2b) [TheClimateGroup - 2008].

11.1.2   Economic Benefits

According to an investigation on the total cost of ownership (TCO), operating expenditure (OPEX) and capital expenditure (CAPEX) account for over 60% and for nearly 40% of the TCO, respectively [Chi - 2011]. Consequently, OPEX is a major factor that the operators need to explore in developing the future RAN. Figure 11.2c illustrates the OPEX analysis of a typical cell site. With reference to the analysis, over 40% of the OPEX is expended on electricity [Chi - 2011].

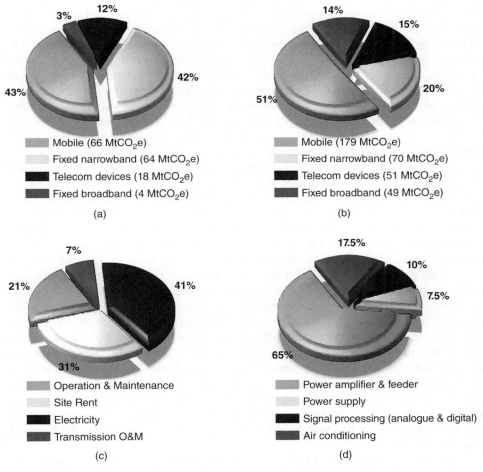

Figure 11.2 Global telecoms emissions %: (a) 2002 (100% = 151 $MtCO_2e$), (b) 2020 (100% = 349 $MtCO_2e$) [TheClimateGroup - 2008], (c) OPEX analysis of a cell site [Chi - 2011], and (d) radio BS components power consumption [Hasan et al. - 2011; Correia et al. - 2010].

Therefore, another reason why consideration is being given to energy-related aspects of the network (apart from the environmental effects) is the need to minimize the usually high energy bills of network providers, since this typically takes a considerable amount of the OPEX. For instance, to operate BSs that are connected to the electrical grid, about \$3000 per year may be required. However, the off-grid BSs that run on diesel power generators may cost approximately ten times more [Hasan et al. - 2011]. A further insight into the components of power consumption of a typical BS is depicted in Figure 11.2d. It shows that about 50% of the power is consumed by the RAN equipment while the remainder is expended on air conditioners and other equipment [Chi - 2011]. Consequently, a reduction in the energy consumption in the ICT sector would really boost the anticipated revenue by the mobile network operators [Humar et al. - 2011; Alsharif et al. - 2013; Alimi et al. - 2017d]. In view of this, the Mobile Virtual Centre of Excellence (VCE) Green Radio project has been established with the major goals of reducing the carbon emissions and OPEX of wireless cellular networks by developing innovative green radio techniques and radio architectures [Han et al. - 2011; Karmokar and Anpalagan - 2013]. In addition, Green-Touch, a consortium of ICT industry, academia, and non-governmental research professionals is determined to enhance the ICT sector EE. Also, the Green-IT project intends to improve the EE standards and energy consumption metrics for networking equipment [Karmokar and Anpalagan - 2013]. Moreover, it has been observed that a proficient approach to saving energy and reducing CO_2 emissions is to reduce the number of BS sites being deployed. Nevertheless, in the traditional RAN, this not only brings about inadequate network coverage but also results in lower capacity and poorer quality of experience (QoE). Consequently, it is important to develop a more cost-effective method of reducing the consumption while concurrently providing suitable network requirements regarding the quality of service (QoS) and QoE [Chi - 2011].

A number of innovative technologies have been proposed in order to reduce the BS power consumption and subsequently lowering the CO_2 emissions and OPEX costs. Among these are software-based solutions that save power by turning off certain carriers during idle hours (e.g. during the night). Another approach is the use of renewable energy-based solutions, such as wind and solar, to power the BS, in line with the local natural conditions. In another development, energy-saving air conditioning technology can be employed. This entails reducing energy consumption of air conditioning equipment by considering the local climate and environment features. Nevertheless, the aforementioned technologies are just ancillary approaches that cannot effectively take care of the associated power consumption fundamental problems, which increase with the number of BSs [Chi - 2011]. Hence, it is very essential to study different paradigm modification schemes like smart grids, efficient BS redesign, energy efficient wireless architectures, and protocols. Similarly, opportunistic network access or cognitive radio, cooperative relaying and small cell-based heterogeneous network deployment need to be considered. In all, the main trade-offs of EE with network performance such as SE, deployment efficiency, bandwidth, and end-to-end service delay have to be analyzed. [Hasan et al. - 2011].

A disruptive approach of addressing the energy consumption challenge is that the mobile operators need to consider EE from the onset of RAN architectural planning for an efficient system design. So, infrastructural change is a crucial approach to deal with the power consumption challenge of the RAN. In order to achieve the aim, centralized

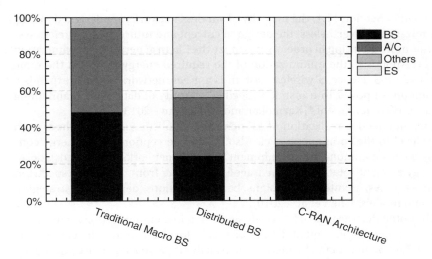

Figure 11.3 BS power consumption comparison. ES: energy saved [Carugi - 2011].

BS schemes can be adopted. A centralized BS-based solution not only reduces the number of BS equipment rooms but also lowers the amount of air conditioning required [Chi - 2011]. Moreover, it exploits resource sharing mechanisms in an attempt to efficiently enhance the BS utilization rate in a dynamic network load condition. Figure 11.3 compares the power consumption between the traditional macro BS, the distributed BS and the C-RAN centralized approaches. Compared to the traditional macro BS, the distributed BS and C-RAN approaches reduce power consumption by 39% and 68%, respectively [Carugi - 2011].

With EE being introduced as one of essential requirements for viable evolution of mobile communication systems, in Section 11.2, we present a comprehensive discussion on EE metric measurement. The green design for energy crunch prevention in the 5G networks is considered in Section 11.3. In this section, schemes, such as hardware solutions, resource allocation, network planning and deployment, and energy harvesting and transfer, that can be employed to improve the EE of the 5G and B5G networks are comprehensively discussed. Fiber based energy efficient networks are discussed in Section 11.4. In section 11.5, we analyze system and power consumption models in which remote unit-based power consumption, centralized unit based power consumption, fronthaul power consumption, and massive MIMO energy efficiency are considered. The obtained simulation results and comprehensive discussions are presented in Section 11.6 and concluding remarks are given in Section 11.7.

11.2 Standardized Energy Efficiency Metric (Green Metric)

In telecommunication networks, *greenness* can be measured by considering factors such as the contributed carbon footprint, battery lifespan extension and economic benefits in view of energy costs. Nevertheless, it is well known that the contributed portion of carbon emissions by telecommunication networks is extremely small. Therefore, estimation of *greenness* appears to be more appropriate with EE [Hasan et al. - 2011].

As mentioned in Section 11.1, the majority of wireless networks are designed mainly for SE maximization. Nevertheless, the design of current and future green wireless systems demands more attention in order to maximize the EE. In general, the maximization of the EE corresponds to the minimization of the required energy to obtain the same throughput and QoS. Hence, a wireless system is a green network when it expends a minimum amount of power to transmit a specified quantity of data while maintaining the stipulated QoS requirements [Karmokar and Anpalagan - 2013].

In addition, an effective realization of EE demands a holistic view from the system architecture level to the component level. Also, a clear perception of the energy consumption by the telecommunication equipment as well as networks is required. Consequently, energy efficient solutions are envisaged to emanate from innovative spectrum management schemes, architectural designs, backhaul options, deployment strategies, EE metrics, and models [Chen et al. - 2010].

It should be noted that the idea of EE is only significant in communications when it can be evaluated. EE metrics offer quantified information for determining the energy efficiency in the network. In general, EE metrics are usually employed to [Chen et al. - 2010]:

- relate energy consumption performance of various systems and components in a similar class;
- set precise development targets on means of reducing energy consumption, as well as long term research on EE;
- determine the EE of a particular configuration in a system and facilitate its adaptation to a more energy efficient configuration.

Moreover, the following institutions and standardization bodies have been attending to the necessity for unified energy metrics [Hamdoun et al. - 2012]:

1. Alliance for Telecommunications Industry Solutions (ATIS).
2. European Telecommunications Standards Institute (ETSI).
3. International Telecommunication Union (ITU).
4. Energy Consumption Rating (ECR) Initiative.

Furthermore, the EE metrics are categorized into absolute and relative metrics. The former specifies the real energy consumed for a performance. In this category, bits/Joule is the most extensively employed metric. This metric indicates how EE is improved, so that the ratio of output and input power/energy can be employed. It is remarkable that the EE metrics requirements at the component, equipment, and system/network level are considerably different [Chen et al. - 2010]. Other optional EE metrics are enumerated in the following subsections.

11.2.1 Power Per Subscriber, Traffic and Distance/Area

The following two energy metrics have been recommended by the ITU [Hamdoun et al. - 2012]:

$$\frac{Power}{Subscriber \cdot Traffic \cdot Distance} \quad [\text{watts bps}^{-1}\ \text{m}^{-1}] \tag{11.1a}$$

$$\frac{Power}{Subscriber \cdot Traffic \cdot Area} \quad [\text{watts bps}^{-1}\text{m}^{-2}]. \tag{11.1b}$$

It is remarkable that the formal metric is meant for use in the wired networks whereas the latter metric is bound for use in the wireless access networks.

11.2.2 Energy Consumption Rating (ECR) Measured in W Gbps^{-1}

The ECR-based metrics can be categorized into:

1. ECR-weighted.
2. ECR over a variable-load cycle.
3. ECR over extended-idle load cycle.

It should be noted that the ECR metric has been developed by an open initiative and has not been standardized. Nevertheless, it is usually employed for energy consumption evaluations. The ECR metric can be expressed as [Hamdoun et al. - 2012]

$$ECR = \frac{P_f}{T_f},$$ (11.2)

where T_f represents the maximum throughput achieved in the measurement and P_f denotes the measured peak power during the test execution.

Furthermore, the energy efficiency rate (EER) is another energy metric that can be derived from the fundamental ECR metric. The EER metric in (bps W^{-1}) is inversely proportional to ECR and can be defined as [Hamdoun et al. - 2012]

$$EER = \frac{1}{ECR}.$$ (11.3)

11.2.3 Telecommunications Energy Efficiency Ratio (TEER)

The ATIS Network Interface, Power and Protection (NIPP) Committee developed the TEER metric. The metric offers a major unified approach for power and energy measurements of the telecommunication equipment as well as the associated power and energy efficiencies. Moreover, apart from the test and measurement methods that have been standardized for the TEER metric, the equipment utilization levels, environmental test conditions, and reporting methods are defined as well. The TEER metric can be expressed as [Hamdoun et al. - 2012]

$$TEER = \frac{\text{Useful work}}{\text{Power}} \quad [?/\text{watt}],$$ (11.4)

where 'useful work' (and its unit) differs with the equipment brand and depends on the equipment function.

11.2.4 Telecommunication Equipment Energy Efficiency Rating (TEEER)

The total power consumption P_{total} using TEEER is expressed as [Hasan et al. - 2011]

$$P_{total} = 0.35P_{max} + 0.4P_{50} + 0.25P_{sleep},$$ (11.5)

where P_{max} denotes power consumption at full rate, P_{50} signifies power consumption at half-rate and P_{sleep} is the power consumption in the sleep mode. The weights are equal to unity and are normally achieved statistically [Hasan et al. - 2011].

11.3 Green Design for Energy Crunch Prevention in 5G Networks

In 5G and B5G networks, prevention of energy crunch demands advanced methods for wireless network design and operation. There has been a general consensus that the thousand fold increase in capacity should be realized at a comparable or lesser power consumption compared to the existing networks. This implies a factor of 1000 or more increase in the efficiency with which each Joule of energy is expended for information transmission [Buzzi et al. - 2016].

In general, the EE of a BS is based on its different basic components as well as the core radio devices such as power amplifiers (PAs) and radio transceivers. Hence, to address the power consumption of BSs so as to achieve a green radio network, the key focus has to be on the access and core networks [Karmokar and Anpalagan - 2013]. Generally, the methods that can be employed to improve the EE of 5G and B5G networks can be classified into several schemes such as hardware solutions, resource allocation, network planning and deployment, and energy harvesting and transfer [Buzzi et al. - 2016]. This is depicted in Figure 11.4 and discussed in the following subsections.

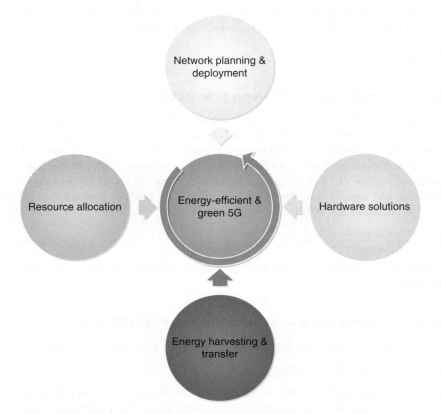

Figure 11.4 Energy-efficient technologies of 5G network [Buzzi et al. - 2016].

11.3.1 Hardware Solutions

The energy efficient hardware solution is based on the wireless system hardware design that explicitly considers energy consumption. It entails green design of the RF chain and employment of simplified transceiver structures. Furthermore, this solution entails adoption of the main architectural changes such as cloud based implementation of the RAN. It also involves the implementation of network function virtualization [Buzzi et al. - 2016].

As mentioned in Section 1.1, PAs consume a substantial amount of power and its EE is subject to modulation, frequency band, as well as operating environment [Buzzi et al. - 2016; Hasan et al. - 2011]. It should be noted that the requirements for PA linearity and high peak-to-average power ratios (PAPR) make BSs inefficient. It should be noted that realization of high linearity in the PAs so as to maintain the radio signals quality requires PAs to be operated considerably below saturation. However, this eventually brings about poor power efficiency. Also, the employed modulation schemes by different communication standards are described by intensely fluctuating signal envelopes with PAPR that is over 10 dB [Hasan et al. - 2011]. Thus, energy efficient design of the PAs has been receiving significant attention. This is not only in the direct circuit design but also in the signal design methods for PAPR reduction [Buzzi et al. - 2016; Hasan et al. - 2011]. For instance, an innovative PA architecture such as digital pre-distorted Doherty architectures and aluminum gallium nitride (GaN) based amplifiers are attractive since they can boost the power efficiency level by more than 50% [Hasan et al. - 2011; Correia et al. - 2010]. Furthermore, another approach to enhance the EE is to shift from the traditional analog RF PAs to switch mode PAs [Hasan et al. - 2011]. The switch mode PAs run cooler and draw a lesser amount of current compared to the traditional PAs. This is achieved by turning the output transistors on and off at an ultrasonic rate while amplifying the signal. Consequently, its total component efficiency is approximately 70%. The development of flexible PA architectures can also be a key contributor for high EE, namely by enabling a better adaptation of the amplifier to the required output power. Furthermore, a number of linearization techniques like Cartesian feedback, digital pre-distortion and feed-forward in conjunction with various type of digital signal processor (DSP) schemes that can lessen the PA linear area requirement can be employed [Hasan et al. - 2011; Claussen et al. - 2008].

In addition, the employment of simplified transceiver architectures that comprises adoption of coarse signal quantization and hybrid analog/digital beamformers can help in increasing the hardware EE by reducing the system complexity and energy consumption. This is more applicable to networks in which massive MIMO and millimeter wave (mmWave) systems are employed [Buzzi et al. - 2016].

As stated earlier, the cloud based implementation of RANs is the main disruptive technology that aids in making the 5G networks more energy efficient. In the cloud-RAN (C-RAN), various functions that are presently executed in the BS are shifted to a remote data center where they are implemented via software. Also, another implementation of C-RAN anticipates light BSs in which just the RF chain as well as the baseband-to-RF conversion stages are present. The light BSs are then connected by means of high capacity links to the data center. Subsequently, all baseband processing and resource allocation algorithms are executed at the data center. This configuration permits high network flexibility and results in considerable savings in both deployment costs and energy consumption. Furthermore, mobile-edge computing is another approach that increases the network flexibility, which results in substantial energy savings [Buzzi et al. - 2016; Alimi et al. - 2018].

11.3.2 Network Planning and Deployment

In an effort to attend to the envisaged amount of devices, a number of advanced technologies have been presented for the planning, deployment, as well as operation of 5G networks [Buzzi et al. - 2016].

11.3.2.1 Dense Networks

Dense networks have been proposed in order to address the huge amount of connected devices through the deployment of more network equipment. For 5G implementation, dense heterogeneous networks and massive MIMO are envisioned to be the main technologies in the network. The massive MIMO implementation entails replacing conventional arrays with a few antennas that are fed by bulky as well as expensive hardware with hundreds of small antennas that are fed by low-cost amplifiers and circuitry. Massive MIMO has the ability to decrease the radiated power by a factor that is proportional to the square root of the amount of deployed antennas without altering the information rate. Nevertheless, this only applies to an ideal, single-cell massive MIMO system without considering the power consumed by the hardware. In addition, self-organizing cells can be employed in dense networks to help in the network deployment in response to traffic conditions. This is due to the fact that self-organizing cells have a salient feature that enables them to autonomously activate/deactivate in response to traffic demands. This can be a key enabler for an enhanced EE in dense networks [Buzzi et al. - 2016].

11.3.2.2 Offloading Techniques

A traffic offloading technique is also a fundamental 5G approach for improving the capacity and EE of the network. The existing user devices have been equipped with multiple radio access technologies such as Bluetooth, cellular, and WiFi to facilitate traffic offloading whenever alternative connection technologies are accessible. This offers additional cellular resources to the users that are not capable of traffic offloading. A number of offloading strategies such as device-to-device communications, visible light communications, local caching, and mmWave cellular are anticipated. In general, traffic offloading techniques have an intense influence on the system EE. This is because direct transmission between neighboring devices can occur at a considerably lower transmit power compared to that required for communication through a BS, which might be remotely located [Buzzi et al. - 2016; Alimi et al. - 2018].

In general, the technique involves deployment of infrastructure nodes so as to maximize the covered area per consumed energy instead of the covered area only. Furthermore, the implementation of BS switch-off/on algorithms (sleep modes), antenna muting methods (deactivation), and self-optimized antenna muting schemes can also help in adapting to the fluctuating traffic conditions, thereby contributing to an additional reduction in energy consumption [Buzzi et al. - 2016; Gandotra et al. - 2017]. It should be noted that, when cells are in sleep mode or switched off, the coverage gaps caused by the sleeping/off cells can be effectively covered by the remaining active cells with intelligent features. The idea of self-organizing networks has been initiated with the intention of improving network performance and flexibility [Hasan et al. - 2011; Alimi et al. - 2018].

11.3.3 Resource Allocation

Since EE happens to be one of the major key performance indicators of 5G networks, efforts are ongoing on a paradigm shift from throughput optimized to EE optimized communication systems. Therefore, the EE of a communication system can be optimized by allocating the system radio resources with the intention of maximizing the EE instead of the throughput. This can be achieved by enhancing the aggregate information that is reliably transmitted per Joule of consumed energy rather than the sole maximization of the amount of information that is reliably transmitted. This method has been shown to offer considerable EE gains at the expense of a slight throughput reduction. There are a number of innovative resource allocation algorithms that have been designed for EE maximization [Buzzi et al. - 2016].

11.3.4 Energy Harvesting (EH) and Transfer

It has been observed that, wireless EH schemes are attractive methods for improving the EE and lowering the entire GHG emissions for wireless communications and networking. Consequently, they has been acknowledged as part of the main technologies for the 5G networks. This can also be attributed to their support for ubiquitous connectivity. Wireless EH can be categorized into dedicated EH and ambient EH [Alimi et al. - 2018; Tabassum et al. - 2015].

11.3.4.1 Dedicated EH

A dedicated EH is a type of EH that involves transmission of energy from dedicated sources to the EH devices on purpose. Owing to its dependence on the deployed dedicated energy sources, it demands extra power consumption. Therefore, an ambient EH is an attractive approach for reducing the grid power consumption in a C-RAN [Alimi et al. - 2018; Ghazanfari et al. - 2016].

11.3.4.2 Ambient EH

Ambient EH is a class of energy harvested from renewable energy sources like wind, solar, electromechanical, and thermoelectric effects. In addition, energy harvested from ambient RF signals like radio networks, BSs, TV towers, and WiFi networks that are detected by EH receivers are also ambient EH [Alimi et al. - 2018; Ghazanfari et al. - 2016]. Therefore, ambient EH facilitates recycling of energy that would otherwise have been wasted [Buzzi et al. - 2016].

MmWave communication with EH potential is one of feasible schemes for achieving the required multi-gigabit data rates while accomplishing green communications. The mmWave band is attractive for wireless EH since it has been envisaged that it will feature large-dimensional antenna arrays with directional beamforming. Also, it is expected to support densely deployed BSs to ensure high EE, SE, and network coverage [Wu et al. - 2017]. A 60 GHz harvester can convert mmWave input power to storable DC power that can be consumed instantaneously or be used to charge storage capacitors or batteries [Nariman et al. - 2017]. The generated power can be employed for wireless charging, and charge coil-free and battery-free solutions for a massive number of low-power wireless devices in the networks. Nevertheless, signal propagation at this band suffers from diffraction and poor penetration. This makes the mmWave band sensitive to blockage by barriers. Consequently, whether mmWave will be more promising for RF EH compared with the conventional band is still debatable.

In general, this scheme helps in operating communication systems by harvesting energy from the environment. The exploited ambient energy by the transmission nodes is converted into a suitable DC signal for powering devices such as cell phones. It should be noted that the energy sources could probably be practical alternatives to regular batteries. Consequently, ambient energy exploitation brings about longer operation lifetime as well as elimination of the related problem of battery replacement under severe situations [Alimi et al. - 2018; Ghazanfari et al. - 2016].

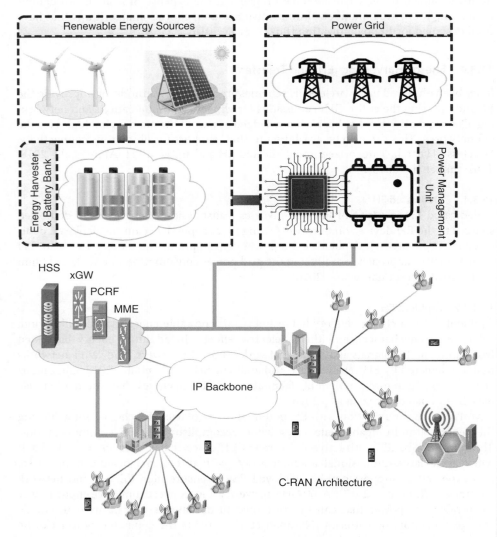

Figure 11.5 Energy efficient hybrid solution for the field deployment of a C-RAN architecture.

EH is becoming more appealing for charging low-power devices wirelessly due to the current technological developments in the electronics industries. Nevertheless, as a result of the erratic nature of the energy influxes over diverse location, time, and weather conditions, EH performance reliability may not be guaranteed [Alimi et al. - 2018; Ghazanfari et al. - 2016]. However, RF based ambient EH is a promising solution for the future power/energy-constrained 5G wireless networks owing to its sustainable nature [Alimi et al. - 2018]. It presents a viable means of reducing the sporadic nature of wireless power sources. The concept entails the combination of EH with wireless power transfer techniques, thereby enabling sharing of energy between the network nodes. One of the major benefits of the scheme is that it enables redistribution of the network energy, thereby extending the lifetime of nodes that are low on battery energy. Moreover, it supports the deployment of dedicated beacons in the network in order to reduce or manage the intermittent nature of the RF energy source. The scheme can also be improved by superimposing energy signals on normal communication signals. Simultaneous wireless information and power transfer is based on this concept [Buzzi et al. - 2016], which also relates to the optical fiber based solution discussed in Section 11.4.

In general, the future cellular network BSs are envisaged to be powered by both on-grid and renewable energy sources. This results into a hybrid energy system, as depicted in Figure 11.5 in which an EH battery is employed to address the erratic nature of renewable energy sources. The main objective of the scheme is to optimize energy utilization in the networks by maximizing the utilization of green (renewable) energy while saving on-grid energy as much as possible. In this scheme, different scheduling techniques are required in order to deal with the presence of multiple energy sources. Scheduling algorithms are also needed for joint control of the transmission power so as to choose the energy source that minimizes the total energy consumption. In addition, power co-operation implementation facilitates the sharing of green power between various BSs on every possible period. This brings about a sustainable and energy efficient wireless network operation [Ismail et al. - 2015].

11.4   Fiber Based Energy Efficient Network

The C-RAN has been considered as an attractive architecture for 5G goal realization by both academia and industry. Nevertheless, for the C-RAN to be a feasible solution, some related issues should be considered carefully. One of these issues is the associated cost and time for the quantity of power lines that will be needed to supply a massive number of remote radio heads (RRHs) in the network. Apart from the fact that the huge energy demanded is unpleasant economically, it is not also eco-friendly. Furthermore, with the required ubiquitous network connectivity, some RRHs in the off-grid sites or places where there is no external power supply have to be considered in order to provide the required coverage. In such a situation, a suitable way of supplying power to the remote RRHs has to be resolved. One of the preferred ways for achieving this purpose is through a C-RAN based on a passive optical network (PON) and power over fiber (PoF) or photonic power [Alimi et al. - 2018].

In the PoF scheme, optical energy is transmitted over fiber to the remotely deployed sites where it is transformed into electrical energy by means of a photovoltaic (PV) converter [Röger et al. - 2008; Wake et al. - 2008]. The PoF system employs an optical fiber to connect a high-power laser dio: de (HPLD) that is located at the central office (CO) to a PV converter that is being deployed at the remote location. In addition, there are two main approaches that can be employed in the PoF scheme. The approaches are categorized based on whether power sources are required in the remote antenna units (RAUs) or not [Alimi et al. - 2018; Röger et al. - 2008; Wake et al. - 2008].

11.4.1 Zero Power RAU PoF Network

A zero power RAU PoF system does not require an electrical power supply or any form of power source in the RAU. Similarly, integration of either coin battery or EH devices to the RAU for effective functioning is not needed. Therefore, this initiates the concept of energy autonomous RAU systems. Furthermore, a zero power RAU PoF system is mainly based on the application of an HPLD for delivering power to remotely deployed equipment like RRHs. It is noteworthy that the PoF system supports transmission of both data and optical power supply on a shared optical fiber infrastructure. The PV power converters are normally used at the remote site for transforming the optical power into electrical power [Alimi et al. - 2018; Wake et al. - 2008; Werthen et al. - 2005].

11.4.2 Battery Powered RRH PoF Network

An optical splitter is normally used in the PON with PoF architecture so that multiple RRHs can be served with a single optical fiber cable. Furthermore, the required electrical power can be supplied to the RRHs via the optical line terminal (OLT) that aggregates them. The PoF scheme is implemented in the OLT to realize this goal. The broadcast communication nature of the PON system wherein the OLT broadcasts the data received from the CO to all RRHs is exploited by the scheme. As a result of this, a number of RRHs receive data that are not intended for them. The received redundant data are subsequently transformed into electrical power with PoF technology [Alimi et al. - 2018; Miyanabe et al. - 2015].

It is noteworthy that the battery, the optical network unit (ONU), and the antenna modules are the key components of the RRH. The idea of the approach is that, at any moment that the RRH's ONU module receives the desired signal from the OLT, it relays it to the antenna module for onward transmission. However, in a situation wherein redundant data are received by the ONU module, the module will transform the optical signal into electrical power. Subsequently, the electrical power produced by the module will be conveyed to the battery module for storage. In the process, the stored electrical power can be employed to drive the ONU and antenna modules. It is remarkable that RRH energy consumption can be lowered by implementing the sleep mode of operation for turning off specific modules. Nevertheless, to facilitate continuous generation and storage of electrical power in the system, the battery modules and the ONU receive unit of each RRH should always be active [Alimi et al. - 2018; Miyanabe et al. - 2015].

11.5 System and Power Consumption Model

Consider a C-RAN cellular network comprising N RRHs and M baseband units (BBUs) as illustrated in Figure 11.6. Then, a set of all RRHs can be denoted by $\mathbb{N} = \{1, 2, \ldots, N\}$. Also, assume that $\mathbb{U} = \{1, 2, \ldots, U\}$ represents a set of users in the C-RAN and U_l signifies a set of users served by the lth RRH. Also, it is assumed that the BBU pool

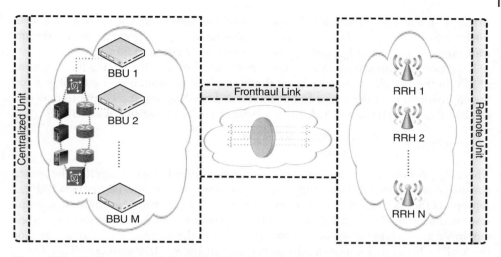

Figure 11.6 C-RAN system model considered for the power consumption analysis.

has a number of processors that are capable of processing the signal from RRHs. Then, the processors in the pool can be represented as $\mathbb{C} = \{1, 2, \ldots, C\}$. Furthermore, in a traditional cellular network transmission design, the transmit power at each BS can be expressed as [Dasi and Yu - 2016]

$$P_{l,tx} = \mathbb{E}[|x_l|^2] \le P_l, \quad l \in \mathbb{L} = \{1, 2, \ldots, L\}, \tag{11.6}$$

where P_l, denotes the BS transmit power budget and x_l represents the transmit signal at the lth BS.

It is noteworthy that for an effective and full description of the BS power consumption, the efficiency of the PA as well as other power consuming modules, such as the cooling system and the BBU pool, should be taken into consideration. Furthermore, the power consumption of the fronthaul link in C-RAN architectures should be considered as well.

The power consumption of the BS components like PA, RF small signal transceiver, main supply (MS), baseband engine (BB), direct current (DC)–DC converter, and active cooling (CO) is contingent on factors such as number of radio chains/antennas, transmission bandwidth and power. Therefore, the power consuming components in a BS is contingent on the BS design and vary from BS to BS. For instance, the BS maximum power consumption can be expressed as [Auer et al. - 2011; Sabella et al. - 2014]

$$P_{BS} = \frac{P_{BB} + P_{RF} + P_{PA}}{(1 - \sigma_{DC})(1 - \sigma_{MS})(1 - \sigma_{CO})} \tag{11.7}$$

where σ_{CO}, σ_{DC}, σ_{MS}, correspond to the loss factors for the CO, DC, and MS, respectively, and P_{RF}, P_{BB}, and P_{PA} represent the power consumption for the RF, BB, and PA, respectively, and are given by [Holtkamp et al. - 2013]

$$P_{RF} = D\frac{W}{10 \text{ MHz}}P'_{RF}, \tag{11.8a}$$

$$P_{BB} = D\frac{W}{10 \text{ MHz}}P'_{BB}, \tag{11.8b}$$

$$P_{PA} = \frac{P_{max}}{D\eta_{PA}(1 - \sigma_{feed})}. \tag{11.8c}$$

where W denotes the bandwidth, D represents the number of BS antennas, η_{PA} signifies the PA efficiency, σ_{feed} indicates the feeder cable losses, P_{max} represents the maximum transmission power, and P'_{BB} and P'_{RF} denote some basic consumption.

The power consumption can be defined by a unified model that is valid for a number of BSs. In the model, a piecewise linear function of the transmit power is employed for the approximation of the evolved node B (eNodeB or eNB) power consumption, which can be modeled as [Dasi and Yu - 2016; Sabella et al. - 2014; Sigwele et al. - 2017]

$$P_l^{eNB} = \begin{cases} P_0 + y \cdot \Delta_p \cdot P_{max} & \text{if } 0 < y < 1 \\ P_{sleep} & \text{if } y = 0 \end{cases} \qquad (11.9)$$

where P_{max} denotes the maximum transmission power (full load), P_0 signifies the power consumption at the minimum non-zero output power (idle mode, i.e. zero load), P_{sleep} represents the BS sleep state (sleep mode, i.e. $P_{sleep} < P_0$) power consumption, Δ_p represents the slope of the load dependent linear power model that is subject to a given employed server, and the scaling parameter y denotes the normalized cell traffic load of the lth BS where $y = 1$ implies a fully loaded system and $y = 0$ denotes an idle system.

It should be noted that in Equation (11.6), there is no provision for the sleep mode as offered in Equation (11.9). In addition, with reference to Equation (11.9), an eNB can be operated in sleep mode whenever it does not have any data to receive or transmit. Since P_{sleep} is usually less than P_0, operating the eNB in sleep mode will help in energy saving. This is due to the fact that sleep mode permits some of the eNB main units to be turned off under a no load condition.

In addition, it is noteworthy that the long term evolution (LTE) model presented in Equation (11.9) cannot be implemented directly on the C-RAN. This is due to the fact that, unlike the traditional architecture, the C-RAN architecture centralizes and shares some components. This requires a modified power consumption model for the C-RAN. The model has to account for the power consumption of the major parts of the C-RAN such as the BBU pool, fronthaul, and RRHs. Therefore, the power consumption model for the C-RAN can be written as

$$\sum_{\substack{i \in I \\ j \in J \\ k \in K}} (P_i^{RU} + P_j^{CU} + P_k^{FH}) \qquad (11.10)$$

$$\sum_{i \in I} \sum_{j \in J} \sum_{k \in K} (P_i^{RU} + P_j^{CU} + P_k^{rmFH}) \qquad (11.11)$$

where P_i^{RU}, P_j^{CU} and P_k^{FH} are the remote unit, centralized unit, and fronthaul power consumptions, respectively.

In the following subsection, we present the power consumption model considering the RRH, BBU, and fronthaul links for the C-RAN system.

11.5.1 Remote Unit Power Consumption

In the C-RAN architecture, the RRHs are low complexity as well as low processing nodes in which RF operations are exclusively performed. It is also remarkable that there are no cooling losses on the RRH, because the cooling is usually realized by natural air. Furthermore, the RRHs depend on self-backhauling. This implies that the backhaul also shares the same wireless channel resources. Consequently, $P_{BB} = 0$ in the expression given in Equation (11.7). In this scenario, the remote unit based power consumption can be expressed as

$$P_i^{RU} = \sum_{n=1}^{N_{RRH}} (P_0^{RRH} + y_{RRH}\Delta_p^{RRH}P_{max}^{RRH}) \tag{11.12}$$

where all the parameters are as defined in Equation (11.9) but in terms of RRH.

11.5.2 Centralized Unit Power Consumption

The traditional DSPs or general purpose processors (GPPs) can be employed as the centralized BBU processors. However, GPPs are preferred due to features such as (re)programmability, affordability, as well as high processing capabilities. A standard GPP power consumption can be modeled as [Sabella et al. - 2014; Sigwele et al. - 2017]

$$P_j^{CU} = P_0^{GPP_j} + y_{GPP_j}\Delta_p^{GPP_j}P_{max}^{GPP_j} \tag{11.13}$$

where all the parameters are as defined in Equation (11.9) but in terms of a specific central processing unit utilization of GPP$_j$.

11.5.3 Fronthaul Power Consumption

The fronthaul network connects the RRHs to the BBU pool in the C-RAN. As discussed in [Alimi et al. - 2018], there are a number of fronthaul technologies that can be employed in the C-RAN. Consequently, fronthaul power consumption depends on the adopted technology. The fronthaul network power consumption can be modeled by assuming a set of parallel communication channels. Then, the fronthaul power consumption can be defined in accordance with the capacity C_k of each fronthaul and the power dissipation $P_{k,max}^{FH}$ as [Dasi and Yu - 2016]

$$P_k^{FH} = \frac{R_k^{FH}}{C_k}P_{k,max}^{FH} = \rho_k R_k^{FH}, \tag{11.14}$$

where $\rho_k = P_{k,max}^{FH}/C_k$ denotes a constant scaling factor and R_k^{FH} represents the fronthaul traffic between the RRH$_l$ and the BBU.

It is remarkable that although Equation (11.14) applies mainly to the microwave fronthaul links, it can also be generalized to other fronthaul technologies like fiber-based ethernet and passive optical networks. In a fiber based connection, the loss per connector and loss per splice also have to be considered [Sigwele et al. - 2017].

In general, the expression for power consumption of C-RAN shows that there are different potential means by which its EE can be enhanced. Apart from reducing the transmit power, the EE can be improved by setting the BSs in sleep mode. However, from other perspective, deactivating some BSs implies reduced proficiency for interference mitigation between the BSs that are active. This approach brings about the need for extra transmit power so as to sustain the QoS and QoE [Dasi and Yu - 2016].

Another notable approach is by decreasing the backhaul traffic. Nevertheless, higher fronthaul rate permits more user information to be shared among the BSs. This facilitates improved cooperation between the BSs and hence, interference mitigation can be realized. This may result in lower transmit power. A joint design is essential so that parameters/features such as BS activation, transmit power, and backhaul traffic rate can be effectively balanced for enhanced EE [Dasi and Yu - 2016]. In the following subsection, we present a means of EE maximization for C-RAN architectures.

11.5.4 Massive MIMO Energy Efficiency

The EE is the ratio of the total system throughput and the overall transmit power. Also, it is the quantity of information/number of bits that can be reliably conveyed per Joule of expended energy and it is expressed as [Prasad et al. - 2017; Buzzi et al. - 2016; Karmokar and Anpalagan - 2013]

$$EE = R/P, \tag{11.15}$$

where R represents the system throughput and P denotes the power expended in attaining R.

Furthermore, the massive MIMO scheme presents multiple orders of SE and EE gains over the current LTE technologies. Consequently, it has been considered as one of the main enablers for 5G systems [Prasad et al. - 2017] A massive MIMO is a multi-user MIMO (MU-MIMO) scheme in which service provisioning for the K user equipment (UE) is offered on the same time frequency resource by a BS that has been equipped with M antennas. In this scenario, $M \gg K$. Moreover, when a massive number of antennas are deployed at the BS, a favorable propagation scenario can be achieved. In this situation, the effects of intra-cell interference, small-scale fading, and uncorrelated noise vanish asymptotically. In addition, when M and K are increased, huge array and multiplexing gains can be realized [Prasad et al. - 2017].

Let us consider zero-forcing (ZF) and maximum ratio combining (MRC) for uplink (UL) detection, the UL linear receive combining matrix can be expressed as [Björnson et al. - 2015]

$$\mathbf{G} = \begin{cases} \mathbf{H} & \text{for MRC,} \\ \mathbf{H}\,(\mathbf{H}^H\mathbf{H})^{-1} & \text{for ZF,} \end{cases} \tag{11.16}$$

where $\mathbf{H}$ is the channel state matrix. Also, consider precoding schemes such as ZF and maximum ratio transmission (MRT) for downlink (DL) transmissions, then the precoding matrix can be defined as

$$\mathbf{V} = \begin{cases} \mathbf{H} & \text{for MRT,} \\ \mathbf{H}\,(\mathbf{H}^H\mathbf{H})^{-1} & \text{for ZF.} \end{cases} \tag{11.17}$$

Moreover, let us assume fractions of UL and DL transmission represented by $\zeta^{(\text{ul})}$ and $\zeta^{(\text{dl})}$, respectively, whose total is equal to unity (i.e. $\zeta^{(\text{ul})} + \zeta^{(\text{dl})} = 1$). Then, the achievable UL, $R_k^{(\text{ul})}$ and DL, $R_k^{(\text{dl})}$ rates (bits s^{-1}) of the kth UE with linear processing can be expressed respectively as [Björnson et al. - 2015]

$$R_k^{(\text{ul})} = \zeta^{(\text{ul})} \left(1 - \frac{\tau^{(\text{ul})}K}{U\zeta^{(\text{ul})}} \right) \overline{R}_k^{(\text{ul})} \tag{11.18a}$$

$$R_k^{(\text{dl})} = \zeta^{(\text{dl})} \left(1 - \frac{\tau^{(\text{dl})}K}{U\zeta^{(\text{dl})}} \right) \overline{R}_k^{(\text{dl})}, \tag{11.18b}$$

where $\tau^{(\cdot)}$ denotes the pilot sequence length, U represents coherence block and $\overline{R}_k^{(\cdot)}$ denotes the gross rate (bits s^{-1}) from the kth UE.

From (11.15), the total EE of the UL and DL transmission can be written as

$$
\text{EE} = \frac{\sum\limits_{k=1}^{K} \left(\mathbb{E}\left\{ R_k^{(\text{ul})} \right\} + \mathbb{E}\left\{ R_k^{(\text{dl})} \right\} \right)}{P}
\tag{11.19}
$$

where $\mathbb{E}$ represents an expectation operator, $P = P_{\text{TX}}^{(\text{ul})} + P_{\text{TX}}^{(\text{dl})} + P_{\text{CP}}$, P_{CP} denotes the circuit power consumption, $P_{\text{TX}}^{(\text{ul})}$ and $P_{\text{TX}}^{(\text{dl})}$, are the average UL and DL PA power consumption, respectively, and are expressed as [Björnson et al. - 2015]

$$
P_{\text{TX}}^{(\text{ul})} = \sigma^2 \frac{B\zeta^{(\text{ul})}}{\eta^{(\text{ul})}} \, \mathbb{E}\left\{ \mathbf{1}_K^T \left(\mathbf{D}^{(\text{ul})} \right)^{-1} \mathbf{1}_K \right\}
\tag{11.20a}
$$

$$
P_{\text{TX}}^{(\text{dl})} = \sigma^2 \frac{B\zeta^{(\text{dl})}}{\eta^{(\text{dl})}} \, \mathbb{E}\left\{ \mathbf{1}_K^T \left(\mathbf{D}^{(\text{dl})} \right)^{-1} \mathbf{1}_K \right\},
\tag{11.20b}
$$

where $\mathbf{1}_K$ denotes the K-dimensional unit vector, B signifies the operating bandwidth of the MIMO system, σ^2 represents the noise variance (in Joule/symbol), $\eta^{(\cdot)}$ represents the PA efficiency at the UEs and $\mathbf{D} \in \mathbb{C}^{K \times K}$.

In addition, under a realistic circuit power consumption model, P_{CP} can be defined as

$$
P_{\text{CP}} = P_{\text{FIX}} + P_{\text{TC}} + P_{\text{CE}} + P_{\text{C/D}} + P_{\text{BH}} + P_{\text{LP}}
\tag{11.21}
$$

where P_{CE} represents the power consumption that is due to the channel estimation process, P_{BH} denotes the load dependent backhaul power consumption, P_{LP} is the power consumption due to linear processing at the BS, $P_{\text{C/D}}$ is the power consumption for channel coding and decoding units, P_{TC} denotes the power consumption of the transceiver chains, and a constant parameter P_{FIX} represents the fixed power consumption.

Moreover, the sum asymptotic SE of a massive MIMO system can be expressed as [Wang et al. - 2017]

$$
C_{\text{sum}} = C^{\text{ul}} + C^{\text{dl}},
\tag{11.22}
$$

where C^{ul} and C^{dl} are the UL and DL capacities, respectively, and are given by [Wang et al. - 2017]

$$
C^{\text{ul}} = \frac{T_{\text{C}} - \tau}{T_{\text{C}}} \sum_{k=1}^{K} \log_2(1 + \gamma^{\text{ul}}),
\tag{11.23a}
$$

$$
C^{\text{dl}} = \frac{T_{\text{C}} - \tau}{T_{\text{C}}} \sum_{k=1}^{K} \log_2(1 + \gamma^{\text{dl}}),
\tag{11.23b}
$$

where T_{C} denotes the channel coherence time and $\gamma^{(\cdot)}$ signifies the asymptotic signal-to-interference-plus-noise ratio.

11.6 Simulation Results and Discussions

We assume a single cell scenario that is operating at the 2 GHz band with the number of users $K = 10$ UEs and a pilot length $\tau = 2K$. A transmission bandwidth of 20 MHz is also assumed for a channel coherence bandwidth of 180 kHz and channel coherence time of length $T_C = 300$ (symbols). Fractions of both UL and DL transmissions are assumed to be 0.5 and the power required for the fronthaul traffic is 0.25 W Gbps^{-1} while that required for coding and decoding is 0.2 W Gbps^{-1} and 0.7 W Gbps^{-1}). Further information on parameter selection can be found in [Björnson et al. - 2015].

Figure 11.7 depicts the SE and maximal EE for a number of BS antennas considering ZF and MRT/MRC as processing schemes under a single-cell scenario. It is observed that the SE increases with an increase in the number of BS antennas, as shown in Figure 11.7a.

In addition, with reference to Figures 11.7a and b, it is observed that the performance of the ZF with perfect channel state information obtained from the UL pilots is better than that of imperfect conditions. Furthermore, with ZF precoding, the transmitter can dynamically null out (suppress) multi-user interference signals at affordable complexity. The ZF precoder also performs even better under high SNR conditions compared with

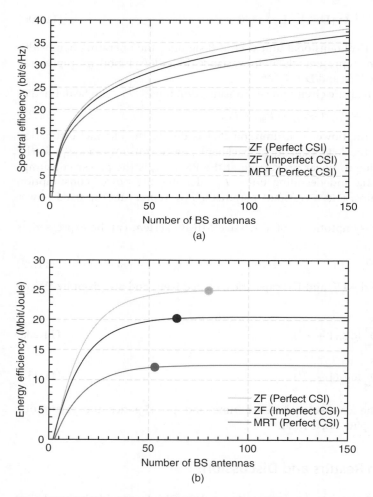

Figure 11.7 (a) SE and (b) maximal EE for a number of BS antennas with ZF and MRT processing under a single-cell scenario.

the MRT processing that demands an infinite M condition (i.e. $M \gg K$, this condition tends to increase the circuit/computational power significantly whereas the EE is not improved).

11.7 Conclusion

In this chapter, we have presented a comprehensive review and discussion on techniques that not only aid in enhancing the 5G EE gains but also help in reducing the OPEX and CO_2 emissions in green wireless cellular networks. Moreover, the related opportunities of massive MIMO implementations in the C-RAN have been comprehensively discussed. The potential advantages of mmWave for EE maximization through EH have been discussed. For instance, with an EH, mmWave input power can be converted into storable DC power. Furthermore, a number of cost-effective wired and wireless schemes for supplying the required power to the network entities in order to realize ubiquitous network connectivity have been presented. In addition, analytical expressions and numerical results have been presented to demonstrate that the EE of the 5G and B5G networks can be improved significantly. We also discuss research challenges and open-ended issues on ways of achieving the requirement of 5G networks in terms of EE in an economical manner.

Acknowledgments

This work is supported by the Fundação para a Ciência e a Tecnologia (FCT) under PhD grant PD/BD/52590/2014. Also, it is supported by the European Regional Development Fund (FEDER), through the Regional Operational Programme of Lisbon (POR LISBOA 2020) and the Competitiveness and Internationalization Operational Programme (COMPETE 2020) of the Portugal 2020 framework, Project 5G (POCI-01-0247-FEDER-024539), ORCIP (CENTRO-01-0145-FEDER-022141) and SOCA (CENTRO-01-0145-FEDER-000010). It is also funded by FCT through national funds under the project COMPRESS - PTDC/EEI-TEL/7163/2014 and by the European Regional Development Fund (FEDER), through the Regional Operational Program of Centre (CENTRO 2020) of the Portugal 2020 framework [Project HeatIT with Nr. 017942 (CENTRO-01-0247-FEDER-017942)]. Our gratitude is also extended to the following funding body Ocean12-H2020-ECSEL-2017-1-783127.

Bibliography

I. A. Alimi, A. L. Teixeira, and P. P. Monteiro. Toward an Efficient C-RAN Optical Fronthaul for the Future Networks: A Tutorial on Technologies, Requirements, Challenges, and Solutions. *IEEE Communications Surveys Tutorials*, 20(1): 708–769, Firstquarter 2018. doi: 10.1109/COMST.2017.2773462.

Isiaka Alimi, Ali Shahpari, Vítor Ribeiro, Artur Sousa, Paulo Monteiro, and António Teixeira. Channel characterization and empirical model for ergodic capacity of free-space optical communication link. *Optics Communications*, 390:123–129, 2017a. ISSN 0030-4018. doi: https://doi.org/10.1016/j.optcom.2017.01.001. URL http://www .sciencedirect.com/science/article/pii/S0030401817300019.

Isiaka Alimi, Ali Shahpari, Artur Sousa, Ricardo Ferreira, Paulo Monteiro, and António Teixeira. Challenges and Opportunities of Optical Wireless Communication Technologies. In Pedro Pinho, editor, *Optical Communication Technology*, chapter 02, pages 5–44. InTech, Rijeka, 2017b. ISBN 978-953-51-3418-3. doi: 10.5772/intechopen.69113. URL https://cdn.intechopen.com/pdfs-wm/55559.pdf.

Isiaka A. Alimi, Paulo P. Monteiro, and António L. Teixeira. Analysis of multiuser mixed RF/FSO relay networks for performance improvements in cloud computing-based radio access networks (CC-RANs). *Optics Communications*, 402:653–661, 2017c. ISSN 0030-4018. doi: http://dx.doi.org/10.1016/j.optcom.2017.06.097. URL http://www .sciencedirect.com/science/article/pii/S0030401817305734.

Isiaka A. Alimi, Paulo P. Monteiro, and António L. Teixeira. Outage probability of multiuser mixed RF/FSO relay schemes for heterogeneous cloud radio access networks (H-CRANs). *Wireless Personal Communications*, 95(1): 27–41, Jul 2017d. ISSN 1572-834X. doi: 10.1007/s11277-017-4413-y. URL https://doi.org/10.1007/s11277-017-4413-y.

Mohammed H. Alsharif, Rosdiadee Nordin, and Mahamod Ismail. Survey of green radio communications networks: Techniques and recent advances. *Journal of Computer Networks and Communications*, 2013:1–13, 2013. ISSN 2090-7141. doi: 10.1155/2013/453893.

G. Auer, V. Giannini, C. Desset, I. Godor, P. Skillermark, M. Olsson, M. A. Imran, D. Sabella, M. J. Gonzalez, O. Blume, and A. Fehske. How much energy is needed to run a wireless network? *IEEE Wireless Communications*, 18(5): 40–49, October 2011. ISSN 1536-1284. doi: 10.1109/MWC.2011.6056691.

E. Björnson, L. Sanguinetti, J. Hoydis, and M. Debbah. Optimal design of energy-efficient multi-user mimo systems: Is massive mimo the answer? *IEEE Transactions on Wireless Communications*, 14(6):3059–3075, June 2015. ISSN 1536-1276. doi: 10.1109/TWC.2015.2400437.

S. Buzzi, C. L. I, T. E. Klein, H. V. Poor, C. Yang, and A. Zappone. A Survey of Energy-Efficient Techniques for 5G Networks and Challenges Ahead. *IEEE Journal on Selected Areas in Communications*, 34(4):697–709, April 2016. ISSN 0733-8716. doi: 10.1109/JSAC.2016.2550338.

Marco Carugi. *C-RAN: innovative solution for "green" radio access networks*. ZTE Corporation, September 2011. white paper, [Online]. Available: https://www.slideshare .net/Netmanias/201109c-ran-innovative-solution-for-green-radio-access-networks.

T. Chen, H. Kim, and Y. Yang. Energy efficiency metrics for green wireless communications. In *2010 International Conference on Wireless Communications Signal Processing (WCSP)*, pages 1–6, Oct 2010. doi: 10.1109/WCSP.2010.5633634.

C-RAN The Road Towards Green RAN. China Mobile Research Institute, October 2011. white paper, [Online]. Available: https://pdfs.semanticscholar.org/eaa3/ca62c9d5653e4f2318aed9ddb8992a505d3c.pdf.

H. Claussen, L. T. W. Ho, and F. Pivit. Effects of joint macrocell and residential picocell deployment on the network energy efficiency. In *2008 IEEE 19th International Symposium on Personal, Indoor and Mobile Radio Communications*, pages 1–6, Sept 2008. doi: 10.1109/PIMRC.2008.4699844.

L. M. Correia, D. Zeller, O. Blume, D. Ferling, Y. Jading, I. Gódor, G. Auer, and L. V. Der Perre. Challenges and enabling technologies for energy aware mobile radio: networks. *IEEE Communications Magazine*, 48(11):66–72, November 2010. ISSN 0163-6804. doi: 10.1109/MCOM.2010.5621969.

B. Dasi and W. Yu. Energy Efficiency of Downlink Transmission Strategies for Cloud Radio: Access Networks. *IEEE Journal on Selected Areas in Communications*, 34(4):1037–1050, April 2016. ISSN 0733-8716. doi: 10.1109/JSAC.2016.2544459.

P. Gandotra, R. K. Jha, and S. Jain. Green Communication in Next Generation Cellular Networks: A Survey. *IEEE Access*, 5:11727–11758, 2017. doi: 10.1109/ACCESS.2017.2711784.

A. Ghazanfari, H. Tabassum, and E. Hossain. Ambient RF energy harvesting in ultra-dense small cell networks: performance and trade-offs. *IEEE Wireless Communications*, 23(2):38–45, April 2016. ISSN 1536-1284. doi: 10.1109/MWC.2016.7462483.

Hassan Hamdoun, Pavel Loskot, Timothy O'Farrell, and Jianhua He. Survey and applications of standardized energy metrics to mobile networks. *annals of telecommunications - annales des télécommunications*, 67(3):113–123, Apr 2012. ISSN 1958-9395. doi: 10.1007/s12243-012-0285-z. URL https://doi.org/10.1007/s12243-012-0285-z.

C. Han, T. Harrold, S. Armour, I. Krikidis, S. Videv, P. M. Grant, H. Haas, J. S. Thompson, I. Ku, C. X. Wang, T. A. Le, M. R. Nakhai, J. Zhang, and L. Hanzo. Green radio: radio techniques to enable energy-efficient wireless networks. *IEEE Communications Magazine*, 49(6):46–54, June 2011. ISSN 0163-6804. doi: 10.1109/MCOM.2011.5783984.

Z. Hasan, H. Boostanimehr, and V. K. Bhargava. Green Cellular Networks: A Survey, Some Research Issues and Challenges. *IEEE Communications Surveys Tutorials*, 13(4):524–540, Fourth 2011. ISSN 1553-877X. doi: 10.1109/SURV.2011.092311.00031.

H. Holtkamp, G. Auer, V. Giannini, and H. Haas. A Parameterized Base Station Power Model. *IEEE Communications Letters*, 17(11):2033–2035, November 2013. ISSN 1089-7798. doi: 10.1109/LCOMM.2013.091213.131042.

I. Humar, X. Ge, L. Xiang, M. Jo, M. Chen, and J. Zhang. Rethinking energy efficiency models of cellular networks with embodied energy. *IEEE Network*, 25(2):40–49, March 2011. ISSN 0890-8044. doi: 10.1109/MNET.2011.5730527.

M. Ismail, W. Zhuang, E. Serpedin, and K. Qaraqe. A Survey on Green Mobile Networking: From The Perspectives of Network Operators and Mobile Users. *IEEE Communications Surveys Tutorials*, 17(3):1535–1556, thirdquarter 2015. ISSN 1553-877X. doi: 10.1109/COMST.2014.2367592.

A. Karmokar and A. Anpalagan. Green Computing and Communication Techniques for Future Wireless Systems and Networks. *IEEE Potentials*, 32 (4):38–42, July 2013. ISSN 0278-6648. doi: 10.1109/MPOT.2013.2245946.

K. Miyanabe, K. Suto, Z. M. Fadlullah, H. Nishiyama, N. Kato, H. Ujikawa, and K. i. Suzuki. A cloud radio: access network with power over fiber toward 5G networks: QoE-guaranteed design and operation. *IEEE Wireless Communications*, 22(4):58–64, August 2015. ISSN 1536-1284. doi: 10.1109/MWC.2015.7224728.

M. Nariman, F. Shirinfar, S. Pamarti, A. Rofougaran, and F. De Flaviis. High-Efficiency Millimeter-Wave Energy-Harvesting Systems With Milliwatt-Level Output Power. *IEEE Transactions on Circuits and Systems II: Express Briefs*, 64(6):605–609, June 2017. ISSN 1549-7747. doi: 10.1109/TCSII.2016.2591543.

K. N. R. S. V. Prasad, E. Hossain, and V. K. Bhargava. Energy efficiency in massive mimo-based 5g networks: Opportunities and challenges. *IEEE Wireless Communications*, 24(3):86–94, 2017. ISSN 1536-1284. doi: 10.1109/MWC.2016.1500374WC.

M. Röger, G. Böttger, M. Dreschmann, C. Klamouris, M. Huebner, A. W. Bett, J. Becker, W. Freude, and J. Leuthold. Optically powered fiber networks. *Opt. Express*, 16(26):21821–21834, Dec 2008. doi: 10.1364/OE.16.021821. URL http://www.opticsexpress.org/abstract.cfm?URI=oe-16-26-21821.

D. Sabella, A. de Domenico, E. Katranaras, M. A. Imran, M. di Girolamo, U. Salim, M. Lalam, K. Samdanis, and A. Maeder. Energy Efficiency Benefits of RAN-as-a-Service Concept for a Cloud-Based 5G Mobile Network Infrastructure. *IEEE Access*, 2:1586–1597, 2014. ISSN 2169-3536. doi: 10.1109/ACCESS.2014.2381215.

Tshiamo Sigwele, Atm S. Alam, Prashant Pillai, and Yim F. Hu. Energy-efficient cloud radio access networks by cloud based workload consolidation for 5G. *Journal of Network and Computer Applications*, 78:1–8, 2017. ISSN 1084-8045. doi: https://doi.org/10.1016/j.jnca.2016.11.005. URL http://www.sciencedirect.com/science/article/pii/S1084804516302740.

H. Tabassum, E. Hossain, A. Ogundipe, and D. I. Kim. Wireless-powered cellular networks: key challenges and solution techniques. *IEEE Communications Magazine*, 53 (6):63–71, June 2015. ISSN 0163-6804. doi: 10.1109/MCOM.2015.7120019.

TheClimateGroup. SMART 2020: Enabling the low carbon economy in the information age. Technical report, The Climate Group/ Global eSustainability Initiative (GeSI), November 2008. URL https://www.theclimategroup.org/sites/default/files/archive/files/Smart2020Report.pdf.

D. Wake, A. Nkansah, N. J. Gomes, C. Lethien, C. Sion, and J. P. Vilcot. Optically powered remote units for radio-over-fiber systems. *Journal of Lightwave Technology*, 26(15):2484–2491, Aug 2008. ISSN 0733-8724. doi: 10.1109/JLT.2008.927171.

Ximing Wang, Dongmei Zhang, Kui Xu, and Wenfeng Ma. On the energy/spectral efficiency of multi-user full-duplex massive MIMO systems with power control. *EURASIP Journal on Wireless Communications and Networking*, 2017 (1):82, May 2017. ISSN 1687-1499. doi: 10.1186/s13638-017-0864-9. URL https://doi.org/10.1186/s13638-017-0864-9.

J.G. Werthen, S. Widjaja, T.C. Wu, and J. Liu. Power over fiber: a review of replacing copper by fiber in critical applications. In *Proc. SPIE*, volume 5871, pages 58710C–58710C–6, 2005.

Y. Wu, Q. Yang, and K. S. Kwak. Energy Efficiency Maximization for Energy Harvesting Millimeter Wave Systems at High SNR. *IEEE Wireless Communications Letters*, 6(5):698–701, Oct 2017. ISSN 2162-2337. doi: 10.1109/LWC.2017.2734087.

12

Fog Computing Enhanced Fiber-Wireless Access Networks in the 5G Era

Bhaskar Prasad Rimal[1] and Martin Maier[*,2]

[1] Electrical and Computer Engineering Department, University of New Mexico, Albuquerque, NM 87131, USA
[2] Optical Zeitgeist Laboratory, Institut National de la Recherche Scientifique (INRS), Montréal, Canada

12.1 Background and Motivation

12.1.1 Next-Generation PON and Beyond

Passive optical networks (PONs) are considered to be an appealing wired access solution due to their high network capacity, reliability, low-cost, increased reach, and relative ease of deployment stemming from their all-passive network infrastructure and point-to-multi-point topology [Kramer and Pesavento - 2002; Kramer et al. - 2012]. PON-based fiber access networks are vital for enabling fast and ultra-fast broadband services [Effenberger et al. - 2007]. Since future access network technologies will need to support very high capacity to cope with the massive growth in traffic demand, optical fiber technology is the de facto future-proof technology for broadband access.

PON technology appears to be the best choice among various wired access technologies for triple-play service delivery (e.g., data, video, and voice) to support ever-growing bandwidth-intensive applications and services (e.g., high definition (HD) TV, 8K ultra-HDTV, high quality video conferencing, 3D displays, holographic imaging, augmented and virtual reality, immersive reality for telemedicine distance learning) due to its reliability, high capacity, and low cost.

Over the past few years, different flavors of PON technology (e.g., gigabit PON (G-PON), ethernet PON (EPON)) have been developed by the ITU-T and IEEE standard bodies to facilitate broadband access. For illustration, the evolution and capacity trend of PONs are shown in Figure 12.1. The ITU-T/FSAN (Full Service Access Network) G.989 series has defined NG-PON2 [ITU-T - 2013; Nesset - 2015] as the state-of-the-art PON technology. This architecture relies on a time and wavelength division multiplexing (TWDM) approach, which multiplexes four wavelengths (optionally eight) in a coordinated manner onto a single fiber, with each wavelength delivering 10 Gb s^{-1} downstream and 2.5 Gb s^{-1} upstream line rates, thereby reaching a total bandwidth of 40 Gb s^{-1} downstream and 10 Gb s^{-1} upstream. TWDM PONs are

* Corresponding Author:Martin Maier maier@emt.inrs.ca

Optical and Wireless Convergence for 5G Networks, First Edition.
Edited by Abdelgader M. Abdalla, Jonathan Rodriguez, Issa Elfergani, and Antonio Teixeira.

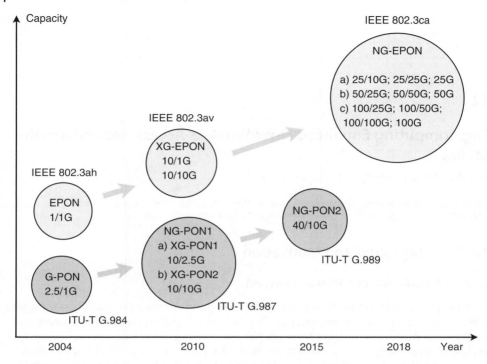

Figure 12.1 Evolution and capacity trends of PONs (different options of downstream and upstream capacity shown for the IEEE 802.3ca NG-EPON are still under discussion).

still considered complex (e.g., coordinating wavelengths together with time slots adds complexity to NG-PON2 dynamic bandwidth allocation algorithms) and is expensive for widespread deployment at present [Lam - 2016]. On the other hand, similar to the TWDM approach, the IEEE 802.3 working group formed the P802.3ca task force [Group - 2016] in November 2015 to develop objectives for the next generation of EPON (NG-EPON) that aims at supporting optical network units (ONUs) with one, two, or four wavelengths, providing system capacities of 25/50/100 Gb s^{-1} as well as coexistence with IEEE 802.3av 10G-EPON. The IEEE 802.3ca standard is expected to be released by October 2018.

Given the recent trends in ethernet based communication systems such as the Common Public Radio Interface (CPRI)[1] over ethernet [Gomes et al. - 2015] in a cloud radio access network [China Mobile Research Institute - 2011], CPRI over ethernet based TDM-PONs [Shibata et al. - 2015], ethernet based transport for mobile fronthaul (IEEE P1914.1), and passive optical local area network (POL)[2] technology for enterprise networks [Network Strategy Partners - 2009; Van-Etter - 2015; Nok - 2016], this chapter

---

1 CPRI is an industry standard specification. It details the key internal interface of radio base stations between the radio equipment control and radio equipment. Please refer to [CPRI - 2015] for more information about CPRI.

2 Passive optical LAN (POL) solutions are based on PON technologies (e.g., EPON, G-PON) and use the same principles as PONs in fiber-to-the-home networks. POLs have been optimized for enterprise LAN environments, for example, enterprises, governments, healthcare, education institutions, and hospitality providers. POL is suitable for larger LAN deployments and long-term operational benefits of the solutions,

considers the most popular and widely deployed PON system to date, that is, TDM EPON (IEEE 802.3ah EPON/IEEE 802.3av 10G-EPON) among many variations of PON technologies, for example, WDM-PON, TWDM-PON [Koonen - 2006]. EPON is based on the ethernet standard, which is simple, easy to manage with a capacity growth driven by the enterprise connectivity, and also provides economies of scale.

While in recent years both the ITU-T and IEEE standards for PON have put all their effort into achieving higher PON capacity (see Figure 12.1), those systems are now mature [Maier - 2014b]. For example, very recently, BT demonstrated high-speed terabit optical technologies and successful trials of a 3Tb s^{-1} optical superchannel [Smith and Zhou - 2016]. This enables operators to scale core optical transmission networks gracefully to satisfy the expected increasing needs of bandwidth intensive applications in 5G networks and beyond. However, instead of just continually expanding the capacity of NG-PONs, the research focus of broadband access networks should shift towards improving efficiency and supporting emerging services and applications (e.g., cloud computing and fog computing, mobile backhauling and fronthauling, virtualization). For instance, in the light of the emerging trends in network function virtualization (NFV) and software-defined networking, very recently the flexible access system architecture (FASA) concept [NTT Corporation - 2016] was introduced by the Nippon Telegraph and Telephone Corporation (NTT). FASA aims to provide greater flexibility in optical access equipment through the use of NFV – a virtual optical line terminal (OLT) for next-generation PONs with the combination of modularized function (software components such as wavelength control, bandwidth control, multi-cast, OAM (operation, administration, and maintenance)) rather than relying on building equipment for a specific purpose.

Generally, it may not be feasible to deploy optical fiber everywhere due to geographical constraints or when mobility is a necessity. Existing wireless access technologies (e.g., wireless fidelity (WiFi), 4G Long Term Evolution (LTE) and LTE-Advanced (LTE-A)), on the other hand, can provide user mobility but require a reliable high-capacity backhaul to satisfy the bandwidth requirements of emerging bandwidth hungry applications, such as 3D displays, HDTV, and high-quality video conferencing. In addition, holographic imaging, immersive experiences such as virtual and augmented reality, telemedicine, distance learning, and other high-bandwidth applications will continue to increase the demand for high speed wireline connectivity. This trend is also in line with *Edholm's law of bandwidth* [Cherry - 2004], which states that a unified (or convergent) optical wireline and wireless network is required to provide both fixed and mobile services to the end users. In the light of this, the convergence of fiber and wireless networks, also known as *fiber-wireless (FiWi) broadband access networks* [Martin et al. - 2008; Maier and Rimal - 2015], is widely viewed as a prominent solution to eliminate the aforementioned shortcomings and to meet the ever growing bandwidth demand, as discussed next.

12.1.2 FiWi Broadband Access Networks

The access network is called the "last mile" of telecommunication networks that connects the central office with end users. Access technologies can be categorized into two groups: (i) wired, for example, digital subscriber line (xDSL) like ITU G.9701 G.Fast,

---

whereby scalable and immediate cost savings (up to 50% savings in operating costs) are best realized [Van-Etter - 2015; Nok - 2016].

fiber-to-the-x (FTTx, x: building/home/node) and (ii) wireless, for example, WiFi (IEEE 802.11b/g/n/ac), WiMax (IEEE 802.16e), 4G LTE/LTE-A. An integrated FiWi broadband access network combines both wired and wireless access technologies. More specifically, FiWi is an architecture that combines features of a wireless network (i.e. ubiquity, flexibility, and cost savings) with an optical fiber network (i.e. reliability, robustness, and high capacity). Further, FiWi networks support emerging applications and provide broadband services to not only fixed subscribers but also mobile users that help foster innovation, generate revenue, and enhance the quality of our everyday lives [Maier - 2014a; Maier and Rimal - 2015].

The IEEE Technical Subcommittee on Fiber-Wireless (TSC-FiWi) Integration [Fiber-Wireless Integration Technical Subcommittee - 2016] defines what FiWi does as follows: "The subcommittee on Fiber-Wireless Integration addresses architectures, techniques, and interfaces for the *integration of fiber and wireless network segments in a unified wired-wireless infrastructure*. It *does not address* architectures or techniques specific to individual optical or wireless networks."

Most radio-and-fiber (R&F) based FiWi networks are composed of a cascaded TDM IEEE 802.3ah EPON in the optical backhaul segment and an IEEE 802.11 a/b/g/n/s WLAN mesh in the wireless front-end network segment. Aside from NG-PONs, for example, IEEE 802.3av 10G-EPON or WDM-PON, optical technologies such as tunable lasers play a vital role in the design of a flexible and cost-effective optical backhaul for FiWi networks [Martin et al. - 2008]. Besides a WiFi front-end, FiWi networks may consist of a cellular front-end such as 4G LTE/LTE-A. It is important to note that according to Aptilo (a leading carrier-grade WiFi supplier), WiFi and LTE are better considered complementary technologies than competing ones. At the same time, there is a strong "WiFi first" trend among users and service providers [Apt - 2017]. On the other hand, optical access technologies consistently offer considerably higher capacity than current leading edge wireless and cellular access technologies (see Figure 12.2 and [Maier - 2015; Rimal et al. - 2017c]), in particular next generation WiFi and cellular mobile radio technologies enabling the future 5G vision. Further, it is important to

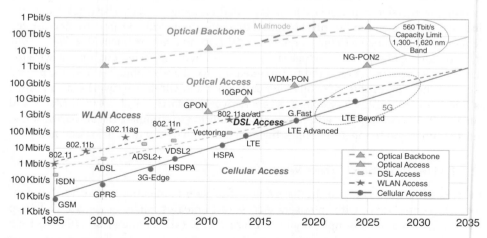

Figure 12.2 Evolution of wired and wireless technologies with projections. Source: [Décina - 2014], based on data by Bell Labs, [Fettweis and Alamouti - 2014].

note that WLAN is seemingly a strong nominee for low-cost high-speed mobile data offloading since it has been invariably able to yield data rates 100 times higher than cellular networks (see Figure 12.2 and [Maier - 2015; Rimal et al. - 2017c]). In addition, it is worth mentioning that IEEE 802.11ax-2019, the next high-throughput WLAN amendment, aims for a four-fold throughput increase compared to IEEE 802.11ac-2013 [Leadership - 2017; Bellalta - 2016]. In the light of these trends and facts, we consider a WiFi front-end in FiWi access networks to meet the low latency and high-capacity requirements of future 5G mobile networks.

12.1.3 Role of Fog Computing

Fog computing is a new computing paradigm that is a highly virtualized platform. Fog computing extends centralized cloud computing (e.g., Amazon EC2) to the edge of networks, thereby enabling applications and services on billions of connected devices and sensors, particularly, in the the context of the Internet of Things (IoT) [Bonomi et al. - 2012]. Fog computing may contain a cached state from the cloud and exploits the store-
and-forward principle.

Fog computing infrastructures differ from centralized clouds in that they are decentralized entities. Fog nodes are deployed at the edge of the internet, they are just one or multiple wireless hops away from corresponding edge devices (e.g., mobile phones, sensors, actuators), and are periodically synchronized with clouds, thus enabling the support of a wide range of applications. For example, compute intensive and latency sensitive applications such as augmented and virtual reality. The unified cloud–fog architecture can leverage both centralized and distributed cloud resources and services. Some of the advantages of fog computing include a reduction in backhaul traffic, improved network reliability by distributing content between the edge and centralized data centers, cost reductions by decomposing and disaggregating access network functions, optimization of central office infrastructures, and delivery of innovative services at the edge of networks.

There are a few industrial applications available for fog computing. Among others, IOx is Cisco's implementation of fog computing. It is an application framework (a router operating system) for fog computing that combines IoT application execution within the fog, secure connectivity with an industry-leading Internetwork Operating System (IOS), and reliable integration with IoT sensors and the cloud [Cisco Systems Inc. - 2017]. In summary, fog computing brings cloud computing capabilities (e.g., computing, storage, caching) closer to edge devices/things. In general, any edge devices that have computing, caching, and storage capability as well as network connectivity can be considered as fog nodes. For instance, industrial controllers, switches, routers, and video surveillance cameras. Fog nodes are best realized by relying on R&F networks and deployed at the edge of FiWi networks, whereby the integrated ONU access point (AP) or ONU mesh portal point (MPP) serve as the *edge of FiWi networks* in order to provide high bandwidth and low latency.

12.1.4 Computation Offloading

Computation offloading is a method by which the execution efficiency of a distributed system is enhanced by shifting compute intensive task(s) from a less capable system to a more capable one. This capability can take the form of computing power, memory, system load, and battery life [Wolski et al. - 2008]. It is also known as surrogate computing or cyber foraging [Satyanarayanan - 2001; Balan et al. - 2007]. Computation offloading is also similar to the concept of SETI@home [Anderson et al. - 2002], where computation tasks are sent to surrogates in order to perform computation. However, note that SETI@home is large-scale public resource computing, whereas computation offloading is typically small-scale computing.

Offloading decisions are typically made based on many parameters, including bandwidth, server speed, memory, the load on the server, and the amount of data interchanged between mobile clients and servers. Several methods can be implemented to get the solutions, such as partitioning programs and predicting parametric variations in the behavior of the application and execution environment. Offloading can be performed at various granularities, e.g., fine-grained, coarse-grained at the process, method/function, class, component, task, service, and application/program levels. Offloading decisions depend on multiple factors such as *why* to offload (e.g., improve energy efficiency), *how* to decide on offloading (i.e. static versus dynamic), *what* mobile systems use offloading (e.g., cellphone, wearables, robots), types of applications, and infrastructure for offloading (e.g., cloud computing) [Kumar et al. - 2013]. There are the following four major steps involved in computation offloading:

- Application modeling. Three types of application model, namely, procedure call, service invocation, and dataflow can be found in the state-of-the-art literature. Among them procedure call is widely used. In a procedure call model, an application can be described with a set of functions, whereby each function in turn calls other functions. Recent studies, for example, MAUI [Cuervo et al. - 2010], CloneCloud [Chun et al. - 2011], and ThinkAir [Kosta et al. - 2012] employ this approach. An application comprises a set of services in the case of service invocation. In both approaches, a graph or tree can be used to represent the application, whereby a node represents the procedure/service and an edge represents the call relationship/services.

- Profiling. Device and network profiling involve collecting device and network information (e.g., CPU, memory state, network bandwidth). This information is used to develop a cost model for the application. For instance, MAUI [Cuervo et al. - 2010] profiles the energy consumption of each part of the application. A unification of static analysis and dynamic profiling was used in CloneCloud to split applications automatically at a fine granularity [Chun et al. - 2011], whereas [Odessa Ra et al. - 2011] does not profile the application, device, and network independently. Instead, it profiles them in runtime and data transmission time.

- Optimization. A mathematical optimizer is used to optimize the objective (e.g., total execution time, energy consumption) for a given application and cost model. The cost model includes completion time, data processing throughput, energy consumption, or any combination thereof. For instance, MAUI [Cuervo et al. - 2010] optimizes the energy consumption of devices, CloneCloud [Chun et al. - 2011] and ThinkAir [Kosta et al. - 2012] optimizes the execution time or energy consumption depending on the programmer's choice, whereas [Odessa Ra et al. - 2011] optimizes the makespan for data streaming applications.

- Implementation. The implementation applies a suitable approach to determine which tasks of an application are remotely executed on the cloud infrastructure.

Three approaches are widely used, namely client-server, VM migration, and mobile agents. Unlike in the traditional client–server architecture, client applications can be partitioned at different levels of granularity (e.g., class, method, task, thread). More specifically, a mobile client requests the server to execute a particular method with given arguments and the server returns the result of the method execution using remote procedure call or remote method invocation protocols. Interesting examples of client-server based computation offloading systems include Spectra [Flinn et al. - 2002; [Chroma Balan et al. - 2007; Cuckoo Kemp et al. - 2010; MAUI Cuervo et al. - 2010; Odessa Ra et al. - 2011]. In the VM migration approach, clients prepare an image of their devices and transmit it to the servers for execution. The VM migration approach is used in Slingshot [Su and Flinn - 2005], CloneCloud [Chun et al. - 2011], ThinkAir [Kosta et al. - 2012], cloudlet [Ha et al. - 2013], cloudlet enhanced FiWi [Rimal et al. - 2017c], and mobile edge computing empowered FiWi [Rimal et al. - 2016, 2017a,b]. In the mobile agent approach, the computation is migrated from mobile devices to servers, whereby mobile agents are autonomous programs that control their movement from machine to machine [Kumar et al. - 2013].

12.1.5 Key Issues and Contributions

Over recent years, mobile data traffic has grown dramatically, mainly due to video services. Figure 12.3 shows the Cisco visual networking index (VNI) global mobile data traffic (only cellular traffic) forecast for 2015–2020. We observe from Figure 12.3 that the amount of mobile video traffic is higher than 50% of all mobile data traffic. This will further increase 11-fold between 2015 and 2020, accounting for 75% of the world's mobile data traffic by the end of 2020. Also, it should be noted that mobile data traffic by 2020 will reach up to 120 times the volume of global mobile traffic ten years earlier (2010) [Cisco Systems Inc. - 2016]. Importantly, a large increase in connections is foreseen due to the phenomenal growth of IoT/machine-to-machine (M2M) devices. In particular, M2M traffic will increase 21 times from 2015 to 2020 and reach 2.1 exabytes per month by 2020 [Cisco Systems Inc. - 2016]. These spiraling growth trends in the number of connected devices and data usage will require high capacity and a high degree of scalability.

Besides the aforementioned mobile data usage trends, another trend stems from new business opportunities. According to the OECD, the digital economy has given rise to a number of new business models, including online payment services, e-commerce, cloud computing, app stores, high-frequency trading, online advertising, and participative networked platforms [OECD Publishing - 2014]. Since these new business models support vertical markets at substantially greater scale, they require new network capabilities and better network performance. Given these trends and evolving new business models, current 4G technologies may limit the growth of mobile services, especially when considering the anticipated needs of 2020 and beyond.

To cope with the trends and challenges mentioned above, future 5G mobile networks will play an important role. The anticipated strict requirements of 5G applications such as ultra-high reliability and ultra-low latency will drive the need for extremely localized services at the edge of networks [Rimal et al. - 2017a]. Given the growing interest in the integration of wired/wireless networks and decentralized networks (e.g., M2M/D2D communications, small cell) in 5G, FiWi networks present an auspicious architecture

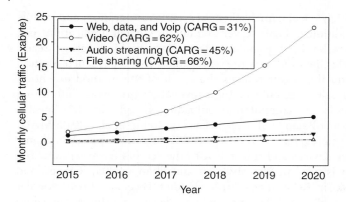

Figure 12.3 Application usage trends: Cisco VNI global mobile data traffic (cellular traffic only) forecast (2015–2020). (Source: adopted from [Cisco Systems, Inc. - 2016]). CARG: compound annual growth rate.

that serves both conventional cloud computing and the emerging decentralized fog computing. However, the integration of fog computing in FiWi networks (termed FC-FiWi henceforth) is not trivial and poses several architectural and resource management issues. A conventional cloud computing offers high storage and processing facilities. On the downside, however, it has large latency and real-time processing issues [Rimal and Maier - 2017; Rimal and Lumb - 2017]. Conversely, fog computing offers lower latency, but on the downside, it has restricted computing and storage capacity compared to a centralized cloud. Therefore, cloud computing and fog computing coexist, thereby supporting a variety of looming services and applications such as mission critical and delay tolerant in 5G networks.

There are two types of coexistence in FC-FiWi networks, namely, cloud/FiWi coexistence and the coexistence of fog and FiWi traffic (e.g., triple-play traffic). Maintaining stringent QoS requirements of each traffic without negatively affecting each other and supporting miscellaneous services and applications in the 5G era in such a network is not trivial, giving rise to diversified communication characteristics and requirements [Rimal et al. - 2017a]. Furthermore, providing ultra-high reliability (99.999 percent availability) at the FiWi network edge is one of the vital concerns in FC-FiWi networks, which creates complexity in designing highly optimized protocols [Rimal et al. - 2017a]. For example, fiber cuts may happen in the optical fiber backhaul segment and faults may occur in fog computing infrastructures. Also, link failures to/from access points may happen. As a result, the introduced network may experience widespread service outages.

Toward this end, in this chapter, we mainly concentrate on the 5G key attributes of high reliability with substantially enhanced connectivity and very low latency and study how and to what extent they can be obtained in FC-FiWi networks. The objective of this chapter is to design an architecture and develop a novel unified (one solution for both network segments to simplify the management of the integrated architecture) resource management scheme as well as to study their achievable performance gains. It should be noted that carrier-sense multiple access with collision avoidance as random medium access control protocol was used on FiWi networks in most of the preceding studies (e.g., [Martin et al. - 2008; Maier - 2014a,b]. However, to meet the real-time, low latency, and low energy consumption requirements of FC-FiWi networks, we need a deterministic access mechanism. Therefore, we develop access protocols based on time division multiple access (TDMA). More specifically, *from the resource management perspective*, we design a unified and decentralized resource management scheme based on two-layered TDMA, where fog traffic is scheduled outside the timeslot of primary FiWi traffic to maintain the coexistence of both cloud and FiWi traffic. *From the modeling point of view*, a thorough analysis of survivability and end-to-end-delay performance is presented. *From the validation and measurements perspective*, an experimental testbed is developed, and a proof of concept is conducted to validate and measure the performance of the proposed network under different performance metrics. The performance of the proposed solution is evaluated through an experiment and analytical results and discussed in detail.

The rest of the chapter is organized as follows. Section 12.2 describes the proposed fog computing enhanced FiWi networks and unified resource management scheme. Section 12.3 presents the detailed analysis of survivability and end-to-end packet delay performance in greater detail. Section 12.4 discusses the implementation and validation, including the concept validation approach, experimental testbed, and obtained results. Finally, Section 12.5 draws conclusions and provides an outlook for future research.

12.2 Fog Computing Enhanced FiWi Networks

12.2.1 Network Architecture

The architecture of FC-FiWi networks is illustrated in Figure 12.4, whereby the optical fiber backhaul is composed of an extensively used IEEE 802.3ah EPON. The extended network reach between the remote optical network units (ONUs), and an OLT can be up to 100 km. The broadband and cloud services are provided to the end uses via the optical backhaul. The OLT connects to the various types of ONUs in a tree-and-branch topology. Some ONUs are stationed at the places of business and/or residential subscribers to provide FTTx (e.g., fiber-to-the-office/home) services to wired subscribers.

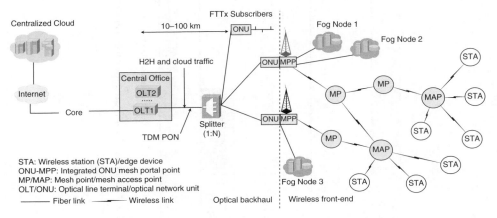

Figure 12.4 Network architecture of fog computing enhanced FiWi networks. (Source: adopted and customized from [Rimal et al. - 2017b]) © 2018 IEEE. Reprinted with permission, from [Rimal et al. - 2017b].

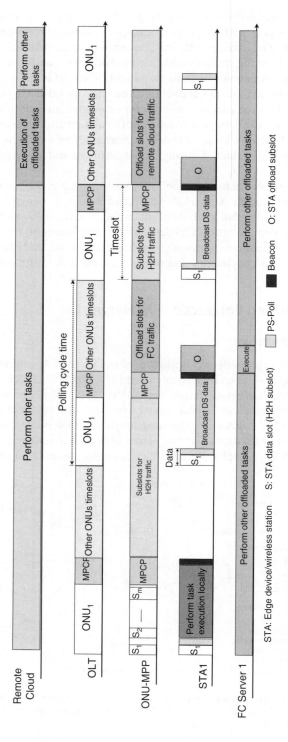

Figure 12.5 Two-layered TDMA based resource management scheme for FC-FiWi networks.

Also, some ONUs are equipped with a wireless MPP to interface with the wireless mesh network. Note that mesh points (MPs) are relay nodes that forward packets to the other MP node(s) and, depending on the configuration parameters, MPs support multiple wireless mesh services. One or more fog servers can be connected to the ONU-MPP through fiber links to provide cloud services at the proximity of subscribers. The envisioned FC-FiWi network offers a full-fledged communications infrastructure.

12.2.2 Protocol Description

The operation of the proposed scheme is illustrated in Figure 12.5. The system is designed through two TDMA layers with polling. The first TDMA tier is intended for the optical backhaul, where the OLT allocates bandwidth and schedules an upstream timeslot for each ONU-AP/MPP, while broadcasting downstream frames to all ONU-APs/MPPs [Rimal et al. - 2017b]. Each ONU-AP/MPP discards received downstream frames that are not intended for it. In the second layer TDMA, the ONU-AP allocates bandwidth in sub-slots and schedules transmissions of both FiWi and cloud traffic (fog and centralized cloud) of all edge devices destined to it. During its assigned timeslot, an ONU-AP/MPP sends its upstream data frames (e.g., fog/cloud offload/FiWi data frame) to the OLT and the ONU-AP/MPP receives downstream data frames (e.g., computation result, downstream FiWi data traffic) from the OLT and immediately broadcasts them to its corresponding edge devices.

Cloud and fog traffic are scheduled outside the ONU-AP timeslot within a PON polling cycle time. This allows the coexistence of fog, cloud, and FiWi traffic without degrading the performance of the FiWi network. Upon receiving a PS-Poll from an edge device, the ONU-AP decides where (fog or cloud) to offload the computation task based on the given service level agreement or QoS requirements. In the cloud offloading scenario, this includes the computation offload request in the REPORT message sent to the OLT at the end of its timeslot [Rimal et al. - 2017b]. The start and duration of computation offload sub-slots are conveyed using a Beacon broadcast to edge devices. Upon receiving offloaded tasks from the ONU-AP, the OLT sends it to the conventional cloud. Similar to the cloud offloading scenario, the transmission of fog traffic is scheduled outside the ONU-AP timeslot within a PON polling cycle time to allow the coexistence of FiWi and fog traffic. In this scenario, the ONU-AP decides fog sub-slots without notifying the OLT.

The transmission sub-slots of STAs/edge devices are allocated by the ONU-AP via Beacon and PS-Poll messages. The ONU-AP broadcasts a Beacon to its STA containing an uplink FiWi subslot map, whereby each STA sends a PS-Poll at the end of its FiWi subslot. The STA sends offloaded traffic and receives computation results from the fog server or cloud. For further technical details, interested readers are referred to [Rimal et al. - 2017b].

12.3 Analysis

12.3.1 Survivability Analysis

The capability of a network to provide continuous services in the case of link or node failures is known as survivability [Rimal et al. - 2017b]. Network elements of our considered

FC-FiWi networks are reliable since they are passive (not powered). However, note that due to fiber cuts, FC-FiWi networks may suffer from link failures. Eventually, the network becomes unreliable, and a single or multiple ONUs and their corresponding subscribers may disconnect from the OLT. Furthermore, a fog computing server may go down due to power failure or server crash. Providing highly reliable fog computing services is very important, especially for delay-critical applications. Different from existing studies, we consider different types of fiber backhaul redundancy schemes and fog server protection schemes.

Note that the ITU-T Recommendation G.983.1 does not cover direct inter-ONU communications to enhance the reliability of PONs. First, interconnection fibers (IFs) are deployed to interconnect neighboring ONUs pairwise (see ONU_{N-1} and ONU_N in Figure 12.6) [Rimal et al. - 2017b]. Each pair of ONUs can mutually protect themselves by using the corresponding IF in the case of a distribution fiber cut of either ONU, at the cost of the extra optical switch at an IF-upgraded ONU. It should be noted that IFs are a promising approach to obtain high connection availability between ONUs and a OLT for sparse population densities in a cost-effective way [Maier - 2012]. Second, we deploy redundant backup fiber links between each node pair of each link, as shown in Figure 12.6. Third, our fog computing (FC) protection scheme includes a redundant FC server and a backup fiber connection between the ONU-MPP and an FC server, as shown in Figure 12.6). An analysis of survivability schemes of FC, fiber backhaul, and a wireless mesh network is presented below.

(1) Failures during packet transmission and task execution, fiber cuts between ONU_k and the FC server, and FC server failure are considered in our FC survivability analysis [Rimal et al. - 2017b]. Let STA = $\{1, 2 \dots N_{st}\}$, ONU= $\{1, 2 \dots N_{on}\}$, FC = $\{1, 2 \dots N_{fc}\}$, and OLT = $\{1, 2 \dots N_{ol}\}$ represent the sets of STAs, ONUs, OLTs, respectively, where the last element of each set indicates its number of elements. Let $P_{f(ij)}$ represent the failure probability of the connection between i and j, where $i, j \in$ (STA $\cup$ ONU $\cup$ OLT $\cup$

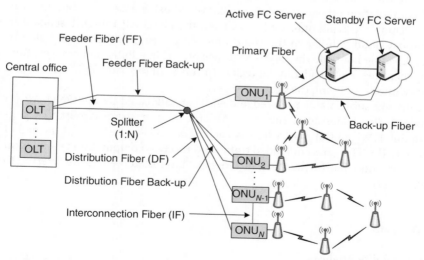

Figure 12.6 Protection schemes for the optical backhaul, fog computing, and wireless protection. (Source: adopted and customized from [Rimal et al. - 2017b].) © 2018 IEEE. Reprinted with permission, from [Rimal et al. - 2017b].

FC). The FiWi failure probability between ONU and FC server is calculated as follows [Rimal et al. - 2017b]:

$$P_{f(ij)}^{ONU_n} = 1 - \prod_{n=1}^{N}(1 - P_{C(ij)})(1 - P'_{C(ij)}),$$

(12.1)

where $P_{C(ij)}$ and $P'_{C(ij)}$ denote the fiber cut probability of the primary and back-up fibers between ONUs and the FC server, respectively.

The reliability of an FC server is given by [Rimal et al. - 2017b]:

$$\mathcal{R}^{FC} = \prod \left[exp\left\{-F_{tx}^{FC} \cdot \left(\frac{P^{tx}}{C_{FC}}\right)\right\} \cdot exp\left\{-F_{ofl}^{FC} \cdot \left(\frac{C'_{fc}}{S_{fc}}\right)\right\} \right] \cdot \prod(1 - P_f^a)(1 - P_f^s),$$

(12.2)

where F_{tx}^{FC}, F_{ofl}^{FC}, P_f^a, and P_f^s represent the transmission failure rate, offloaded packet execution failure rate, failure probability of the active FC server, and failure probability of the standby FC server, respectively. The FiWi connectivity probability of ONU_k with a reliable FC service is computed from Equations (12.1) and (12.2).

$$P_{C(ij)}^{ONU_k} = \prod_{n=1}^{N}(1 - P_{f(ij)}^{ONU_n}) \cdot \mathcal{R}^{FC}.$$

(12.3)

(2) Backhaul segment. The FiWi failure probability between ONU_i and the OLT using the considered fiber redundancy schemes is calculated as follows [Rimal et al. - 2017b]:

$$P_{f(ij)}^{ONU_n} = 1 - \prod_{n=1}^{N}(1 - P_{C(is)})(1 - P'_{C(is)}) \cdot (1 - P_{C(sj)})(1 - P'_{C(sj)}),$$

(12.4)

where $P_{C(is)}$, $P'_{C(is)}$, $P_{C(sj)}$, and $P'_{C(sj)}$ represent the fiber cut probability of the distribution, backup distribution, feeder, and back-up feeder fibers, respectively, whereby s represents the splitter in the PON.

(3) FC-FiWi front-end segment. The co-located ONU-MPPs use the wireless mesh network to route traffic wirelessly from an ONU to another ONU. It then optically forwards the traffic to the OLT. Let $P_f^{\text{Path}_i^{w(y,x)}}$ be the probability that the ith wireless path $\text{Path}_i^{w(y,x)}$ between ONU_x and ONU_y fails. Further, let NW_i MPs be wirelessly connected to MPP_y and MPP_x via $\text{Path}_i^{w(y,x)}$ and MP_{iz_k} is the MP that is k wireless hops away. Overall, for a given failure probability of ONU, MPP, MP, MAP, and FC, the FiWi connectivity probability of STA_k in FC-FiWi networks is computed as follows [Rimal et al. - 2017b]:

$$P_{C(ij)}^{STA_k} = (1 - P_f^{MAP_q})(1 - P_f^{MP_1}) \cdots (1 - P_f^{MP_l})(1 - P_f^{MPP_y})$$

$$\left(1 - \prod_{\forall | x \leftrightarrow y}\left[1 - \left(1 - \prod_{i=1}^{N_{WP(y,x)}} P_f^{\text{Path}_i^{w(y,x)}}\right)(1 - P_f^{ONU_x})\right]\right) \cdot$$

(12.5)

$$\left(\prod_{n=1}^{N}(1 - P_{C(ij)})(1 - P'_{C(ij)})\right) \cdot \left(\prod_{n=1}^{N}(1 - P_{f(ij)}^{ONU_n}) \cdot \mathcal{R}^{FC}\right),$$

where STA_k is connected to MAP_q, which is consecutively connected to MPP_y through a number of intermediate MPs, that is $MP_1, \ldots, MP_l$, whereby the values in the last term are substituted from Equations (12.1) and (12.2).

12.3.2 End-to-End Delay Analysis

Computation offloading should be performed if the time to execute a task on the edge device locally is longer than the response time of offloading that task onto a fog server. We define this response time difference as offload gain [Rimal et al. - 2017a]. The response time efficiency is defined as the ratio of the offload gain and the response time of a task that is locally executed on edge devices [Rimal et al. - 2017a]. Further, let the packet delay denote the time a packet waits in a data buffer. The mean packet delay from the edge device to the OLT (i.e. end-to-end) is the sum of the propagation delay and the mean end-to-end packet delay of FiWi traffic. The performance of the proposed scheme is examined by considering a polling system with M/G/1 queues [Rimal et al. - 2017b]. Specifically, we assume that the MP/MAP serve packets in a first-come-first-serve fashion and each MP/MAP in the wireless mesh network is modeled as an M/M/1 queue. Hence, the overall average end-to-end packet delays for FC, centralized cloud, and FiWi are given by [Rimal et al. - 2017b]:

$$
D_{\text{FC}} = \begin{cases} \dfrac{T_c}{2N}(3N + 1 + \rho^{\text{FC}}), & \text{for single hop} \\[2ex] \left[\dfrac{T_c}{2N}(3N + 1 + \rho^{\text{FC}})\right] + \displaystyle\sum_{i=1}^{\mathcal{N}-1} \left[\dfrac{1}{\mu C_i} + \dfrac{1}{2\mu C_i} + \dfrac{\rho_i}{\mu C_i - \lambda_i}\right], & \text{for multi-hop} \end{cases}
$$

$$(12.6)$$

$$
D_{\text{cloud}} = \begin{cases} \left(\dfrac{T_c}{2N}(3N + 1 + \rho^{\text{cloud}}) + T_{\text{prop3}}, & \text{for single-hop} \\[2ex] \left(\dfrac{T_c}{2N}(3N + 1 + \rho^{\text{cloud}}) + T_{\text{prop3}} + \\[2ex] \displaystyle\sum_{i=1}^{\mathcal{N}-1} \left[\dfrac{1}{\mu C_i} + \dfrac{1}{2\mu C_i} + \dfrac{\rho_i}{\mu C_i - \lambda_i}\right], & \text{for multi-hop} \end{cases}
$$

$$(12.7)$$

Table 12.1 System parameters and default values.

Parameter	Description	Value
C, C_{cloud}, C_{FC}	ONU-AP, cloud, Fog transmission capacity	6900 Mbits^{-1}, 10 Gbs^{-1}, 10 Gbs^{-1}
N	Number of ONUs in the FC-FiWi network	32, 64
M	Number of wireless STAs	8–100
T_{prop1}, T_{prop2}, T_{prop3}, T_{prop4}	Air propagation delay between the edge device/STA and ONU-AP, propagation delay between ONU-AP and OLT, fiber propagation delay between OLT and conventional cloud, fiber propagation delay between the ONU-AP and FC server	0.00033 ms, 0.05 ms, 50 ms, 0.01 ms
T_g, T_c	Guard time between two consecutive slots, PON polling cycle time	1µs, ms (variable)
$T_{\text{pon}}^{\text{ms}}$, $T_{\text{wl}}^{\text{ms}}$	Transmission of MPCP message and STA PS-Poll	0.512µs, 0.12µs
P_f^{MPP}, P_f^{MP}, P_f^{MAP}	Failure probability of an MPP, MP, and MAP	10^{-7}

$$
\mathcal{D}_{\text{fiwi}} = \begin{cases} \dfrac{\lambda \overline{X^2}}{2(1-\rho^{\text{fiwi}})} + \dfrac{(3N-\rho^{\text{fiwi}})\overline{V}}{2(1-\rho^{\text{fiwi}})} + \dfrac{\sigma_v^2}{2\overline{V}} + \overline{X} + 2T_{\text{prop2}} + \\[4mm] \dfrac{(\text{P}^{tx}+\text{P}^{rx})}{r_d} + 2T_{\text{prop1}}, \qquad\qquad\qquad \text{for single hop} \\[5mm] \dfrac{\lambda \overline{X^2}}{2(1-\rho^{\text{fiwi}})} + \dfrac{(3N-\rho^{\text{fiwi}})\overline{V}}{2(1-\rho^{\text{fiwi}})} + \dfrac{\sigma_v^2}{2\overline{V}} + \overline{X} + 2T_{\text{prop2}} + \dfrac{(\text{P}^{tx}+\text{P}^{rx})}{r_d} + \\[4mm] 2T_{\text{prop1}} + \displaystyle\sum_{i=1}^{\mathcal{N}-1} \left[\dfrac{1}{\mu C_i} + \dfrac{1}{2\mu C_i} + \dfrac{\rho_i}{\mu C_i - \lambda_i} \right], \qquad \text{for multi-hop} \end{cases}
$$

$$(12.8)$$

where σ_v^2 represents the variance of each reservation time and r_d is the maximum bandwidth that a wireless communication link can offer defined by the well known Shannon capacity. Further, $\overline{X}$ and $\overline{X^2}$ represent the first and second moments of packet service time of ONU-AP$_i$, $\overline{V}$ represents the first moment of reservation time, and $\rho^{\text{fiwi}} = \lambda\overline{X}$ is the aggregated FiWi traffic load. The remaining parameters are defined in Table 12.1. Further details on the analysis of each quantity in the above equations can be found in [Rimal et al. - 2017b, 2016].

12.4 Implementation and Validation

The performance of our proposed scheme is assessed by means of numerical results and a proof-of-concept demonstration. The overall conceptual validation approach is illustrated in Figure 12.7. We have conducted a proof-of-concept demonstration to validate and measure the performance of our scheme in terms of different performance metrics. A more detailed description of our experimental testbed is presented in the following.

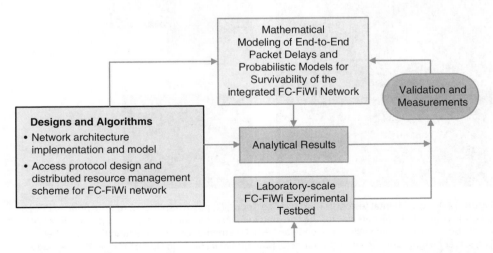

Figure 12.7 Conceptual validation approach: modeling, numerical simulations, and experimental testbed.

12.4.1 Experimental Testbed

The experimental testbed consists of an EPON (Sun Telecom 4×ONU) stationed at a network reach of 20 km from the OLT and a WLAN (with a capacity of 1 Gb s^{-1}). Those ONUs connect the fog server (Dell OptiPlex 9020 desktop machine) to the multiple edge devices. We consider an FC-FiWi network that comprises an integrated ONU-AP associated with one edge device (Dell Inspiron 3521 notebook). To develop a fog computing infrastructure, the OpenStack++ open source cloud platform is deployed at the FC-FiWi network edge, where we deploy a VM running on the fog server with four cores virtual CPUs with a clock speed of 3.6 GHz. The VM instance is allocated a 50 GB disk and 10 GB of RAM. To emulate a centralized cloud, the Amazon Elastic Compute Cloud (Amazon EC2) was used as the public cloud provider. Figure 12.8 shows the experimental testbed of our fog computing enhanced FiWi network along with a list of the deployed networking equipment.

12.4.2 Results

This section presents our obtained results for various network configurations. The IEEE 802.11ac very high throughput WLAN operates at a line rate of 6900 Mbit s^{-1}. We vary the FiWi traffic from 0.3 to 0.9. The traffic load at an integrated ONU-AP/MPP is normalized and the maximum traffic does not exceed 1.20 bytes of the PS-Poll frame. The data load of a computation task is divided into packets, while the application is

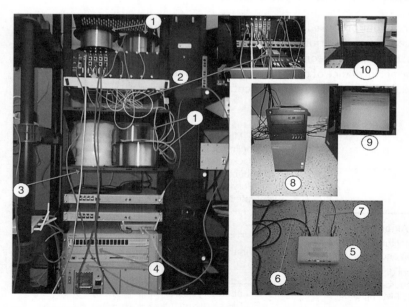

Figure 12.8 Experimental testbed: (1) optical fiber loops; (2) passive splitter at remote node of EPON; (3) cable connecting the gateway for the core network; (4) cable connecting access points; (5) WLAN access point; (6) ethernet cable connecting an ONU; (7) ethernet cable connecting the fog server; (8) Fog server hosting OpenStack++ platform; (9) running VM instance in OpenStack++; 10) STA/edge device running edge application. (Source: adopted from [Rimal et al. - 2017c].) © 2018 IEEE. Reprinted with permission, from [Rimal et al. - 2017c].

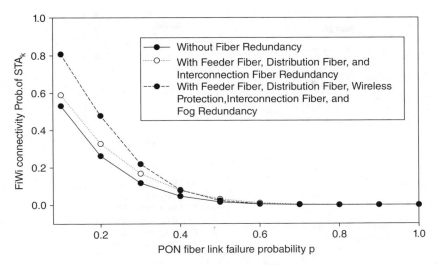

Figure 12.9 FiWi connectivity probability of an edge device versus optical backhaul failure probability.

split into fine-grained tasks, similar to [Cuervo et al. - 2010]. The computation capability of the centralized cloud is assumed to be 1000 times higher than that of the fog. For illustration, a face detection application is considered for computation offloading, though it should be noted that the proposed scheme is applicable to any application. The remaining parameter settings are summarized in Table 12.1.

Recall from above that survivability is one of the critical issues to be tackled in FC-FiWi networks. Figure 12.9 illustrates the beneficial impact of our survivability schemes, where an ONU wirelessly connects to its neighboring ONUs through three-hop wireless mesh paths in the wireless protection scheme. From the obtained result in Figure 12.9 we observe that by jointly employing feed fiber, distribution fiber, interconnection fiber, wireless protection, and fog redundancy schemes, the FiWi connectivity is significantly improved for various PON fiber link failure probabilities compared to the scenario without fiber redundancy. For example, a FiWi connectivity probability of 0.81 for STAs is achieved for a typical scenario with a PON fiber link failure probability of 0.1. Clearly, this shows that our proposed scheme is able to provide highly reliable broadband and fog computing services.

Delay depends on the traffic load and for increasing FiWi traffic loads both D_{FC} D_{cloud} increase. A similar tendency is found in the scenario of FiWi packet delay as well. We notice that $D_{cloud} > D_{FC}$, as depicted in Figure 12.10. For instance, in the scenario of 32 ONU-APs/MPPs, D_{FC} and D_{FC} are below 133.5 ms and 233.50 ms, respectively, for all FiWi traffic loads. According to Equations 12.6, 12.7, and 12.8, the delay and not only the function of the PON polling cycle time is also considerably influenced by the propagation delay and aggregate traffic load at an ONU-AP/MPP. For example, for a given threshold of 44.5 ms (as an example) of the fog traffic delay, the upper bound of acceptable aggregate FiWi traffic load of 32 ONU-APs/MPPs must not exceed 0.7. To decide where to offload computation tasks is another important concern in FC-FiWi networks, for which Figure 12.10 provides valuable insights. For a typical case of 32 ONU-APs/MPPs, at a FiWi traffic load of less than or equal to 0.5, fog experiences a

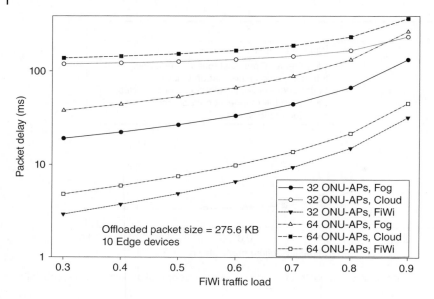

Figure 12.10 Delay performance.

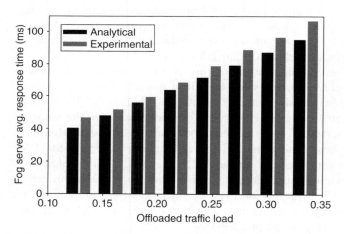

Figure 12.11 Experimental versus analytical results: impact of traffic load on the average response time of the fog server.

delay of less than 26.74 ms, compared to 126.7 ms in the case of cloud. This means that many delay-sensitive applications can be offloaded onto FC-FiWi rather than onto the centralized cloud.

To verify the accuracy of our analytical models, we have run experiments in a real-world setting. Figure 12.11 confirms the right match between our obtained experimental and analytical results. Specifically, Figure 12.11 shows that the fog's average response times achieved from the experiments are somewhat higher than the corresponding analytical values with a maximum FiWi traffic load of 0.95 and polling cycle time of 5 ms. For example, the average response time of a fog server of 68.7 ms was obtained at an offloaded traffic load of 0.22, compared to the corresponding analytical value of 63.6 ms.

12.5 Conclusions and Outlook

12.5.1 Conclusions

In this chapter, we introduced FC-FiWi networks based on the integration of EPON, WLAN, and fog computing for the creation of multi-purpose assets. For the efficient operation of the integrated network, we developed a two-layer TDMA based unified resource management scheme to schedule FiWi, fog, and cloud traffic at the same time. While the proposed network is resilient against fiber cuts, the central OLT forms a single point of failure. To estimate survivability for both FiWi and fog computing segments, different protection schemes were proposed and evaluated. We developed an analytical framework to examine the packet delay performance of both FiWi and cloud traffic. The obtained results reveal the effectiveness of implementing fog computing in FiWi networks. More specifically, the proposed scheme offers low end-to-end offload delay and high reliability at the edge of the FC-FiWi network without negatively influencing the network performance of FiWi broadband traffic. The obtained results reveal that for a typical scenario of 32 ONU-APs/MPPs and a FiWi traffic load of 0.6, a fog and cloud packet delay of 33.41 ms and 133.41 ms can be obtained, respectively, without degrading the network performance for FiWi traffic.

12.5.2 Outlook

Given the growing progress on, and interest in, 5G networks, FiWi access networks have great potential to enable the emergence of a variety of novel applications and creating new exciting opportunities. Consequently, the contributions made in this chapter offer opportunities for further improvements. Some of the potential directions for future research that may build upon the work in this chapter are as follows:

- FC-FiWi over blockchain networks. Blockchain is the distributed and transparent public ledger that records all transaction data, allowing all non-trusting peers to interact with each other without a third-party intermediary. Blockchain technology consists of cryptography, mathematics, algorithms, and economic models, combining peer-to-peer networks using a distributed consensus algorithm [Gervais et al. - 2014]. Among many blockchain technology platforms, Bitcoin, created by Satoshi Nakamoto in 2009 [Nakamoto - 2008], is one of the most popular decentralized cryptocurrencies based on blockchain technology. The basis of blockchain technology is the decentralized consensus mechanisms (i.e. all blockchain nodes agree on the same state of a system), which has been studied and applied in the area of distributed computing [Chen et al. - 1992]. However, a public blockchain network faces scalability issues, especially in terms of network bandwidth, size, and storage requirements, and exhibits a long-tail distribution of the information propagation delay [Tschorsch and Scheuermann - 2016], on which blockchain's security heavily depends. However, longer delays increase the possibility of cyber attacks. In their current design, blockchains are not suitable for handling heavy computations. To address these issues, FC-FiWi networks would be an interesting topic.

- Homogeneous versus heterogeneous machines. The current implementation of the algorithms, models, and host configuration assume homogeneity of all machines in the fog and remote cloud. However, previous research studies showed that the cloud environment is mostly a heterogeneous environment with machines of the same category having different capabilities. Similarly, a fog may be deployed using multiple servers with a certain heterogeneity. For future work, the proposed models and results may be extended by taking machines with heterogeneous capabilities into account.
- Fast moving versus quasi-stationary users. The proposed models and algorithms do not take mobility into account. It is evident that computation offloading onto the fog is challenging because of the intermittent connections between fog and mobile users. An interesting extension of the proposed solutions would be to incorporate a broad range of realistic mobility scenarios with connectivity predictions. More specifically, the impact of fast-moving and quasi-stationary users on the performance of FC-FiWi networks would be an attractive topic for future work.

Bibliography

D. P. Anderson, J. Cobb, E. Korpela, M. Lebofsky, and D. Werthimer. Seti@home: An experiment in public-resource computing. *Communications of the ACM*, 45(11):56–61, Nov. 2002.

High Efficiency (HE) Wireless LAN Task Group. Aptilo Networks, Dec.5 2017. URL http://www.anpdm.com/newsletterweb/44475F447248435D4474424A59/494A514279494B5B417341445E43.

R. K. Balan, D. Gergle, M. Satyanarayanan, and J. Herbsleb. Simplifying cyber foraging for mobile devices. In *Proc. 5th International Conference on Mobile Systems, Applications and Services (MobiSys)*, pages 272–285, June 2007.

B. Bellalta. IEEE 802.11ax: High-efficiency WLANs. *IEEE Wireless Communications*, 23(1):38–46, 2016.

F. Bonomi, R. Milito, J. Zhu, and S. Addepalli. Fog computing and its role in the Internet of Things. In *Proc., First Edition of the MCC Workshop on Mobile Cloud Computing*, pages 13–16, Aug. 2012.

M-S Chen, K-L Wu, and Philip S. Yu. Efficient decentralized consensus protocols in a distributed computing system. In *Proc. IEEE International Conference on Distributed Computing Systems*, pages 426–433, 1992.

S. Cherry. Telecom: Edholm's law of bandwidth. *IEEE Spectrum*, 41(7):58–60, July 2004.

China Mobile Research Institute. C-RAN: The road towards green RAN. *White Paper*, pages 1–48, Oct. 2011.

B.G. Chun, S. Ihm, P. Maniatis, M. Naik, and A. Patti. Elastic execution between mobile device and cloud. In *Proc. ACM EuroSys*, pages 301–314, 2011.

Cisco Systems, Inc. Cisco visual networking index: Global mobile data traffic forecast update (2015-2020). *White Paper*, pages 1–39, 2016.

Cisco Systems Inc. Cisco Visual Networking Index: Mobile Forecast Highlights (2015-2020). Technical report, Cisco Systems Inc., Oct.20 2016. URL http://www.cisco.com/assets/sol/sp/vni/forecast_highlights_mobile/.

Cisco Systems Inc. Cisco IOx. URL https://developer.cisco.com/site/iox/. Accessed on Dec. 20, 2017.

CPRI. Common Public Radio Interface (CPRI); Interface Specification, V 7.0. Technical report, CPRI, 2015.

E. Cuervo, A. Balasubramanian, D.-Ki Cho, A. Wolman, S. Saroiu, R. Chandra, and P. Bahl. MAUI: Making smartphones last longer with code offload. In *Proc. ACMMobiSys*, pages 49–62, 2010.

M. Décina. Future of networks. In *Proc. IEEE Technology Time Machine (TTM)*, Oct. 2014.

F. Effenberger, D. Cleary, O. Haran, G. Kramer, R. D. Li, M. Oron, and T. Pfeiffer. An introduction to PON technologies. *IEEE Communications Magazine*, 45 (3):S17–S25, 2007.

G. Fettweis and S. Alamouti. 5G: Personal mobile Internet beyond what cellular did to telephony. *IEEE Communications Magazine*, 52(2):140–145, 2014.

Fiber-Wireless Integration Technical Subcommittee, 2016. URL http://ifwt.committees .comsoc.org/. Accessed on Feb. 14, 2018.

J. Flinn, S. Park, and M. Satyanarayanan. Balancing performance, energy, and quality in pervasive computing. In *Proc. 22nd International Conference on Distributed Computing Systems*, ICDCS 02, pages 217–226, July 2002.

A. Gervais, G. Karame, S. Capkun, and V. Capkun. Is bitcoin a decentralized currency? *IEEE security & privacy*, 12(3):54–60, 2014.

N. J. Gomes, P. Chanclou, P. Turnbull, A. Magee, and V. Jungnickel. Fronthaul evolution: From CPRI to Ethernet (invited paper). *Optical Fiber Technology*, 26:50–58, 2015.

IEEE 802.3 Next Generation Ethernet Passive Optical Network (NG-EPON) Study Group, 2016. URL http://www.ieee802.org/3/NGEPONSG/. Accessed on Oct. 25, 2016.

K. Ha, P. Pillai, W. Richter, Y. Abe, and M. Satyanarayanan. Just-in-time provisioning for cyber foraging. In *Proc. 11th Annual International Conference on Mobile Systems, Applications, and Services*, pages 153–166, June 2013.

ITU-T. 40-Gigabit-capable passive optical networks (NG-PON2): General requirements. *ITU-T G.989.1 Recommendation*, pages 1–26, Mar. 2013.

R. Kemp, N. Palmer, T. Kielmann, and H. Bal. Cuckoo: A computation offloading framework for smartphones. In *Proc. 2nd International Conference on Mobile Computing, Applications, and Services (MobiCASE)*, pages 59–79, Oct. 2010.

T. Koonen. Fiber to the home/fiber to the premises: What, where, and when? *Proceedings of the IEEE*, 94(5):911–934, 2006.

S. Kosta, A. Aucinas, P. Hui, R. Mortier, and X. Zhang. Dynamic resource allocation and parallel execution in the cloud for mobile code offloading. In *Proc., IEEE INFOCOM*, pages 945–953, 2012.

G. Kramer and G. Pesavento. Ethernet passive optical network (EPON): Building a next-generation optical access network. *IEEE Communications Magazine*, 40(2):66–73, 2002.

G. Kramer, M. De Andrade, R. Roy, and P. Chowdhury. Evolution of optical access networks: Architectures and capacity upgrades. *Proceedings of the IEEE*, 100(5):1188–1196, 2012.

K. Kumar, J. Liu, Y.-H. Lu, and B. Bhargava. A survey of computation offloading for mobile systems. *Mobile Networks and Applications*, 18(1):129–140, 2013.

C. F Lam. Fiber to the home: Getting beyond 10 Gb/s. *OSA Optics and Photonics News*, 27(3):22–29, 2016.

Task Group Leadership. High efficiency (he) wireless lan task group. Technical report, IEEE, Dec.9 2017. URL http://www.ieee802.org/11/Reports/tgax_update.htm.

M. Maier. Survivability techniques for NG-PONs and FiWi access networks. In *Proc. IEEE ICC*, pages 6214–6219, 2012.

M. Maier. FiWi access networks: Future research challenges and moonshot perspectives. In *Proc. IEEE ICC, Workshop on Fiber-Wireless Integrated Technologies, Systems and Networks*, pages 371–375, 2014a.

M. Maier. The escape of sisyphus or what post NG-PON2 should do apart from neverending capacity upgrades. *Photonics*, 1(1):47–66, Mar. 2014b.

M. Maier and B. P Rimal. The audacity of Fiber-Wireless (FiWi) networks: Revisited for clouds and cloudlets (invited paper). *China Communications*, 12(8):33–45, Aug. 2015.

M. Maier. Towards 5G: Decentralized routing in fiwi enhanced LTE-A hetnets. In *Proc. IEEE International Conference on High Performance Switching and Routing (HPSR)*, pages 1–6, July 2015.

M. Martin, N. Ghazisaidi, and M. Reisslein. The audacity of fiber-wireless (FiWi) networks (invited paper). In *Proc. Third International Conference on Access Networks (AccessNets)*, pages 1–10, Oct. 2008.

S. Nakamoto. Bitcoin: A peer-to-peer electronic cash system, 2008.

D. Nesset. NG-PON2 technology and standards. *IEEE/OSA Journal of Lightwave Technology*, 33(5):1136–1143, 2015.

LLC (NSP) Network Strategy Partners. Transformation of the enterprise network using passive optical LAN. *White Paper*, pages 1–13, May 2009.

Nokia Passive Optical LAN. Nokia, Aug. 25 2016. URL https://networks.nokia.com/solutions/passive-optical-lan.

NTT Corporation. Flexible access system architecture (FASA). *White Paper, Ver. 1.0*, pages 1–29, June 2016.

OECD Publishing. The digital economy, new business models and key features. *Addressing the Tax Challenges of the Digital Economy*, Sept. 2014.

M.-R. Ra, A. Sheth, L. Mummert, P. Pillai, D. Wetherall, and R. Govindan. Enabling interactive perception applications on mobile devices. In *Proc. 9th international conference on Mobile systems, applications, and services*, MobiSys 11, pages 43–56, June 2011.

B. P. Rimal and I. Lumb. The rise of cloud computing in the era of emerging networked society. In *Cloud Computing: Principles, Systems and Applications*, pages 3–25. Springer, 2017.

B. P. Rimal and M. Maier. Workflow scheduling in multi-tenant cloud computing environments. *IEEE Transactions on Parallel and Distributed Systems*, 28(1): 290–304, 2017.

B. P. Rimal, D. Pham Van, and M. Maier. Mobile-edge computing vs. centralized cloud computing in fiber-wireless access networks. In *2016 IEEE Conference on Computer Communications Workshops (INFOCOM WKSHPS)*, pages 991–996, 2016.

B. P. Rimal, D. P. Van, and M. Maier. Mobile edge computing empowered fiber-wireless access networks in the 5G Era. *IEEE Communications Magazine*, 55(2):192–200, 2017a.

B. P. Rimal, D. Pham Van, and M. Maier. Mobile-edge computing versus centralized cloud computing over a converged fiwi access network. *IEEE Transactions on Network and Service Management*, 14(3):498–513, 2017b.

B. P. Rimal, D. Pham Van, and M. Maier. Cloudlet enhanced fiber-wireless access networks for mobile-edge computing. *IEEE Transactions on Wireless Communications*, 16(6):3601–3618, 2017c.

M. Satyanarayanan. Pervasive computing: Vision and challenges. *IEEE Personal Communications*, 8(4):10–17, 2001.

N. Shibata, T. Tashiro, S. Kuwano, N. Yuki, Y. Fukada, J. Terada, and A. Otaka. Performance evaluation of mobile front-haul employing Ethernet-based TDM-PON with IQ data compression [invited]. *IEEE/OSA Journal of Optical Communications and Networking*, 7(11):B16–B22, 2015.

K. Smith and Y. R. Zhou. Optical transmission innovation for long term broadband internet growth. *IEEE ComSoc Technology News (CTN) [Online]*, Aug. 2016. URL https://www .comsoc.org/ctn/strengthening-backbone-5g-and-beyond.

Y-Y. Su and J. Flinn. Slingshot: Deploying stateful services in wireless hotspots. In *Proc. 3rd International Conference on Mobile Systems, Applications, and Services*, pages 79–92, June 2005.

F. Tschorsch and B. Scheuermann. Bitcoin and beyond: A technical survey on decentralized digital currencies. *IEEE Communications Surveys & Tutorials*, 18(3):2084–2123, 2016.

L. L. Van-Etter. Design and installation challenges and solutions for passive optical LANs. *White Paper, 3M Communication Markets Division*, pages 1–9, 2015.

R. Wolski, S. Gurun, C. Krintz, and D. Nurmi. Using bandwidth data to make computation offloading decisions. In *Proc. IEEE International Symposium on Parallel and Distributed Processing*, pages 1–8, Apr. 2008.

13

Techno-economic and Business Feasibility Analysis of 5G Transport Networks

Forough Yaghoubi[1], Mozhgan Mahloo[2], Lena Wosinska[1,3], Paolo Monti[1,3], Fabricio S. Farias[4], Joao C. W. A. Costa[4], and Jiajia Chen*,[1,3]

[1] School of Electrical Engineering and Computer Science, KTH Royal Institute of Technology, Kista, Sweden
[2] ÅF Technology, Digital Solutions Division, Stockholm, Sweden
[3] Department of Electrical Engineering, Chalmers University of Technology, Gothenburg, Sweden
[4] School of Electrical and Computer Engineering, Institute of Technology of the Federal University of Pará, Belém, Brazil

13.1   Introduction

The exponential growth of mobile data traffic, mainly driven by multimedia services and an increase in the number of connected devices, brings new challenges for mobile network operators (MNOs) [see Cisco - 2015; UMTS Forum - 2011; Osseiran et al. - 2013; Zander and Mähönen - 2013]. Traditionally this capacity growth has been addressed by looking for new spectrum opportunities, enhancing spectrum efficiency, and/or adding macro cell sites. However, the spectrum is not an infinite resource, and its efficiency is improving at a much slower pace than the increase in the capacity demand [see Femto forum - 2010]. It is also complicated to acquire new base station (BS) sites in urban areas, not to mention the inefficiency of using macro sites to serve users that mainly reside indoors [see Norman - 2010; Ericsson AB - 2013]. A promising way to solve this capacity crunch is to deploy heterogeneous networks (HetNets), where high-power macro cells provide coverage, and less expensive outdoor/indoor small cells are deployed close to the end user to ensure that capacity is provided only where it is needed.

The benefits of HetNets over homogeneous deployments (i.e. macro cells only) have been demonstrated in a number of studies in terms of both cost and power consumption [see Markendahl et al. - 2008; Markendahl and Mäkitalo - 2010; Aleksic et al. - 2013; Khirallah and Rashvand - 2011; Claussen et al. - 2008]. The authors of [Markendahl et al. - 2008] show that using only small cells, especially for indoor coverage, can decrease the deployment cost of radio access networks (RANs) by nearly a factor of five. The study in [Markendahl and Mäkitalo - 2010] emphasized that small cell deployments are a cost efficient alternative to macro densification, especially in scenarios where the capacity demand is high (e.g., beyond 100 Mbps km^{-2}). The work in [Aleksic et al. - 2013] evaluates the benefits of indoor small cells on improving the capacity available for residential and enterprise users, while [Khirallah and Rashvand - 2011] discuss the energy reduction

---

* Corresponding Author: Jiajia Chen; jiajiac@kth.se

Optical and Wireless Convergence for 5G Networks, First Edition.
Edited by Abdelgader M. Abdalla, Jonathan Rodriguez, Issa Elfergani, and Antonio Teixeira.
© 2020 John Wiley & Sons Ltd. Published 2020 by John Wiley & Sons Ltd.

that mobile operators can achieve thanks to HetNet deployments. Finally, the work in [Claussen et al. - 2008] shows that with HetNet deployments it is possible to achieve energy saving up to 60% in urban areas. On the other hand, the introduction of small cells has an impact on the backhaul network [see Farias et al. - 2013a; Tombaz et al. - 2011], which is responsible for collecting data traffic at the BSs (i.e. after baseband processing) and for sending it to the metro/aggregation segment. With HetNets the backhaul segment tends to become more complex compared to homogeneous wireless deployment solutions [see Juniper Networks - 2011]. This is because a large number of links are required to aggregate the data traffic generated by all the small cells, each one working at a peak rate at least tens of Mbps. These rates can hardly be guaranteed with legacy copper based infrastructures. in particular not over a long distance [see Farias et al. - 2013b]. Such a scenario is likely to force mobile operators to upgrade their backhaul networks to avoid potential bottlenecks in terms of capacity.

In a competitive telecom market, where new services and technologies are frequently appearing, and the revenue margin is constantly reducing [see Giles et al. - 2004], for each new deployment/upgrade all the possible cost drivers need to be carefully considered. Knowing that the cost of the backhaul segment is already a not negligible part of the total cost of ownership (TCO) in homogeneous wireless networks [see Geitner - 2005], it can be expected that with an increasing number of small cells (i.e. as in a HetNet deployment) the impact of the backhaul segment on the TCO of RANs will become even more critical [see Skyfiber - 2013]. Therefore, finding cost efficient backhaul solutions have become an important challenge in recent years. The rationale is simple: if not correctly predicted, the extra backhaul cost might reduce the benefits brought by the deployment of small cells.

On the other hand, restricting the analysis of the impact of backhaul only to TCO assessment is risky. TCO can be used to compare the various backhaul alternatives, but it provides estimates only about the cost aspects of a given technology/architecture. This means that from a TCO analysis it is not possible to extrapolate any information about the profitability of a specific backhaul solution, (i.e. the amount of money gained with respect to the investment). These considerations include many other factors such as yearly revenues, user penetration rates, the number of competitors in the area, and regulations, just to name a few. As a result, profitability can only be assessed with a comprehensive economic feasibility framework that in addition to TCO is also able to provide estimates about the net present value (NPV) of a backhaul deployment [see Cid et al. - 2010].

There are some attempts in the literature to provide cost modeling of mobile networks. [Frias and Pérez - 2012] numerically studied the cost savings achievable with the deployment of indoor small cells in regards to RAN deployments. The work in [Ahmed et al. - 2013] studies various wireless architectures (both homogeneous and heterogeneous) and tries to assess the impact of backhaul on the entire TCO. An author in [Frias and Pérez - 2012] proposed a comprehensive methodology to analyze the TCO of a number of backhaul network options based on fiber, copper, and microwave. The use case under examination considers a European urban scenario with both outdoor and indoor users. The latter is served by a layer of femto cells deployed inside the buildings where they reside, while the former are catered by macro BSs. However, many details are ignored that might affect the results in terms of cost saving. The authors in [Soh et al. - 2003; Kuo et al. - 2010] compared different microwave based backhaul topologies (including mesh and tree) with respect to their total cost. They reached the conclusion that mesh structures are a cost efficient option for homogeneous wireless deployment, but they did not consider HetNet deployments.

The study presented in [Senza fili - 2011] compares the deployment cost of backhauling in a long term evolution (LTE) homogeneous wireless network using fiber and microwave technologies. This study concludes that for scenarios with low cell density, the microwave is the cheapest option. However, only a homogeneous wireless deployment was evaluated in this work. The works in [Monti et al. - 2012; Tombaz et al. - 2014] compare different technology and topology options for backhauling a HetNet deployment, but their main focus is only on energy consumption. The work in [Mahloo et al. - 2014b] introduces for the first time a general TCO evaluation methodology of mobile backhaul networks, including a detailed breakdown of the capital expenditure (CAPEX) and operational expenditure (OPEX). On the other hand, no NPV analysis is provided. From the list of research efforts described so far, it becomes clear that there is a need for an assessment framework that in addition to mere cost evaluation also estimates the NPV of a given backhaul deployment. This issue has been addressed in [Yaghoubi et al. - 2018] where the authors provided complete economic feasibility evaluation considering TCO and NPV calculation.

This chapter introduces a techno-economic framework that provides a complete market analysis of the various business actors for any type of mobile access network deployments (including both the homogeneous and the heterogeneous cases). The aim of the proposed model is to advise operators on what type of backhaul investment should be made, at which point in time, and based on what technology, in order to maximize their profits. This is done by providing a complete cash flow and NPV evaluation on top of a detailed TCO estimation. It should be noted that in techno-economics, the NPV is the most important criteria for understanding whether a network deployment is profitable or not. Finally, a sensitivity analysis of several important cost factors is also included, showing the influence of uncertainty on the assumptions and input values used in the case studies.

13.2 Mobile Backhaul Technologies

The backhaul network is responsible for aggregating the users' traffic from the wireless access (i.e. BSs) to the metro/backbone segment of the network. Operators may select from more than one technology for their backhaul network(s) according to their needs in terms of capacity, reliability, cost, and expected deployment duration. In general, backhaul networks may be based on copper, fiber, microwave, or a combination of all these technologies (i.e. hybrid backhaul) [see Ercisson AB - 2014]. Recent backhaul technologies such as millimeter wave [see Nie et al. - 2013] and free space optic [see Feng et al. - 2014] have been evaluated in the literature and were introduced to the market, but they are not yet mature enough for deployment on a massive scale.

Currently, microwave represents nearly 50% of all backhaul deployments, and it is expected that it will maintain this share of the market in the years to come [see Ercisson AB - 2014]. This is mainly due to the moderate installation cost and its relatively short time to deploy. With the most recent technological development, it is possible to have microwave links operating at 1 Gbps up to few kilometers [see Ercisson AB - 2014;

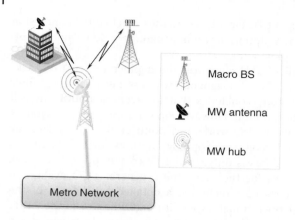

Figure 13.1 Microwave based backhaul architecture.

Macro BS

MW antenna

MW hub

Metro Network

Coldrey et al. - 2013]. Regardless of the topology in which it is organized (i.e. mesh, tree, ring, star, or any combination of them), a microwave based backhaul consists of several point-to-point (P2P) or point-to-multipoint (P2MP) links, each one requiring antennas at every endpoint. In the case of a P2P link, which is the major focus of this chapter, one antenna is located at the BS (or on top of a building), while the other one is connected either to a switch at the first aggregation point of the backhaul infrastructure (in the case of a multi-stage backhaul) or directly to a switch at the metro/aggregation node. Where several microwave antennas are co-located in one place, a tower mast (i.e. also referred to microwave hub) needs to be installed. When possible microwave antennas can be also installed on the same tower as the macro base station antennas. Figure 13.1 shows an example of a simple microwave based backhaul architecture.

Copper-based backhaul segments amount approximately to 20% of all current existing backhaul deployments [see Tipmongkolsilp et al. - 2011; Ercisson AB - 2017]. Most likely, they will be gradually replaced by other technologies due to their limited ability to provide high capacities over long distances (i.e. more than 100 Mbps can be guaranteed only up to 300 m [see Farias et al. - 2013b]). However, there is some new advancement in copper technology that may lead to its survival for several more years before it is replaced totally by other technologies. One example is the G.fast standard that is currently developed by ITU-T and aims at achieving bit rates up to 1 Gbps but still over short distances [see Lins et al. - 2013; Nokia - 2018].

Fiber based backhaul can provide ultra-high capacity over long distances. However, it is relatively time-consuming and expensive to deploy a fiber infrastructure, especially if it is done from scratch (i.e. greenfield deployment). On the other hand, in places where a cable infrastructure (e.g., ducts from a previous copper installation) is already available, faster deployment of fiber is possible [see Farias et al. - 2016].

Fiber access networks can be deployed as P2MP topology or as P2P interconnection. In the latter case, one optical line terminal (OLT), a device corresponding to the service provider endpoint, located in the central office (CO) is connected to an optical network unit (ONU), the endpoint at the user/wireless network side, via a dedicated fiber link (Figure 13.2a). In a P2MP architecture (Figure 13.2b), each OLT is connected to several ONUs via a passive splitter (i.e. in the case of passive optical networks (PONs)),

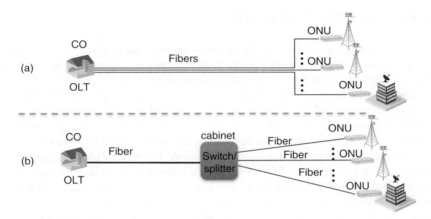

Figure 13.2 Fiber based backhaul architecture.

or through an ethernet switch (i.e. in the case of active optical networks). Splitters and ethernet switches are located in a node that is referred to as the remote node. Thanks to the ability of fiber to deliver high capacity over long distances (i.e. in the order of several tens of kilometers), a backhaul network based on the optical transmission makes it possible to connect cells directly to the mobile core network, without any intermediate stages. This is not the case for a microwave based backhaul network where the connectivity between the microwave hub and the mobile core network typically requires a fiber link (Figure 13.1). Fiber cables are often installed inside ducts that are buried under the ground. This process is referred to as digging (trenching) and it represents the most expensive part of a fiber network deployment.

To achieve better utilization of the radio resources some operators prefer to use a centralized radio access network (C-RAN) architecture [see Chih-Lin et al. - 2014]. Unlike traditional RANs, where the radio unit and the baseband processing unit (BBU) are co-located at the base station site, in a C-RAN architecture, BBUs are decoupled from the BSs and centralized into one or more BBU hotel(s). In a C-RAN architecture, the transport network is divided into two parts: fronthaul and backhaul. The fronthaul segment is responsible for transporting the traffic between the remote radio units (RRUs) and the BBU hotels, while the backhaul part provides transport connectivity between the BBU hotels and the mobile core network. The communication between RRUs and BBUs, which can be based on, e.g., the common public radio interface (CPRI) protocol [see CPRI - 2013], normally requires capacities in the order of tens of Gbps, making optical transmissions (i.e. in particular, radio over fiber techniques) the perfect candidate for fronthaul.

In summary, there is a consensus that fiber and microwave are the two main candidates for backhauling current and future mobile network deployments. For this reason, the case study carried out in this chapter focuses on these two technologies. In addition, the framework proposed in this chapter addresses specifically only the backhaul segment (i.e. no C-RAN), but it is general enough to also be applied to the other 5G transport solutions (e.g., fronthaul).

13.3 Techno-economic Framework

This section first discusses the role played by business viability assessment in network deployments, and then it presents a framework that can be used by mobile network operators to estimate the profitability of a given backhaul deployment.

Figure 13.3 presents the life cycle of a communication network, which typically consists of four phases: planning, initial installation, operational phase (e.g., providing connections to customers, keeping a network up and running), and teardown. The planning phase takes place before any new deployment. This is the most crucial step for understanding if a deployment is feasible or not, and to reduce the risk of the investment not bringing profit. The reason is the following. Even if a technology is already mature enough to be deployed, the market may not be ready for it, e.g., the user penetration might be too low, or potential users may not be willing to pay extra for a specific service. All these aspects need to be assessed via a comprehensive techno-economic and risk analyses framework in order to validate the economic viability of a new deployment. More specifically, it is crucial to quantify the total expenses required during the network lifetime, as well as the estimated revenues and cash flows.

This information should then be used to estimate the payback period, i.e. the time required for the return of the investment. If the payback period is too long, or the total cash flow is negative, it is not advisable (from a purely economic point of view) to carry out the project.

If the results from the planning phase show that a given deployment is profitable, then the initial installation phase starts. During this phase, operators sustain an upfront cost, money that is typically considered as part of the CAPEX in the TCO calculations. Once deployed, the network needs to be kept up and running (i.e. the operational phase). All the expenses that occur in this phase are considered part of the OPEX.

Finally, when the current network needs to be replaced/upgraded (e.g., new technology is ready to be deployed), the customers will gradually be subscribed to the new network. This period is referred to as the teardown phase, and it ends when the migration to the new service(s) is completed. The expenses related to the network teardown phase include mainly labor cost, and they typically represent a relatively small part of the TCO compared to the money spent during the installation and the operational phase. For this reason, the impact of the teardown phase can be, in a first approximation, ignored in most cases.

As it was already explained, the profitability of a given deployment is a crucial point in deciding whether or not a new deployment should take place. For this, a complete

Figure 13.3 Network life cycle.

Figure 13.4 Techno-economic framework.

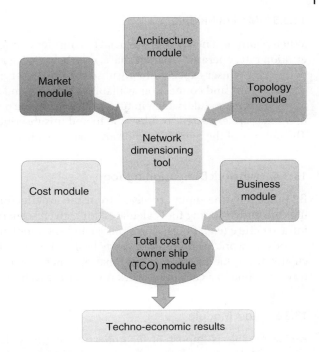

techno-economic framework analysis is required. Figure 13.4 presents an assessment methodology (consisting of a techno-economic market and risk analysis) that can be used by mobile operators to analyze the business feasibility of a given backhaul deployment. This framework consists of several modules, each one presented in the following sections where the required input and the expected output will be described in detail.

13.3.1 Architecture Module

The objective of this module is to define the technology used in the backhaul segment together with the type of components to be installed in each location. For example, in the case of a microwave-based backhaul, antennas are required in both sides of the microwave links, while in the case of fiber-based backhaul components such as OLT, splitting devices, and ONUs need to be installed at central offices, remote nodes, and user premises, respectively.

13.3.2 Topology Module

The network topology defines the way in which the various components of a given architecture are interconnected. Examples of network topologies are ring, star, tree, and mesh. Another important parameter included in the topology module is the demographical data of the region that is under study. The number of buildings, user density, size of the geographical region, existing infrastructure (e.g., available ducts) are also an input for the topology module. In terms of output, the topology module can compute the number of network nodes (e.g., COs, remote nodes, cabinets), their locations, the distances between different nodes, and the equipment type that should be installed in each location. These parameters are then provided to the network dimensioning tool.

13.3.3   Market Module

When planning a network it is crucial to consider market related data such as user penetration rate, operator's market share, user behavior, service prices, user churn rate (i.e. percentage of user expected to unsubscribe from a service), area throughput, quality of service (QoS), and connection availability, just to name a few. These parameters are fed to the market module, which in turn can estimate the possible revenues, the number of users that are expected to subscribe to and unsubscribe from the service under exam. The outputs of the market module are then provided to the dimensioning tool.

13.3.4   Network Dimensioning Tool

By processing the inputs received from the architecture, topology, and market modules the dimensioning tool calculates, for a given scenario, the amount of required new infrastructure (e.g., fibers, ducts, hubs) and the number of components needed in the various network locations on a yearly basis. Moreover, the dimensioning tool also calculates the value of some operational parameters related to the labor activities (e.g., traveling time to a certain network node for reparation).

13.3.5   Cost Module

A cost module is important for understanding how the various cost parameters in a TCO module vary with time. For example the price of a specific component normally decreases as a function of the (increasing) production volumes, the quantity purchased on the market, and the level of maturity of the technology. On the other hand, the expenses related to human resources (e.g., technician salaries) typically increase each year. Therefore, price variation should be considered while calculating the network expenses. Price erosion in time can be calculated via a learning curve that is used in the industry to predict the reduction of the cost of a product [see Verbrugge et al. - 2009]. However, finding the right learning curve is not an easy task. The cost module applied in the proposed framework refers to a widely used linear formula for calculating the cost variation, presented below:

$$P_j = P_0 + \alpha P_{j-1} \tag{13.1}$$

where P_j denotes the price in a year j of network lifetime, and P_0 is the price at the beginning of the project. The coefficient α denotes the cost change factor. This parameter has a negative value when calculating how the price for hardware components varies (i.e. this price normally decrease with time). On the other hand, α has a positive value when calculating how the price varies in time for parameters not related to the hardware, e.g., salaries and energy costs. In reality α might also vary with time, however for simplicity, in this study, α is considered to be constant during the whole network lifetime.

13.3.6   Total Cost of Ownership (TCO) Module

This section presents the TCO module used in the proposed framework. The module covers both the CAPEX and the OPEX aspects of the backhaul segment. More specifically, the module includes all the costs incurred during the backhaul lifetime (i.e. from

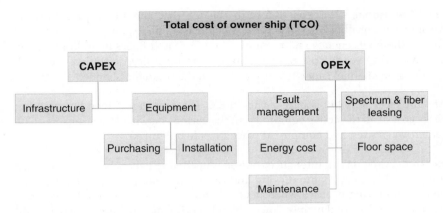

Figure 13.5 Cost classification of a TCO module.

the network deployment phase, when an upfront investment is required, up to all cost aspects related to the operational processes).

Figure 13.5 presents the cost classification according to the proposed cost module. Since, in general, a backhaul segment may comprise more than one technology (i.e. a hybrid architecture), the proposed module accounts for the presence of both fiber and microwave. The details of each part are presented next.

13.3.6.1 Capital Expenditure (CAPEX)

The CAPEX refers to all the expenses related to the backhaul network deployment cost. According to the model in Figure 13.5, CAPEX can be divided into two main parts, i.e. equipment and infrastructure costs. They are described in the following.

Equipment cost The equipment cost is the sum of all expenses related to purchasing the backhaul components, i.e. according to the results of the dimensioning tool, and to install them in their specific locations.

Infrastructure cost The total infrastructure cost of a mobile backhaul segment corresponds to the investment needed to deploy the fiber infrastructure as well as the cost of leasing fibers (when the fiber infrastructure has already been deployed by other providers and is available for leasing). It also includes the expenses needed to install the microwave hubs, i.e. masts and antennas, where needed. The fiber infrastructure cost includes all the expenses related to trenching, purchasing of fiber cables, and pumping fibers into the ducts. Trenching can be defined as placing the optical fibers inside the ducts that are buried under the ground. In many cases, MNOs prefer to lease fibers instead of deploying their infrastructure. In such a case, the infrastructure cost includes an upfront charge per kilometer of leased fiber paid to the infrastructure owner.

13.3.6.2 Operational Expenditure (OPEX)

OPEX refers to the expenses occurred during network operations over a predefined time interval (i.e. the network operational time). The main OPEX components are indicated in Figure 13.5 and they are defined below.

Spectrum and fiber leasing This cost refers to the fee that should be paid to lease microwave spectrum or fiber infrastructure. When leasing fibers, an MNO is charged a yearly fee for the maintenance and reparation of the rented fibers in addition to the upfront expenses. The yearly cost of spectrum leasing for a licensed microwave link varies depending on channel capacity (i.e. class i) and the frequency band.

Energy cost The electricity bill is part of the OPEX. This cost is obtained by summing up the energy cost of all the active equipment in the various backhaul locations (i.e. CO, cabinets, microwave sites, and equipment placed inside buildings).

Maintenance cost A regular maintenance routine is needed to keep a backhaul network up and running. This includes monitoring and testing the equipment, updating software (including renewing licenses when needed), and replacing deteriorated components (e.g., batteries). The total maintenance cost reflects the maintenance cost of central offices, cabinets, and microwave links. In order to ensure that the network and all the services are running as expected, full-time monitoring is also required. This translates into extra monitoring expenses in the total maintenance cost of the network. Operators consider several rounds of maintenance procedures for each central office depending on the number of users and services covered by each one of them. Microwave links also require regular monitoring, because antennas might tilt and lose their line of sight.

Fault management Fault management refers to the expenses related to the reparation of failures that might occur in a backhaul network. The total yearly reparation cost for the backhaul network is defined as the sum of the reparation cost of each failure occurring during the year as well as the penalty paid to the users as specified in the service level agreement (SLA).

The reparation cost depends on the cost of replacing a failed component (if needed) in each year, the mean reparation time of each device, and the time to travel to the location of the failure. The penalty quantifies the fine that operators need to pay to the customers when the service interruption is longer than the threshold defined in the SLA, T_{SLA}. Let t show a period where T_{SLA} needs to be satisfied, which can be a year, a month or even a day. If the mobile backhaul has a failure in this period t, one or more macro cells might be out of service, and potentially a large number of customers can lose connectivity. Therefore, we can consider penalty costs for a backhaul provider when the macro cell backhaul connectivity is lost due to a failure. N_j^{Mac} and $unAv_{ij}$ denote the number of macro cells with high importance in year j and the connection unavailability of the backhaul link to the macro cell i. The penalty rate $P_i^{co/h}$ is agreed in the SLA and is dependent on the importance of service outage for the customer.

$$\text{Penalty} = \sum_{j=1}^{L_n} \sum_{i=1}^{N_j^{Mac}} P_i^{co/h}(unAv_{ij} - T_{SLA}). \tag{13.2}$$

The above equation is valid if the service outage time is larger than T_{SLA}, otherwise the penalty cost is equal to zero.

Floor space cost The floor space cost is a yearly rental fee paid by an operator to house its equipment, i.e. to place components in racks with standard size in various locations.

13.3.7 Business Models and Scenarios

This module accounts for the business-related parameters, i.e. which are the actors in the area, the co-operation models between actors and the various governmental entities (e.g., municipalities). Business actors can be categorized into the following: physical infrastructure provider, a network provider, a service provider, and the MNO. Other important business-related parameters are the market share of each operator, open access models (if any), and regulations with respect to sharing the infrastructure. For example, if inside a commercial building it is not possible to install multiple independent networks, two mobile operators may agree to share the infrastructure. Another possibility is for the incumbent to offer roaming service to any newcomer wanting to support its customers inside that building. Some possible business cases related to backhaul deployments are mentioned below.

- Case 1: an operator provides both mobile and fixed network services to its customers. In this case, the operator could reuse part of the fixed network infrastructure for backhaul services. In this way, the cost of the backhaul infrastructure (which is a relevant part of the backhaul TCO) can be reduced, and the period for the return on the investment will be shorter.
- Case 2: a mobile operator pays to a fixed network provider for backhaul connectivity to each base station.
- Case 3: a mobile operator deploys its backhaul network independently.

In the case of backhaul for small cells deployed indoors, the two main business models considered in the literature are the closed subscriber group and the open subscriber group. In the former case only a closed group of users can access the indoor cells (i.e. it is considered as private network for improving the quality of the service) while in the latter case everyone in the range of the cell can connect [see Frias and Pérez - 2012b], so that the small cells can be considered as part of mobile operator network and offload the macro cell traffic.

13.3.8 Techno-economic Module

The feasibility of any network deployment projects can only be decided using a techno-economic analysis that also includes cash flow (CF) and NPV considerations. By referring only to OPEX and CAPEX it is possible to understand the costs aspects of the project, but it is not possible to assess the feasibility of the project from the cost perspective. Cash flow refers to the amount of money gained (i.e. revenue) and spent during the lifetime of the network. After calculating the cash flow, it is possible to estimate the NPV. Equation 13.3 indicates the extra gain that a project will bring with respect to the invested money [see Nikolikj and Janevski - 2015]. If the NPV is negative, the project will be typically rejected as it does not have any financial benefit.

$$\text{NPV} = \sum_{j=1}^{L_n} \frac{\text{CF}_j}{(1+r)^j} \tag{13.3}$$

where L_n, CF_j, and r, denote the network lifetime, cash flow in a year j, and the discount rate, respectively. The "discount rate is a factor for estimating the present value of the future cash flows by considering the time value of the money and the risk or uncertainties of future incomes.

13.4 Case Study

This section presents a case study where the proposed business feasibility framework is applied. In the case study, we calculate the overall cost to deploy and operate a mobile network including both the backhaul and the RAN segment considering a network lifetime of 10 years.

13.4.1 Application of Methodology/Scenarios

Topology model   A 5×5 km dense urban area representing an average European city with a population density of 3000 users per kilometer [see Auer et al. - 2012] is considered. The area consists of 100 multistory buildings per square kilometer, with five floors per building and two apartments per floor. The buildings are placed according to the Manhattan model [see Marsan et al. - 1991], which is a geometric model widely considered in dense urban areas.

Architecture model   Two options are assumed for the wireless deployment: homogeneous (i.e. using macro BSs only), and heterogeneous (i.e. macro BSs serve outdoor users while small cells are deployed indoors to provide coverage inside for indoor users). The guaranteed bandwidths of 300 Mbps per building and 600 Mbps per macro cell are assumed. Two backhaul technologies are considered in the case study: microwave and fiber. For the microwave only case, P2P microwave links are used to backhaul the data traffic from both the macro BS and the small cells. At each building, a switch gathers the data traffic from all its indoor cells and send them via a rooftop microwave antenna to the closest hub in the area. The hubs then transfer the data traffic to the metro/aggregation network via other P2P microwave links. Traffic from the macro cells is also sent to the backbone via one or more hops of P2P microwave links depending on the distances from the metro node (Figure 13.6).

In the scenario with fiber backhauling, the data traffic from the indoor users is collected using a fiber to the building architecture, i.e. an aggregation switch inside each

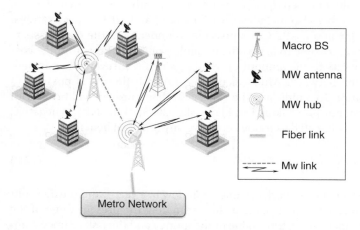

Figure 13.6 Considered microwave based backhaul architecture.

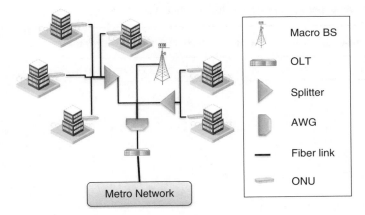

Figure 13.7 Considered fiber based backhaul architecture.

building is co-located with an ONU in the building basement, connected to an OLT using a PON architecture (Figure 13.7). The macro cells are also backhauled following the same way (i.e. one ONU per macro site connects to the OLT in the central office located in the corner of the considered area via an optical distribution network). Since the infrastructure deployment cost is known as the most expensive process of the TCO in case of fiber networks, two scenarios are assessed for the fiber-based backhaul. If no fiber infrastructure is available in the area, the mobile operator needs to deploy (i.e. trench) its infrastructure. However, if there is already an existing fiber infrastructure available, the operator could lease fiber instead of trenching. It is assumed that the fiber backhaul is based on the hybrid time and wavelength division multiplexing PON concept promoted by the full service access network forum as the candidate technology for the future high capacity optical access networks [see FSAN Group - 2018].

In this technology, a four wavelength channel at 10 Gbps is sent from the transceiver array in the OLT to a first remote node where an array wavelength grating routes each wavelength to a separate splitter located in a second remote node [see Mahloo et al. - 2014a]. Each wavelength coming to a splitter is shared among all the connected ONUs using time division multiplexing technology (Figure 13.7). Based on the assumption mentioned above, the following six scenarios are considered for the case study:

- Scenario 1: homogeneous RAN deployment backhauled via microwave links (Ho_MW).
- Scenario 2: microwave backhaul is used to serve a HetNet for RAN (He_MW).
- Scenario 3: operator deploys its fiber infrastructure (i.e. trenching is required) to backhaul its homogeneous RAN network (Ho_Tr).
- Scenario 4: operator leases fiber to backhaul homogeneous RAN network (Ho_Le).
- Scenario 5: operator deploys its fiber infrastructure (i.e. trenching is required) to backhaul its HetNet for RAN (He_Tr).
- Scenario 6: operator leases fiber to backhaul its HetNet for RAN (He_Le).

Market model The main criteria considered for the dimensioning of RAN is the required throughput per square kilometer in each year (shown in Table 13.1) as well as the

Table 13.1 Considered throughput per square kilometer [see Tombaz et al. - 2014].

Year	2014	2016	2018	2020	2022	2024
Throughput (Mbps)	15	30	60	119	235	470

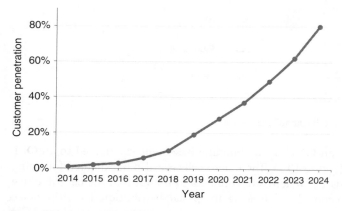

Figure 13.8 User penetration curve.

coverage constraint, which is fulfilled using macro and small cell deployments. We used the model presented in [Tombaz et al. - 2014] to calculate the required number of macro and small cells. Figure 13.8 represents the considered penetration curve related to the mobile customers joining the network in each year.

Table 13.2 summarizes the cost parameters used to calculate the TCO. The fault management cost is calculated based on the values in [see Mahloo et al. - 2013; Telecom India - 2014]. The results of the dimensioning work for the RAN, based on the inputs from the market module, are shown in Table 13.3. We assume to have one small cell per building in the first year in order to have enough indoor coverage. Half of the users are covered using the macro, and the other half is served by the indoor cells.

Business model In order to assess the total required investment cost of an MNO in the worst case scenario, we selected the business model number 3 presented in section 13.3.7. For small cells, we consider the open subscriber group model where indoor cells are managed and owned by the MNO. We assume that the operator has 30% of the market share in the region, i.e. only 30% of the total users are considered for the revenue calculations.

13.4.2 Techno-economic Evaluation Results

The results of the techno-economic evaluation of the scenarios as mentioned above are presented in this section. Figure 13.9 shows the total cost of ownership for the mobile network considering the TCO values for the backhaul and the RAN during ten years of network lifetime. It is evident that HetNet deployments decrease the cost of the RAN dramatically compared to the homogeneous case. However, according to the results, the

Table 13.2 Input values used for cost calculation gathered from [Frias and Pérez - 2012a; Ahmed et al. - 2013; Mahloo et al. - 2014a, 2013; Paolini - 2011; Oughton and Frias - 2017].

Component/parameter	Value
Number of team (β)	1
Cost change factor (salary) (α)	7%
Cost change factor (hardware) (α)	-3%
Discount rate (r)	10%
Subscription fee (λ)	€30
Number of tech./team (Tech$_{te}$)	2
Technician salary/hour (Tech$_j^s$)	€52
Energy cost/kWh	€0.1
Indoor yearly rental fee/m^2	€220
Outdoor yearly rental fee/m^2	€180
Small/large microwave antenna	€500/2000
G-ethernet switch	€1800
Microwave hub + installation	€20000
Ethernet switch	€150
Yearly spectrum leasing/MHz	€5
OLT ($4x$10G array transceiver)	€7000
ONU	€150
Power splitter (1:16/1:32)	€170/340
Fiber/km	€80
Trenching/km	€45000
Leasing upfront fee/km	€800
Yearly fiber leasing fee/km	€200
Macro base station and cell site	€48000
Small indoor base station	€250

Table 13.3 Number of installed cells.

Year	Heterogeneous		Homogeneous
	Macro cell	Small cell	Macro cell
2014	4	2500	8
2016	6	4000	15
2018	9	5750	30
2020	18	7500	60
2022	36	10000	119
2024	72	12500	237

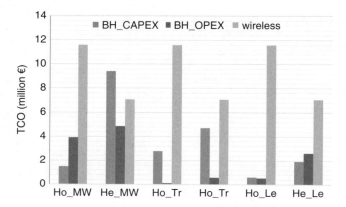

Figure 13.9 Total cost of ownership of a mobile network including both RAN and backhaul.

backhaul for HetNet, in all scenarios, is more than twice as expensive (in terms of TCO) than the corresponding homogeneous case. The results show that HetNet deployment with microwave backhauling is the most expensive scenario. This is due to the component cost and the power consumed by the microwave links, which increase almost linearly with the number of small cells. Moreover, the bigger the area, the higher the possibility to share some infrastructure using a fiber based backhaul compared to the microwave backhaul case. Therefore, fiber based backhauling is more cost efficient, in areas with a high density of small cells, even if an operator needs to deploy its fiber infrastructure. This confirms the claim made earlier in the chapter that it is important to carefully choose a proper backhaul technology in order to minimize the impact on the TCO of a HetNet deployment.

Figure 13.10 presents the cost breakdown to assess the impact of each cost element of the TCO for the backhaul segment. The considered cost elements are: fault management (FM), floor space (FS), spectrum and fiber leasing (Sp&Le), maintenance (M), energy (En.), infrastructure (Infra.), and component cost (Equip.). From Figure 13.10 it becomes evident that each cost item has a different impact on the TCO depending on the

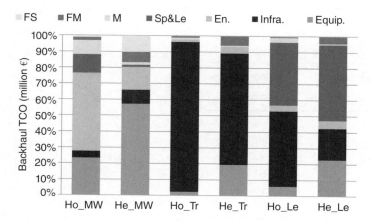

Figure 13.10 Cost breakdown of the backhaul TCO elements.

deployed strategy. For example, energy and equipment cost are the dominant part of a microwave based backhaul while these two elements always amount to less than 30% for a fiber based backhaul. The figure also shows that the impact of FM cost increases in the case of HetNet deployments due to the higher amount of equipment and infrastructure needed in the network.

Figure 13.11 shows the TCO evolution for all the scenarios presenting a TCO breakdown year by year. As can be seen from the figure both the backhaul technology and the RAN deployment strategy influence the distribution of expenses in the time domain during the considered ten years. In the case of HetNet deployments, huge upfront investment is required in the first year to have good indoor coverage backhauling all the small cells (one small cell per building in the first year is considered). For the scenarios with fiber trenching, also a significant amount of money needs to be invested in the first year as most of the trenching needs to be done in the beginning.

Another interesting aspect presented in Figure 13.11 is the proportion of CAPEX and OPEX for the backhaul technologies. In the scenario with microwave backhaul for homogeneous wireless deployment, the OPEX is huge, and it increases considerably with capacity growth. However, the OPEX is a tiny portion of the yearly expenses in case of fiber based backhaul with trenching, where the infrastructure cost is dominant.

Considering an average monthly subscription fee of €30 per user (for voice and data) and a discount rate of 10%, the NPV has been calculated for all the scenarios, and it is presented in Figure 13.12. Except for the case of HetNet deployments with microwave backhaul (i.e. where the backhaul TCO is extremely expensive), all the scenarios have a positive NPV and can be considered economically viable. Another interesting aspect to notice is the following. The He-Tr deployment has the lowest TCO value compared to all three homogeneous scenarios. On the other hand, its NPV (i.e. the total amount of profit at the end of the 10 year network lifetime) is the lowest. This is because most of the investment for both backhaul and RAN needs to be done in the first years of the project. Typically the same amount of money is worth more at present than in the future due to the potential of producing income if invested earlier. This example shows the importance of business viability analyses, i.e. the technology with the lowest TCO value might not be the one economically preferable for a long-term investment project.

13.4.3 Sensitivity Analysis

There are lots of uncertainties in the input parameters considered in the cost study that might change the results dramatically. A thorough sensitivity analysis helps to understand the impact of variations in the input values on the cost results and identify the key cost factors, which give a better view of the possible risks of the project and the reliability of the results.

This section analyses how the variation of some key input parameters of the techno-economic tool might impact the conclusions that were just discussed. Based on the backhaul cost breakdown presented in Figure 13.10, we first identify the most costly elements of each deployment option and then we calculate how the TCO of the backhaul varies when the value of these key elements is changed. In the case of a microwave based backhaul, power consumption and equipment costs are the most expensive elements, while for the fiber scenarios, the infrastructure cost related to the trenching or leasing costs has the highest share in the TCO of the backhaul. To illustrate the impact of each one

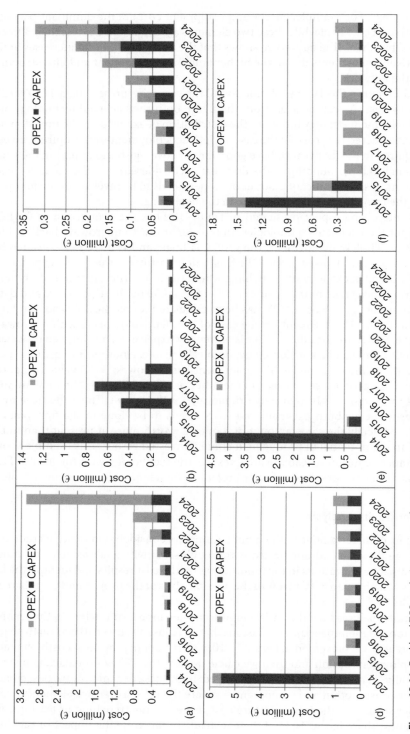

Figure 13.11 Backhaul TCO evolution per year for (a) Ho-MW, (b) Ho-Tr, (c) Ho-Le, (d) He-MW, (e) He-Tr, and (f) He-Le scenarios.

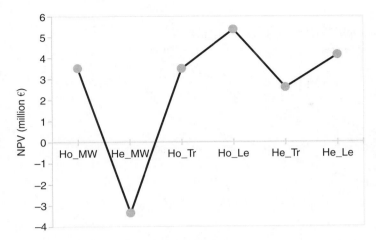

Figure 13.12 NPV at the end of the 10 year network lifetime.

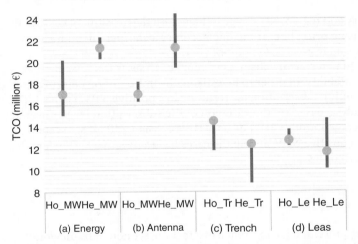

Figure 13.13 Sensitivity analysis of the TCO varying: (a) energy cost (±50%), (b) MW antenna price (±50%), (c) amount of re-usable trenching (from 0% to 100%), (d) price of leasing (±50%).

of these elements, Figure 13.13 shows the fluctuation of the TCO of a mobile network deployment (including both RAN and backhaul).

The grey circles in Figure 13.13, represent the TCO values for each scenario calculated in the previous section, and the green bars demonstrate its variation if the values of corresponding input parameters are changed. It can be seen that even by decreasing the energy price by half or by using antennas costing half the price, microwave based solutions are still the most expensive options among all the scenarios. An interesting finding from Figure 13.13 is the difference between a greenfield scenario where no infrastructure is available (0%) and a brownfield one where a certain amount of fiber infrastructure is available, and only partial trenching is needed (> 0%). If more than half of the required trenching is available, then fiber based backhaul for a HetNet scenario offers the lowest TCO for providing mobile services. The price of leasing fibers, which varies a lot

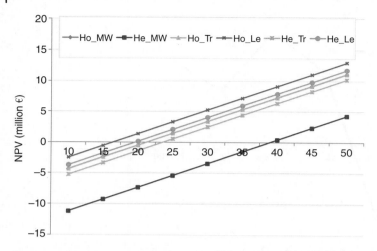

Figure 13.14 Sensitivity analysis of NPV based on the variation of the monthly subscription fee.

depending on the country and region, has a huge influence on the investment cost. With a 50% increase in the leasing price, fiber trenching becomes a cheaper and more attractive option than fiber leasing from a TCO point of view.

Another important parameter in any business viability analyses is the revenue that is related to the project profitability. In order to investigate the impact of revenue per user and to find the lowest monthly subscription fee resulting in a positive NPV, Figure 13.14 presents a sensitivity analysis based on the variation of the subscription fee per user. It is evident that if the average monthly fee per user is more than €40 all six scenarios, even He-Mw, can reach a positive NPV, and all the cases have a negative NPV when the charged monthly fee is €15 or less.

13.5 Conclusion

In summary, this chapter presented a comprehensive techno-economic framework for estimating the TCO of a backhaul network segment as well as for analyzing the business viability of a given wireless network deployment. The chapter focused on two backhaul technologies: microwave and fiber. The proposed framework is applied to a case study to estimate the TCO and the NPV of a heterogeneous and homogeneous wireless deployment. The results showed a considerable increase in the value of the backhaul TCO in the case of a heterogeneous deployment compared to a conventional homogeneous scenario. The results also showed that fiber is the most cost-efficient technology to provide a high capacity backhaul for heterogeneous wireless deployments. The cheapest alternative is to lease fiber connectivity when possible. Our results also highlight the importance of selecting the right backhaul technology in order to keep the economic benefits brought by heterogeneous wireless deployments. The proposed business viability analysis also indicated the importance of having a complete techno-economic assessment instead of having a pure cost calculation. Our NPV results showed that a lower TCO does not always lead to a higher profit because in the long-term project investments done in different time periods affect the total profit of the project.

Bibliography

A. A. W. Ahmed, J. Markendahl, C. Cavdar, and A. Ghanbari. Study on the effects of backhaul solutions on indoor mobile deployment macrocell vs. femtocell. *Proc. of IEEE International conference on Personal, Indoor and Mobile Radio Communications (PIMRC)*, 2013.

M. Aleksic, S. and Deruyck, W. Vereecken, W. Joseph, M. Pickavet, and L. Martens. Energy efficiency of femtocell deployment in combined wireless/optical access networks. *Elsevier Computer Networks*, 57:1217–1233, 2013.

Gunther Auer, Oliver Blume, Vito Giannini, Istvan Godor, Muhammad Ali Imran, Ylva Jading, Efstathios Katranaras, Magnus Olsson (EAB), Dario Sabella, Per Skillermark, and Wieslawa Wajda. Energy efficiency analysis of the reference systems, areas of improvements and target breakdown. *EARTH Deliverable D2.3*, 2012.

I. Chih-Lin, C. Rowell, S. Han, Z. Xu, G. Li, and Z. Pan. Toward green and soft: A 5G perspective. *IEEE Communications Magazine*, 52:66–73, 2014.

C. Cid, M. Ruiz, L. Velasco, and G. Junyent. Costs and revenues models for optical networks architectures comparison. *IX Workshop in G/MPLS Networks*, 2010.

Cisco. Cisco visual networking index: Global mobile data traffic forecast update 2013:2018. *Cisco Inc., USA, C11-738429-00*, 2015.

H. Claussen, L.T.W. Ho, and F. Pivit. Effects of joint macrocell and residential picocell deployment on the network energy efficiency. *Proc. of IEEE International Symposium on Personal, Indoor and Mobile Radio Communications (PIMRC)*, 2008.

J. Coldrey, M. Berg, L. Manholm, C. Larsson, and J. Hansryd. Non-line-of-sight small cell backhauling using microwave technology. *IEEE Communications Magazine*, 51:78–84, 2013.

CPRI. Common public radio interface (CPRI), v6. 0. *CPRI Specification*, 6, 2013.

Ercisson AB. Ericsson microwave towards 2020 report. *Ercisson AB*, 2014.

Ercisson AB. Ericsson microwave outlook. *Ercisson AB*, 13:97–113, 2017.

Ericsson AB. Ericsson radio dot system. *Ericsson AB, white paper*, 2013.

F. S Farias, P. Monti, A. Vastberg, M. Nilson, J. C. W. A. Costa, and L. Wosinska. Green backhauling for heterogeneous mobile access networks: What are the challenges? *Proc. of International Conference on Information, Communications and Signal Processing (ICICS)*, 2013a.

F. S. Farias, M. Fiorani, S. Tombaz, M. Mahloo, L. Wosinska, J. C. W. A. Costa, and P. Monti. Cost-and energy-efficient backhaul options for heterogeneous mobile network deployments. *Photonic Network Communications*, 32(3): 422–437, 2016.

F.S. Farias, Borges G.S., Rodrigues R. M, A. L. Santana, and J.C.W.A. Costa. Real-time noise identification in dsl systems using computational intelligence algorithms. *Proc. of International Conference on Advanced Technologies for Communications (ATC)*, pages 252–255, 2013b.

Femto forum. Femtocells-natural solution for offload. *Femto forum white paper*, 2010.

D. Feng, W. Sun, and W. Hu. Hybrid radio frequency and free space optical communication for 5G backhaul. 2014.

Z. Frias and J. Pérez. Techno-economic analysis of femtocell deployment in long-term evolution networks. *EURASIP Journal on Wireless Communications and Networking*, 2012:1–12, 2012a.

Z. Frias and J. Pérez. Techno-economic analysis of femtocell deployment in long-term evolution networks. *EURASIP Journal on Wireless Communications and Networking*, 2012:1–12, 2012b.

FSAN Group. Full service access network. *http://www.fsan.org*, 2018.

T. Geitner. Vodafone group technology update. *Vodafone*, 2005.

T. Giles, J. Markendahl, J. Zander, P. Zetterberg, P. Karlsson, G. Malmgren, and J. Nilsson. Cost drivers and deployment scenarios for future broadband wireless networks - key research problems and directions for research. *Proc. of IEEE Vehicle Technology Conference (VTC)*, 2004.

Juniper Networks. Mobile backhaul reference architecture. *Juniper Networks white paper*, 2011.

J.S. Khirallah, C. and Thompson and H. Rashvand. Energy and cost impacts of relay and femtocell deployments in long-term-evolution advanced. *IET Communications*, 5:2617–2628, 2011.

F-C. Kuo, F.A. Zdarsky, J. Lessmann, and S. Schmid. Cost efficient wireless mobile backhaul topologies: an analytical study,. *Proc of IEEE Global Telecommunications Conference (GLOBECOM)*, 2010.

S. Lins, P. Figueiredo, and A. Klautau. Requirements and evaluation of copper-based mobile backhaul for small cells LTE networks. *Proc. of IEEE International Microwave and Optoelectronics Conference (IMOC)*, 2013.

M. Mahloo, C. M. Machuca, J. Chen, and L. Wosinska. Protection cost evaluation of wdm-based next generation optical access networks,. *Journal of Optical Switching and Networking (OSN)*, 10:89–99, 2013.

M. Mahloo, J. Chen, L. Wosinska, A. Dixit, B. Lannoo, D. Colle, and C. M. Machuca. Toward reliable hybrid WDM/TDM passive optical networks. *IEEE Communication Magazine*, 52:14–23, 2014a.

M. Mahloo, P. Monti, J. Chen, and L. Wosinska. Cost modeling of backhaul for mobile networks. *Proc. of IEEE International Conference on Communications (ICC)*, 2014b.

J. Markendahl and Ö. Mäkitalo. A comparative study of deployment options, capacity and cost structure for macrocellular and femtocell networks. *Proc. of IEEE International Symposium on Personal Indoor and Mobile Radio Communications (PIMRC)*, 2010.

J. Markendahl, Ö. Mäkitalo, and J. Werding. Analysis of cost structure and business model options for wireless access provisioning using femtocell solutions. *Proc. of European International Telecommunications Society (ITS) Conference*, 2008.

M. A. Marsan, G. Albertengo, A. Francese, and F. Neri. Manhattan topologies for passive all-optical networks. *Proc. of Annual European Fiber Optic Communications and Local Area Network Exposition*, 1991.

P. Monti, S. Tombaz, L. Wosinska, and J. Zander. Mobile backhaul in heterogeneous network deployments: technology options and power consumption. *Proc. of International Conference on Transparent Optical Networks (ICTON)*, 2012.

S. Nie, G. R. MacCartney, S. Sun, and T. S Rappaport. 72 GHz millimeter wave indoor measurements for wireless and backhaul communications,. *Proc. of International Symposium on Personal Indoor and Mobile Radio Communications (PIMRC)*, 2013.

V. Nikolikj and T. Janevski. State-of-the-art business performance evaluation of the advanced wireless heterogeneous networks to be deployed for the "tera age". *Wireless Personal Communications*, 84(3):2241–2270, 2015.

Nokia. G.fast. *https://networks.nokia.com/solutions/g.fast*, 2018.

T. Norman. Wireless network traffic 2010-2015: forecasts and analysis. Analysys Mason, 2010.

A. Osseiran et. al. The foundation of the mobile and wireless communications system for 2020 and beyond: Challenges, enablers and technology solutions. *Proc. Vehicular Technology Conference (VTC Spring)*, pages 1–5, 2013.

E. J. Oughton and Z. Frias. The cost, coverage and rollout implications of 5g infrastructure in britain. *Telecommunications Policy*, 2017.

M. Paolini. An analysis of the total cost of ownership of point-to-point, point-to-multipoint, and fibre options. *White paper on crucial economics for mobile data backhaul*, 2011.

Senza fili. Crucial economics for mobile data backhaul. *Senza fili consulting white paper*, 2011.

Skyfiber. How to meet your backhaul capacity needs while maximizing revenue. *Skyfiber white paper*, 2013.

W-S. Soh, Z. Antoniou, and Hyong S. Kim. Improving restorability in radio access networks. *Proc of IEEE Global Telecommunications Conference (GLOBECOM)*, 2003.

Telecom India. Consultation paper on allocation and pricing of microwave access (mwa) and microwave backbone (mwb) rf carriers. *Telecom regulatory authority of India*, 2014.

O. Tipmongkolsilp, S. Zaghloul, and A. Jukan. The evolution of cellular backhaul technologies: Current issues and future trends. *IEEE Communications Surveys & Tutorials*, 13:97–113, 2011.

S. Tombaz, P. Monti, K. Wang, A. Vastberg, M. Forzati, and J. Zander. Impact of backhauling power consumption on the deployment of heterogeneous mobile networks. *Proc. of IEEE Global Telecommunications Conference (GLOBECOM)*, 2011.

S. Tombaz, P. Monti, F. Farias, M. Fiorani, L. Wosinska, and J. Zander. Is backhaul becoming a bottleneck for green wireless access networks? *Proc. of IEEE International Conference on Communications (ICC)*, 2014.

UMTS Forum. Mobile traffic forecasts 2010-2020. *UMTS Forum, Nokia*, 2011.

S. Verbrugge, K. Casier, J. Van Ooteghem, and B. Lannoo. white paper: Practical steps in techno-economic evaluation of network deployment planning. 2009.

F. Yaghoubi, M. Mahloo, L. Wosinska, P. Monti, F. S. Farias, and J. C. W. A. Costa. A techno-economic framework for 5G transport networks. *IEEE Wireless Communications*, 25: 56–63, 2018.

J. Zander and P. Mähönen. Riding the data tsunami in the cloud: myths and challenges in future wireless access. *IEEE Communications Magazine*, 51: 145–151, 2013.

Index

Optical and Wireless Convergence for 5G Networks, First Edition.
Edited by Abdelgader M. Abdalla, Jonathan Rodriguez, Issa Elfergani, and Antonio Teixeira.
© 2020 John Wiley & Sons Ltd. Published 2020 by John Wiley & Sons Ltd.